7th International Workshop on the Dark Side of The Universe

(DSU 2011)

Journal of Physics: Conference Series Volume 384

Beijing, China
26-30 September 2011

ISBN: 978-1-62276-345-0
ISSN: 1742-6588

Printed from e-media with permission by:

Curran Associates, Inc.
57 Morehouse Lane
Red Hook, NY 12571

Some format issues inherent in the e-media version may also appear in this print version.

Printed by Curran Associates, Inc. (2013)

Institute of Physics
Dirac House, Temple Back
Bristol BS1 6BE UK

Phone: 44 1 17 929 7481
Fax: 44 1 17 920 0979

techtracking@iop.org

Additional copies of this publication are available from:

Curran Associates, Inc.
57 Morehouse Lane
Red Hook, NY 12571 USA
Phone: 845-758-0400
Fax: 845-758-2634
Email: curran@proceedings.com
Web: www.proceedings.com

TABLE OF CONTENTS

PREFACE

The seventh international workshop 'The Dark Side of the Universe (DSU 2011)' was organized by The Kavli Institute for Theoretical Physics China (KITPC), from 26–30 September 2011 in Beijing. DSU 2011 is a continuation of DSU workshops previously held in Seoul (2005), Madrid (2006), Minnesota (2007), Cairo (2008), Melbourne (2009) and Leon (2010). The workshop was associated with the KTIPC program 'Dark matter and new physics', 21 September–6 November, 2011. The topics of the workshop include: dark matter candidates and theory, direct, indirect and accelerator dark matter searches, baryogenesis, new physics beyond the standard model, the origin of dark energy, experimental aspects of dark energy, nonstandard cosmology and astroparticle physics and Ultra hight energy cosmic rays etc.

The workshop attracted more than 130 participants from more than ten countries and regions. The talks presented at the workshop can be downloaded from the website http://kitpc.itp.ac.cn/dsu2011.

The workshop is financially supported by the Chinese Academy of Sciences, the National Science Foundation China (NSFC), and the China Center of Advance Science and Technology (CCAST).

Peer review statement

All papers published in this volume of *Journal of Physics: Conference Series* have been peer reviewed through processes administered by the proceedings Editors. Reviews were conducted by expert referees to the professional and scientific standards expected of a proceedings journal published by IOP Publishing.

Journal of Physics: Conference Series **384** (2012) 012001 doi:10.1088/1742-6596/384/1/012001

Gamma-ray lines from dark matter decay

Alejandro Ibarra

Physik-Department T30d, Technische Universität München
James-Franck-Straße, 85748 Garching, Germany

E-mail: alejandro.ibarra@ph.tum.de

Abstract. No known astrophysical process can generate a monoenergetic gamma-ray with energy in the TeV range, resulting in very stringent constraints on the lifetime of dark matter particles which decay producing gamma-ray lines. We derive in this work constraints on the decay width from observations at current IACTs as well as the estimated sensitivity of the projected CTA. We also discuss the implications of these limits for two dark matter models where the dark matter particle decays at tree level producing gamma-ray lines, namely the gravitino in supersymmetric models without R-parity conservation and a vector of a hidden $SU(2)$ gauge group. We also discuss the constraints on scenarios where the gamma-ray line is generated at the one loop level.

1. Introduction

Despite the many pieces of evidence for the existence of dark matter particles in our Universe [1], very little is known about their properties from the Particle Physics point of view. Namely, nothing is known about the spin or the parity, while the mass, the interaction cross section with nuclei or the lifetime are only very weakly constrained. In this work we aim to derive constraints on the lifetime of a dark matter particle with mass in the TeV range.

Our motivation is twofold. First, no matter particle is guaranteed to be stable and hence the lifetime of the dark matter particle should be determined from observations. This rationale is in complete analogy to the searches for proton decay. Whereas the proton stability is understood in the Standard Model as arising from an accidental baryon number conservation in the renormalizable Lagrangian, higher dimensional operators might violate baryon number inducing the proton decay. This is the case, *e.g.* in Grand Unified models. Similarly, the dark matter particle stability could be attributed to an approximate symmetry in the renormalizable Lagrangian which might be broken, or not, by higher dimensional operators. If the dark matter particle is a fermion which decays via a dimension 6 operator suppressed by a large mass scale M, the dark matter lifetime reads, for $\mathcal{O}(1)$ couplings,

$$\tau_{\rm DM} \sim 10^{26}\,{\rm s}\left(\frac{\rm TeV}{m_{\rm DM}}\right)^5\left(\frac{M}{10^{15}\,{\rm GeV}}\right)^4 . \tag{1}$$

This lifetime, as will argue below, can be probed in cosmic ray and gamma-ray observations if the dark matter mass is ~ 1 TeV and the suppression scale of the dimension six operators is $\sim 10^{15}$ GeV, which is interestingly close to the scale of Grand Unification.

Furthermore, from the experimental side, the PAMELA measurements [2] of the positron fraction and the Fermi-LAT measurements [3] of the electron+positron flux have revealed the

Decay Channel	$m_{\psi_{\rm DM}}$ [GeV]	$\tau_{\rm DM}$ [10^{26}s]
$\psi_{\rm DM} \to \mu^+\mu^-\nu$	3500	1.1
$\psi_{\rm DM} \to \ell^+\ell^-\nu$	2500	1.5
$\phi_{\rm DM} \to \mu^+\mu^-$	2500	1.8
$\phi_{\rm DM} \to \tau^+\tau^-$	5000	0.9
$\psi_{\rm DM} \to W^\pm\mu^\mp$	3000	2.1

Table 1. Decay channels for fermionic and scalar dark matter, $\psi_{\rm DM}$ and $\phi_{\rm DM}$, respectively, that best fit the measurements of the electron flux and the positron fraction by Fermi and PAMELA, respectively, for the MED propagation model and the NFW halo profile (for details, see [5]). The decay mode $\psi_{\rm DM} \to W^\pm\mu^\mp$ is in tension with the PAMELA results on the antiproton-to-proton ratio.

existence of an excess of positrons at energies larger than 7 GeV, and a possible excess of electrons and/or positrons at energies between 600 GeV and 1 TeV. These excesses could be explained by the decay of a dark matter particle with a mass in the TeV range and lifetime of the order of 10^{26} s [4, 5], see Fig. 1. In this work we will also investigate whether this explanation to the electron/positron excesses could be tested in searches for gamma-ray lines.

2. Observational limits on gamma-ray lines

The decay of a dark matter particle with mass $m_{\psi_{\rm DM}}$ in a two-body decay involving a photon and a neutral particle N, $\psi_{\rm DM} \to \gamma N$, produces a gamma-ray with energy

$$E_\gamma = \frac{m_{\psi_{\rm DM}}}{2}\left(1 - \frac{m_N^2}{m_{\psi_{\rm DM}}^2}\right) . \tag{2}$$

When the neutral particle is massless, we will write ν instead of N in the following.

The flux of monochromatic gamma rays from the decay of dark matter in the Milky Way halo is given by a line-of-sight integral over the dark matter distribution [6]. This component of the gamma-ray flux is explicitly given by

$$\frac{dJ_{\rm DM}^{\rm halo}}{dE} = \frac{\Gamma(\psi_{\rm DM} \to \gamma N)}{4\pi m_{\psi_{\rm DM}}}\,\delta\left(E_\gamma - E\right)\int_{\rm l.o.s.} d\vec{l}\,\rho_{\rm DM}^{\rm MW}(\vec{l})\,, \tag{3}$$

where $\Gamma(\psi_{\rm DM} \to \gamma N)$ denotes the partial decay width of the process and $\rho_{\rm DM}^{\rm MW}$ is the Milky Way's dark matter halo density profile. We adopt the NFW profile here, which has the form

$$\rho_{\rm DM}^{\rm MW}(r) = \frac{\rho_{\rm c}}{r/r_{\rm c}\left(1 + r/r_{\rm c}\right)^2}\,, \tag{4}$$

with the parameters $\rho_{\rm c} = 0.35\,{\rm GeV\,cm^{-3}}$ and $r_{\rm c} = 20$ kpc [7], leading to a local dark matter density of 0.4 GeV/cm^3 [8]. The gamma-ray flux from dark matter decay inside the Galactic halo has only a mild angular dependence and can be considered as isotropic for our purposes (for details on anisotropies in the Galactic gamma-ray flux from dark matter decay, see ref. [9]). The extragalactic contribution stemming from the decay of dark matter at cosmological distances is generally fainter than the Galactic flux, and we will neglect this component here.

The Fermi LAT collaboration has conducted a negative search for Galactic gamma-ray lines in the diffuse flux in the energy range from 30 to 200 GeV [10]; the 2σ limits on the partial decay width corresponding to $\psi_{\rm DM} \to \gamma\nu$ are shown in Fig. 1 for the halo profile Eq. (4). This search has been recently extended in [11] for fermionic dark matter particles with masses $2 - 600$ GeV which decay $\psi \to \gamma\nu$.

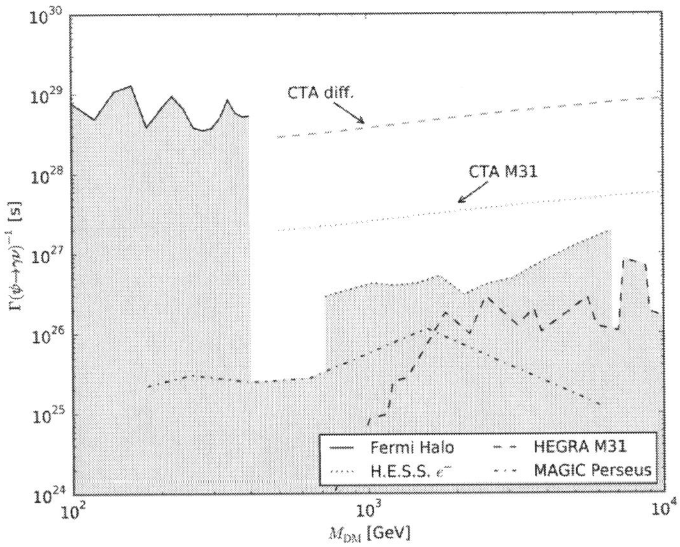

Figure 1. Lower limits on the inverse decay width of a dark matter particle which decays via $\psi_{\rm DM} \to \gamma\nu$ from line searches in M31 by HEGRA, from line searches in the diffuse flux by Fermi LAT and from observations of the Perseus cluster by MAGIC. Further bounds can be derived from the $(\gamma{+})e^-$ observations of H.E.S.S. Our estimates of the reach of the future CTA in measurements of the flux from M31 or spectral variations in the diffuse $\gamma + e^-$ flux are shown as red lines.

For larger dark matter masses, it is possible to derive constraints on the decay width from observations undertaken by Imaging Atmospheric Cherenkov Telescopes (IACTs). The HEGRA collaboration has published upper limits on the gamma-ray line flux from M31 [12]. Besides, the MAGIC collaboration has published upper limits on the gamma-ray flux from the Perseus galaxy cluster [13]. Lastly, the H.E.S.S. collaboration has published measurements of the electron flux at TeV energies [14, 15]. The measured electron flux may be contaminated with diffuse gamma rays by no more than $\approx 50\%$ [14], thus allowing to translate the electron flux into upper bounds on gamma-ray lines from dark matter decay in the Galactic halo. The 99% C.L. limits on the decay width of dark matter into gamma-ray lines from all these measurements are presented Fig. 1 (for further details, see [16]). In the future, the Cherenkov Telescope Array (CTA, see ref. [17] for a recent discussion) will undertake searches of gamma-ray lines with greater sensitivity. In Fig. 1 we also show our estimation for the expected 2σ limit on the decay width from observations of M31 and the diffuse background assuming no signal is observed [16].

3. Tree level decays

The present bounds on the intensity of gamma-ray lines set very stringent constraints on models where the dark matter particle decays at tree level. We will discuss here two well motivated decaying dark matter models which can produce gamma-ray lines with a sizable rate.

3.1. Gravitino in SUSY models without R-parity conservation

Among the most interesting dark matter candidates proposed stands the gravitino [18] which is abundantly produced by thermal scatterings in the very early Universe. If the gravitino is the lightest supersymmetric particle (LSP), a fraction of their initial population will survive until today with a relic density which is calculable in terms of very few parameters, the result being [19]

$$\Omega_{3/2}h^2 \simeq 0.27 \left(\frac{T_R}{10^{10}\,\mathrm{GeV}}\right)\left(\frac{100\,\mathrm{GeV}}{m_{3/2}}\right)\left(\frac{m_{\tilde{g}}}{1\,\mathrm{TeV}}\right)^2 , \tag{5}$$

where T_R is the reheating temperature of the Universe, $m_{3/2}$ is the gravitino mass and $m_{\tilde{g}}$ is the gluino mass. In predicting the relic abundance of gravitinos, the main uncertainty arises from our ignorance of the thermal history of the Universe before Big Bang nucleosynthesis (BBN) and in particular of the reheating temperature after inflation. However, we have strong indications that the Universe was very hot after inflation. Namely, the discovery of neutrino masses provided strong support to leptogenesis as the explanation for the observed baryon asymmetry of the Universe [20]. This mechanism can reproduce the observed baryon asymmetry very naturally if the reheating temperature of the Universe was above 10^9 GeV [21]. Therefore, the abundance of gravitinos can reproduce the dark matter relic density inferred by WMAP for the ΛCDM model, $\Omega_{\rm CDM}h^2 \simeq 0.1$ [22], for natural values of the input parameters in Eq. (5).

Being capable of reproducing the correct relic density is a necessary requirement for any dark matter candidate, but not the only one. Namely, the physical model accounting for the dark matter should not spoil the successful predictions of the standard cosmology. However, in scenarios with gravitino dark matter, when R-parity is conserved, the next-to-LSP (NLSP) is typically present during or after Big Bang nucleosynthesis, jeopardizing the successful predictions of the standard nucleosynthesis scenario [23]. A simple solution which avoids the BBN constraints altogether consists in introducing a small amount of R-parity violation, so that the NLSP decays into two Standard Model particles before the onset of Big Bang nucleosynthesis [24].

When R-parity is not exactly conserved, the gravitino LSP is no longer stable. Nevertheless, the gravitino decay rate is doubly suppressed by the Planck mass and by the small R-parity violation [25], yielding a lifetime which is typically larger than the age of the Universe Nevertheless, gravitinos could be decaying at a rate sufficiently large to allow the detection of the decay products in experiments, thus opening the possibility of the indirect detection of the elusive gravitino dark matter [26, 27, 28, 29, 30].

We show in Fig. 2 the branching ratios of the gravitino in the interesting case of bilinear R-parity violation [29]. Especially for small masses the branching ratio into monoenergetic gamma-rays can be sizable. Hence, we expect very stringent constraints on decaying gravitinos from the negative searches of monoenergetic gamma-rays in the range $E_\gamma = 1 - 300$ GeV. This is illustrated in Fig. 3, right plot, where we show the expected gamma-ray flux of a gravitino with mass 200 GeV and lifetime $\tau = 7 \times 10^{26}$ s [30]. The gamma-ray line is predicted to be fairly intense and is in fact in conflict with the present constraints from the Fermi-LAT.

3.2. Vector of a hidden SU(2) gauge group

This model considers an extension of the Standard Model where the gauge group contains a hidden non-abelian group, $SU(2)_{\rm HS}$, with gauge bosons A^μ. It is assumed that the hidden non-abelian symmetry is spontaneously broken via the vacuum expectation value of a complex scalar field, ϕ. It is also assumed that all the Standard Model particles are singlets under $SU(2)_{\rm HS}$, thus the Standard Model only couples to the hidden sector via the Higgs portal term $|\phi|^2|H|^2$, being H the Standard Model Higgs doublet. With this assumptions, it can be shown that the Lagrangian displays a $SO(3)$ custodial symmetry in the A_i^μ component space, which prevents any decay into $SO(3)$ singlets (such as Standard Model particles) [31]. Consequently, if the model is described just by the renormalizable Lagrangian, the three A_i^μ components are degenerate in mass and are absolutely stable. Furthermore, the parameters of the model can be chosen to produce a relic density of the hidden vector dark matter in agreement with observations [31]. Nevertheless,

Journal of Physics: Conference Series **384** (2012) 012001 doi:10.1088/1742-6596/384/1/012001

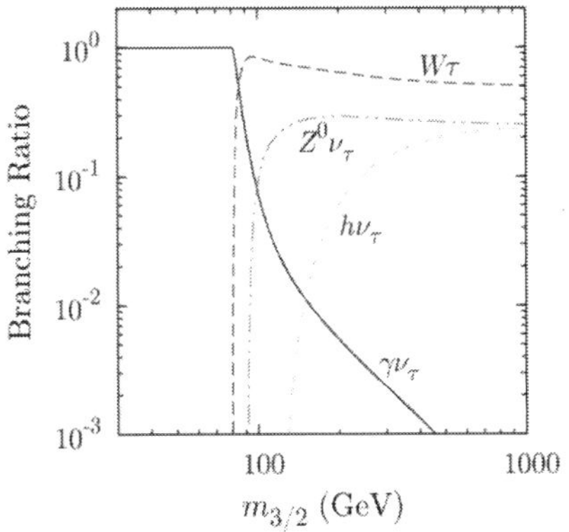

Figure 2. Branching ratios for gravitino decay in a supersymmetric scenario with bilinear R-parity violation.

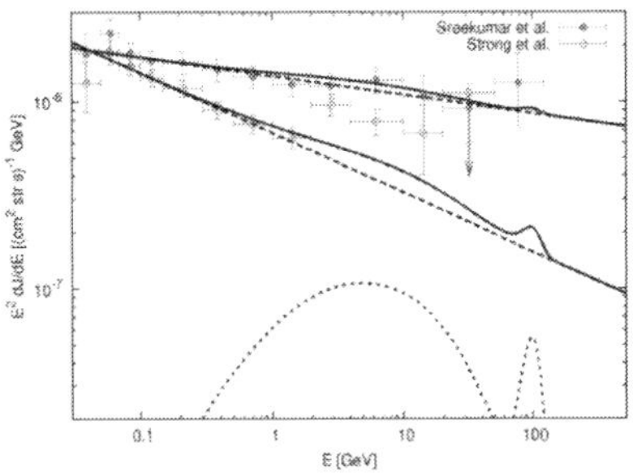

Figure 3. Predicted gamma-ray flux from decaying gravitinos with mass $m = 200$ GeV and lifetime $\tau = 7 \times 10^{26}$ s.

one expects the existence of non-renormalizable operators in the Lagrangian which break the custodial symmetry after the spontaneous symmetry breaking of $SU(2)_{HS}$ and $SU(2)_L \times U(1)_Y$, for example [32].

$$\text{(A)} \qquad \frac{1}{\Lambda^2}\,\mathcal{D}_\mu\phi^\dagger\phi\;\mathcal{D}_\mu H^\dagger H\;, \tag{6}$$

$$\text{(C)} \qquad \frac{1}{\Lambda^2}\,\mathcal{D}_\mu\phi^\dagger\mathcal{D}_\nu\phi\;F^{\mu\nu Y}\;. \tag{7}$$

In turn, the breaking of the custodial symmetry leads to the decay of the dark matter hidden gauge bosons, with a decay rate and branching ratios which depend on the parameters of the model.

We show in Fig. 4, left plot, the expected gamma-ray flux from hidden vector dark matter particles which decay via the operator A in Eq. (6). The dark matter mass is 600 GeV and the various couplings have been chosen in order to reproduce the correct relic density and to be in agreement with direct detection experiments [32]. For the scenario shown in the plot the lifetime is $\tau = 1.1 \times 10^{27}$ s, which is marginally allowed by the Fermi-LAT results. On the other hand, in the right plot we show the predicted gamma ray flux in a scenario where a hidden vector dark matter particle with mass 1550 GeV decays with lifetime 1.6×10^{27} s via the operator C, Eq. (7). In this case the dark matter particle can decay with similar branching ratios into $\gamma\eta$, the hidden sector Higgs, and γh, the visible sector Higgs, resulting in a double gamma-ray line feature [32]. This scenario is allowed by current measurements but could be tested at the CTA.

4. Radiative decays

Searches for gamma-ray lines provide very stringent constraints on the width of fermionic dark matter particles which decay $\psi \to \gamma\nu$. From Fig. 1 it follows that $\Gamma^{-1}(\psi \to \gamma\nu) \gtrsim 5\times 10^{28} - 10^{29}$ s for m_ψ between 1 and 600 GeV, while in the future observations at CTA could reach this same limit for larger masses, between 600 GeV and 10 TeV. These limits are 2-3 orders of magnitude better than the limits set by observations of the cosmic electron/positron flux in that same mass range, see *e.g.* Fig. 1. Therefore, it is interesting to explore whether the constraints on models from one-loop induced gamma-ray lines can be competitive with the constraints from tree-level decays into leptons.

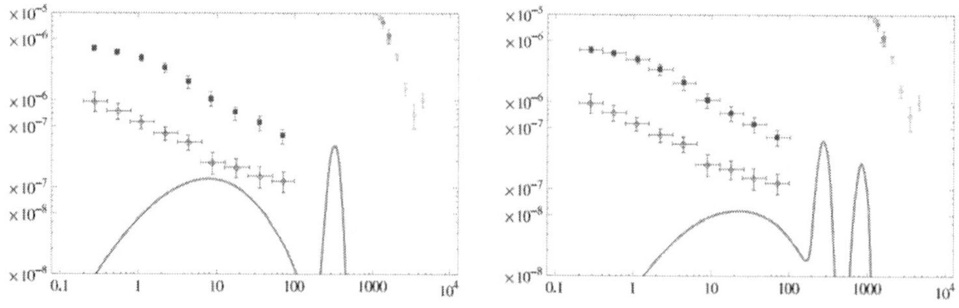

Figure 4. Predicted gamma-ray flux from the decay of a hidden $SU(2)$ gauge boson dark matter particle. In the left plot the decay is induced by the non-renormalizable operator Eq. (6) while in the right plot by Eq. (7). Details are given in the text.

Let us first consider the decay width for the radiatively induced gamma-ray line, $\Gamma(\psi_{\rm DM} \to \gamma\nu)$, when the dark matter decays at tree level into muons of the same chirality. The result reads [16]:

$$\Gamma(\psi_{\rm DM} \to \gamma\nu) \simeq \frac{3\alpha_{\rm em}}{8\pi}\Gamma(\psi_{\rm DM} \to \mu^+_{L,R}\mu^-_{L,R}\nu)\,. \quad (8)$$

In the case that the dark matter decays democratically into all flavours, $\psi_{\rm DM} \to \ell^+_{L,R}\ell^-_{L,R}\nu$, the result is a factor of three larger. On the other hand, if the decay is mediated by a heavy vector, the corresponding decay widths must be multiplied by an additional factor of 9. In especific models, such as when the dark matter is constituted by hidden gauginos of an unbroken $U(1)$, several scalar particles circulate in the loop resulting in an enhancement of the decay rate [33].

We show in Fig. 5 the limits on the inverse decay width of the radiative process $\psi_{\rm DM} \to \gamma\nu$ assuming that the democratic decay $\psi_{\rm DM} \to \ell^+\ell^-\nu$ reproduces the PAMELA positron excess (yellow regions) and the Fermi electron excess (black regions), when the decay is mediated by a heavy scalar (left plot) or by a heavy vector (right plot). It is apparent from the plot that at large masses the CTA observations of the diffuse background will provide better constraints on these models that those stemming from the electron/positron measurements. Furthermore, it is predicted that the CTA will observe a gamma-ray line if these decay channels are the explanation of the electron/positron excesses observed by PAMELA and Fermi [16].

The decay width into gamma-ray lines can be enhanced if the daughter particle N in the decay $\psi \to \gamma N$ is degenerate in mass with the dark matter particle. Considering the tree level decay $\psi_{\rm DM} \to \mu^+\mu^- N$, the radiatively induced decay process $\psi_{\rm DM} \to \gamma N$ has a width [16]:

$$\Gamma(\psi_{\rm DM} \to \gamma N) \simeq \frac{3\alpha_{\rm em}}{8\pi}\begin{cases} \frac{5}{12} & \text{for } m_N \to m_{\psi_{\rm DM}},\ \eta = +1 \\ \frac{20}{(1-m_N^2/m_{\psi_{\rm DM}}^2)^2} & \text{for } m_N \to m_{\psi_{\rm DM}},\ \eta = +1 \end{cases}, \quad (9)$$

with η the relative CP parity between the dark matter particle and the daughter neutral particle. Hence, if these two particles have opposite CP parities, the branching ratio into monochromatic gamma-rays increases as their masses become more and more degenerate, as shown in Fig. 6.

In order to illustrate the enhancement of the branching ratio and the impact from the gamma-line constraints, we consider in Fig. 7 seven benchmark scenarios for which $m_{\psi_{\rm DM}}$ and m_N are of comparable size and the decay generates a gamma line at 170 GeV (benchmarks 1, 2, 3) or at 5 TeV (benchmarks 4, 5, 6, 7). All the benchmark scenarios reproduce the PAMELA positron data, and all except scenarios 1, 2 and 3 additionally reproduce the electron spectrum measured

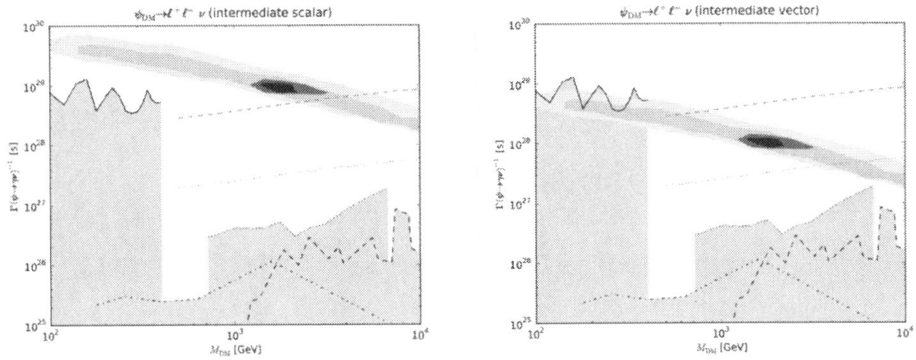

Figure 5. Same as fig. 1. In addition, the yellow and black shaded regions show the parts of the parameter space that are relevant for the decaying dark matter interpretation of the PAMELA/Fermi $e^{\pm}$ anomalies with the decay channel $\psi_{\rm DM} \to \ell^+\ell^-\nu$. The intermediate particle is a scalar in the left plot and a vector in the right plot, and the final fermions are assumed to have the same chirality.

by Fermi. Benchmark points 1, 3, 4 and 6 have $\eta = -1$ and $m_{\psi_{\rm DM}}/m_N$ =0.81, 0.39, 0.95 and 0.58, while 2 and 4 have $\eta = 1$ and $m_{\psi_{\rm DM}}/m_N$ =0.57, 0.58, respectively. It is interesting to note that present experiments are sensitive to dark matter models decaying into neutral daughter particles close in mass to the dark matter particle, when the produced leptons have the same chirality: benchmark points 1 and 4 are excluded by present gamma line constraints, while 6 and perhaps 3 and 5 could be tested in future experiments. In contrast benchmark point 2 and 7 cannot be probed with present or planned experiments (benchmark point 7 corresponds to a decay into leptons of opposite chiralities, and lies out of the figure).

5. Conclusions

We have discussed in this paper the scenario where a dark matter particle with mass in the TeV range decays producing a monoenergetic gamma-ray. We have derived upper limits on the decay width from observations at current ACTs and we have estimated the projected sensitivity of the future CTA. Lastly we have discussed the implications of these constraints on concrete models where the gamma-ray line is produced at tree level or at the one loop level.

Acknowledgments

The author is grateful to the DSU 2011 organizers for their kind invitation to participate in this workshop. The author is also grateful to Chiara Arina, Wilfried Buchmüller, Laura Covi, Michael Grefe, Thomas Hambye, Tetsuo Shindou, Fumihiro Takayama, David Tran and Christoph Weniger for very pleasant collaborations on which this talk is based. This work was partially supported by the DFG cluster of excellence "Origin and Structure of the Universe."

References

[1] See, *e.g.* G. Bertone, D. Hooper and J. Silk, Phys. Rept. **405** (2005) 279 [arXiv:hep-ph/0404175].
[2] O. Adriani *et al.* [PAMELA Collaboration], Nature **458** (2009) 607. [arXiv:0810.4995 [astro-ph]].
[3] A. A. Abdo *et al.* [The Fermi LAT Collaboration], Phys. Rev. Lett. **102**, 181101 (2009). [arXiv:0905.0025 [astro-ph.HE]].
[4] A. Ibarra and D. Tran, JCAP **0902** (2009) 021. [arXiv:0811.1555 [hep-ph]].
[5] A. Ibarra, D. Tran and C. Weniger, JCAP **1001**, 009 (2010). [arXiv:0906.1571 [hep-ph]].
[6] G. Bertone, W. Buchmuller, L. Covi and A. Ibarra, JCAP **0711**, 003 (2007) [arXiv:0709.2299 [astro-ph]].

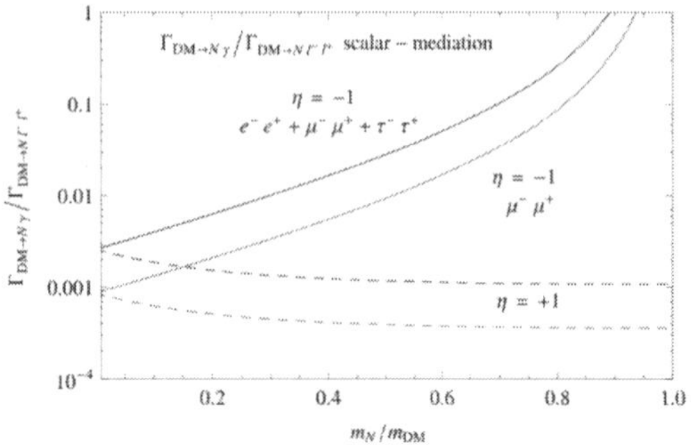

Figure 6. Ratio of the decay rates $\Gamma(\psi_{\rm DM} \to \gamma N)/\sum_\ell \Gamma(\psi_{\rm DM} \to \ell^+\ell^- N)$ when the decay is mediated by a scalar. The four cases correspond to single-flavor decay (red) and democratic decay into all flavors (blue), as well as ψ and N having the same CP parity (dashed, $\eta = +1$) or opposite CP parity (solid, $\eta = -1$)

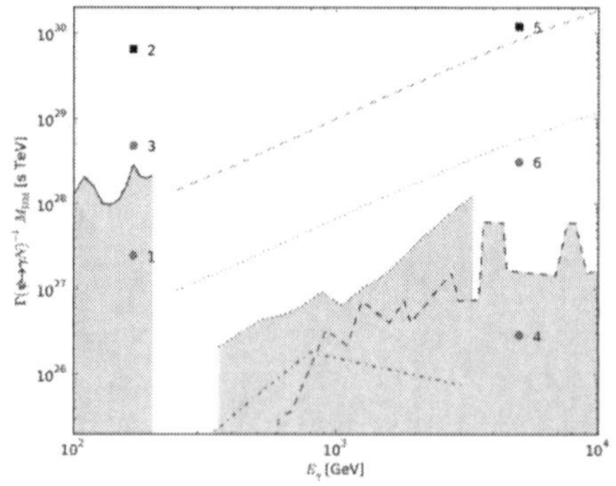

Figure 7. Same as fig. 1, but with different scaling of the axes to allow for non-vanishing m_N. The black squares and red dots show the predictions for the different benchmark scenarios (see text). Black squares correspond to scenarios with $\eta = +1$, while red dots correspond to $\eta = -1$.

[7] J. F. Navarro, C. S. Frenk, S. D. M. White, Astrophys. J. **490** (1997) 493-508. [astro-ph/9611107].
[8] R. Catena, P. Ullio, JCAP **1008**, 004 (2010). [arXiv:0907.0018 [astro-ph.CO]].
[9] A. Ibarra, D. Tran, C. Weniger, Phys. Rev. **D81** (2010) 023529. [arXiv:0909.3514 [hep-ph]].
[10] A. A. Abdo *et al.*, Phys. Rev. Lett. **104**, 091302 (2010) [arXiv:1001.4836 [astro-ph.HE]].
[11] G. Vertongen and C. Weniger, JCAP **1105** (2011) 027 [arXiv:1101.2610 [hep-ph]].
[12] F. A. Aharonian *et al.* [HEGRA Collaboration], Astron. Astrophys. **400**, 153 (2003) [arXiv:astro-ph/0302347].
[13] J. Aleksic *et al.* [The MAGIC Collaboration], Astrophys. J. **710** (2010) 634-647. [arXiv:0909.3267 [astro-ph.HE]].
[14] F. Aharonian *et al.* [H.E.S.S. Collaboration], Phys. Rev. Lett. **101** (2008) 261104 [arXiv:0811.3894 [astro-ph]].
[15] F. Aharonian *et al.* [H.E.S.S. Collaboration], H.E.S.S," Astron. Astrophys. **508** (2009) 561. [arXiv:0905.0105 [astro-ph.HE]].
[16] M. Garny, A. Ibarra, D. Tran and C. Weniger, JCAP **1101** (2011) 032 [arXiv:1011.3786 [hep-ph]].
[17] T. C. Consortium, [arXiv:1008.3703 [astro-ph.IM]].
[18] H. Pagels and J. R. Primack, Phys. Rev. Lett. **48** (1982) 223.
[19] M. Bolz, A. Brandenburg and W. Buchmüller, Nucl. Phys. B **606** (2001) 518. [arXiv:hep-ph/0012052].
[20] M. Fukugita and T. Yanagida, Phys. Lett. B **174** (1986) 45.
[21] S. Davidson and A. Ibarra, Phys. Lett. B **535** (2002) 25; [arXiv:hep-ph/0202239]. W. Buchmüller, P. Di Bari and M. Plümacher, Annals Phys. **315** (2005) 305. [arXiv:hep-ph/0401240].
[22] D. N. Spergel *et al.* [WMAP Collaboration], Astrophys. J. Suppl. **170** (2007) 377 [arXiv:astro-ph/0603449].
[23] For a review, see F. D. Steffen, Eur. Phys. J. C **59** (2009) 557 [arXiv:0811.3347 [hep-ph]].
[24] W. Buchmüller, L. Covi, K. Hamaguchi, A. Ibarra and T. Yanagida, JHEP **0703**, 037 (2007). [arXiv:hep-ph/0702184].
[25] F. Takayama and M. Yamaguchi, Phys. Lett. B **485** (2000) 388. [arXiv:hep-ph/0005214].
[26] A. Ibarra and D. Tran, Phys. Rev. Lett. **100** (2008) 061301. [arXiv:0709.4593 [astro-ph]].
[27] A. Ibarra and D. Tran, JCAP **0807** (2008) 002. [arXiv:0804.4596 [astro-ph]].
[28] K. Ishiwata, S. Matsumoto and T. Moroi, arXiv:0805.1133 [hep-ph].
[29] L. Covi, M. Grefe, A. Ibarra and D. Tran, JCAP **0901** (2009) 029 [arXiv:0809.5030 [hep-ph]].
[30] W. Buchmuller, A. Ibarra, T. Shindou, F. Takayama and D. Tran, JCAP **0909** (2009) 021 [arXiv:0906.1187 [hep-ph]].
[31] T. Hambye, JHEP **0901** (2009) 028 [arXiv:0811.0172 [hep-ph]].
[32] C. Arina, T. Hambye, A. Ibarra and C. Weniger, JCAP **1003** (2010) 024 [arXiv:0912.4496 [hep-ph]].
[33] A. Ibarra, A. Ringwald and C. Weniger, JCAP **0901** (2009) 003 [arXiv:0809.3196 [hep-ph]].

Journal of Physics: Conference Series **384** (2012) 012002 doi:10.1088/1742-6596/384/1/012002

Fermi Large area telescope results: The sky at high energies and the Quest for Dark Matter signals

Aldo Morselli
Via della Ricerca Scientifica 1, 00133 Roma, Italy
E-mail: aldo.morselli@roma2.infn.it

Abstract. Successfully launched in June 2008, the Fermi Gamma-ray Space Telescope, formerly named GLAST, has been observing the high-energy gamma-ray sky with unprecedented sensitivity in the 20MeV 300GeV energy range and electrons + positrons in the 7 GeV- 1 TeV range, opening a new observational window on a wide variety of astrophysical objects.

1. Introduction

The Fermi Observatory carries two instruments on-board: the Gamma-ray Burst Monitor (GBM) [1] and the Large Area Telescope (LAT) [2]. The GBM, sensitive in the energy range between 8 keV and 40 MeV, is designed to observe the full unocculted sky with rough directional capabilities (at the level of one to a few degrees) for the study of transient sources, particularly Gamma-Ray Bursts (GRBs). The LAT is a pair conversion telescope for photons above 20 MeV up to a few hundreds of GeV. The operation of the instrument through the first three years of the mission was smooth at a level which is probably beyond the more optimistic pre-launch expectations. The LAT has been collecting science data for more than 99% of the time spent outside the South Atlantic Anomaly (SAA). The remaining tiny fractional down-time accounts for both hardware issues and detector calibrations. More than 650 million gamma-ray candidates (i.e. events passing the background rejection selection) were made public and distributed to the Community through the Fermi Science Support Center (FSSC) [1]. Over the first three years of mission the LAT collaboration has put a considerable effort toward a better understanding of the instrument and of the environment in which it operates. In addition to that a continuous effort was made to in order to make the advances public as soon as possible. In August 2011 the first new event classification (Pass 7) since launch was released, along with the corresponding Instrument Response Functions. Compared with the pre-launch (Pass 6) classification, it features a greater and more uniform exposure, with a significance enhancement in acceptance below 100 MeV.

2. The Second Fermi-LAT catalog

The high-energy gamma-ray sky is dominated by diffuse emission: more than 70% of the photons detected by the LAT are produced in the interstellar space of our Galaxy by interactions of high-energy cosmic rays with matter and low-energy radiation fields. An additional diffuse component with an almost-isotropic distribution (and therefore thought to be extragalactic in

[1] The FSSC is available at *http://fermi.gsfc.nasa.gov/ssc*

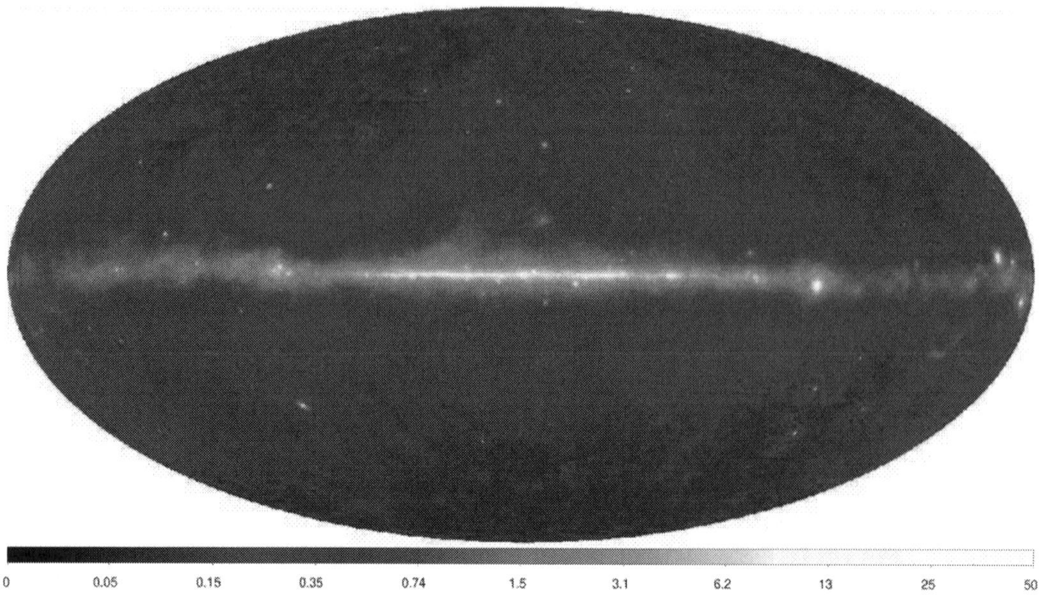

Figure 1. Sky map of the energy flux derived from 24 months of observation. The image shows γ-ray energy flux for energies between 100 MeV and 10 GeV, in units of 10^{-7} erg cm^{-2} s^{-1} sr^{-1}.

origin) accounts for another significant fraction of the LAT photon sample. The rest consists of various different types of point-like or extended sources: Active Galactic Nuclei (AGN) and normal galaxies, pulsars and their relativistic wind nebulae, globular clusters, binary systems, shock-waves remaining from supernova explosions and nearby solar-system bodies like the Sun and the Moon.

The Second Fermi-LAT catalog (2FGL) [3] is the deepest catalog ever produced in the energy band between 100 MeV and 100 GeV. Compared to the First Fermi-LAT (1FGL) [4] , it features several significant improvements: it is based on data from 24 (vs. 11) months of observation and makes use of the new Pass 7 event selection. The energy flux map is shown in figure 1 and the sky-distribution of the 1873 sources is shown in figure 2. It is interesting to note that 127 sources are firmly identified, based either on periodic variability (e.g. pulsars) or on spatial morphology or on correlated variability. In addition to that 1170 are reliably associated with sources known at other wavelengths, while 576 (i.e. 31% of the total number of entries in the catalog) are still unassociated.

3. Indirect Dark Matter searches

One of the major open issues in our understanding of the Universe is the existence of an extremely-weakly interacting form of matter, the Dark Matter (DM), supported by a wide range of observations including large scale structures, the cosmic microwave background and the isotopic abundances resulting from the primordial nucleosynthesis. Complementary to direct searches being carried out in underground facilities and at accelerators, the indirect search for DM is one of the main items in the broad Fermi Science menu. The word indirect denotes here the search for signatures of Weakly Interactive Massive Particle (WIMP) annihilation or decay processes through the final products (gamma-rays, electrons and positrons, antiprotons) of such processes. Among many other ground-based and space-borne instruments, the LAT plays a

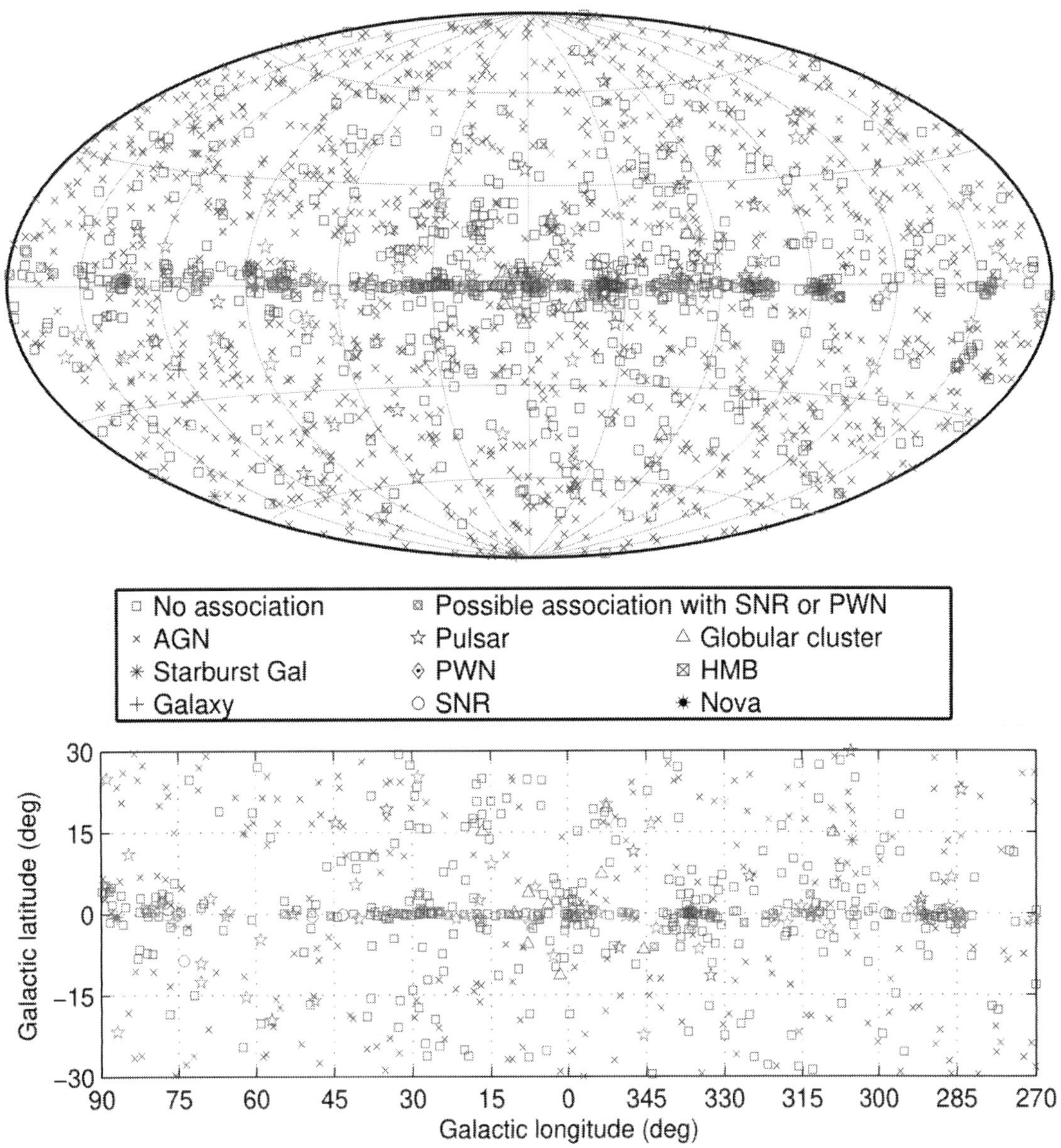

Figure 2. Full sky map (top) and blow-up of the inner Galactic region (bottom) showing sources by source class. Identified sources are shown with a red symbol, associated sources in blue.

prominent role in this search through a variety of distinct search targets: gamma-ray lines, Galactic and isotropic diffuse gamma-ray emission, dwarf satellites, CR electrons and positrons.

3.1. Galactic center

The Galactic center (GC) is expected to be the strongest source of γ-rays from DM annihilation, due to its coincidence with the cusped part of the DM halo density profile [5], [6]. A preliminary analysis of the data, taken during the first 11 months of the Fermi satellite operations is presented in [8], [9]. The diffuse gamma-ray backgrounds and discrete sources, as we know them today, can

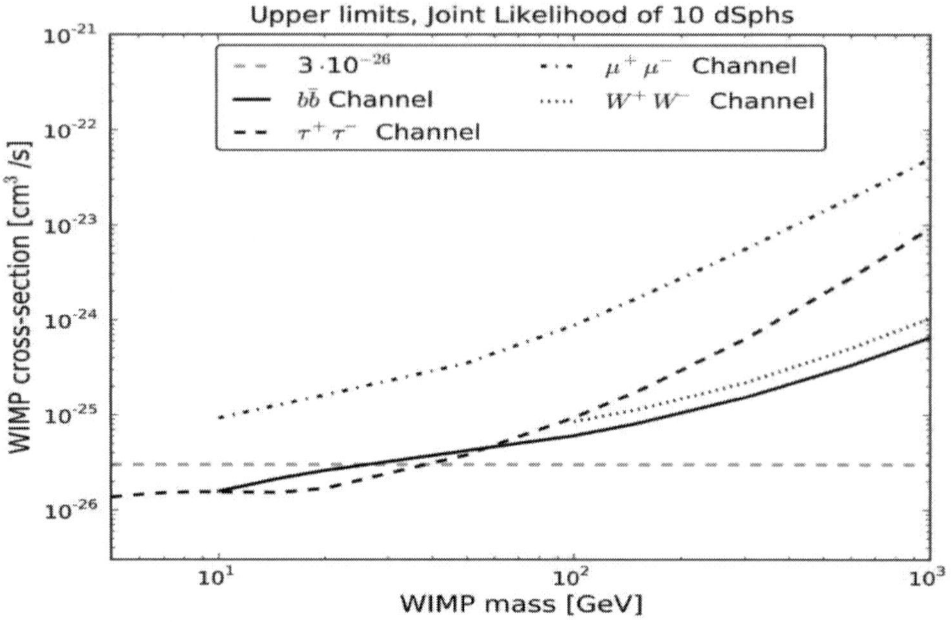

Figure 3. Derived 95% C.L. upper limits on WIMP annihilation cross sections for different channels

account for the large majority of the detected gamma-ray emission from the Galactic Center. Nevertheless a residual emission is left, not accounted for by the above models [8], [9].

Improved modeling of the Galactic diffuse model as well as the potential contribution from other astrophysical sources (for instance unresolved point sources) could provide a better description of the data. Analyses are underway to investigate these possibilities.

3.2. Dwarf galaxies

Dwarf satellites of the Milky Way are among the cleanest targets for indirect dark matter searches in gamma-rays. They are systems with a very large mass/luminosity ratio (i.e. systems which are largely DM dominated). The LAT detected no significant emission from any of such systems and the upper limits on the γ-ray flux allowed us to put very stringent constraints on the parameter space of well motivated WIMP models [11].

A combined likelihood analysis of the 10 most promising dwarf galaxies, based on 24 months of data and pushing the limits below the thermal WIMP cross section for low DM masses (below a few tens of GeV), has been recently performed [12].

The derived 95% C.L. upper limits on WIMP annihilation cross sections for different channels are shown in figure 3. The most generic cross section ($\sim 3 \cdot 10^{-26} cm^3 s^{-1}$ for a purely s-wave cross section) is plotted as a reference. These results are obtained for NFW profiles but for cored dark matter profile the J-factors for most of the dSphs would either increase or not change much so these results includes J-factor uncertainties [12]. With the present data we are able to rule out large parts of the parameter space where the thermal relic density is below the observed cosmological dark matter density and WIMPs are dominantly produced non-thermally, e.g. in models where supersymmetry breaking occurs via anomaly mediation. These γ-ray limits also constrain some WIMP models proposed to explain the *Fermi* LAT and PAMELA e^+e^- data, including low-mass wino-like neutralinos and models with TeV masses pair-annihilating into muon-antimuon pairs.

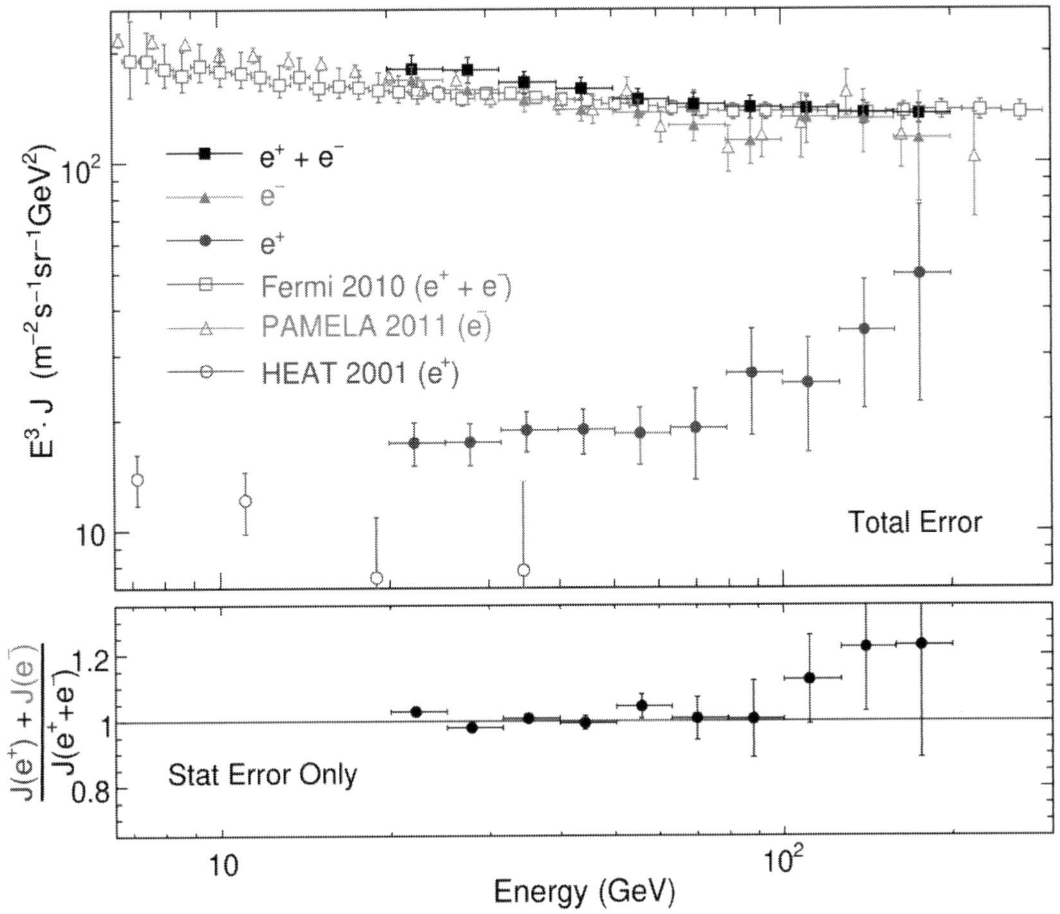

Figure 4. Energy spectra for e^+, e^-, and $e^+ + e^-$ (control region). In the control region where both species are allowed, this analysis reproduces the Fermi LAT results reported previously for the total electron plus positron spectrum [13, 14] (gray). Previous results form HEAT [16] and PAMELA [15] are shown for reference. The bottom panel shows that the ratio between the sum and the control flux is consistent with 1 as expected.

Future improvements (apart from increased amount of data) will include an improved event selection with a larger effective area and photon energy range, and the inclusion of more satellite galaxies.

3.3. Gamma-ray lines

A line at the WIMP mass, due to the 2γ production channel, could be observed as a feature in the astrophysical source spectrum [7]. Such an observation would be a "smoking gun" for WIMP DM as it is difficult to explain by a process other than WIMP annihilation or decay and the presence of a feature due to annihilation into γZ in addition would be even more convincing.

Up to now however no significant evidence of gamma-ray line(s) has been found in the first 11 months of data, between 30 and 200 GeV [10] and work is ongoing to extend the energy range of the analysis and include more data.

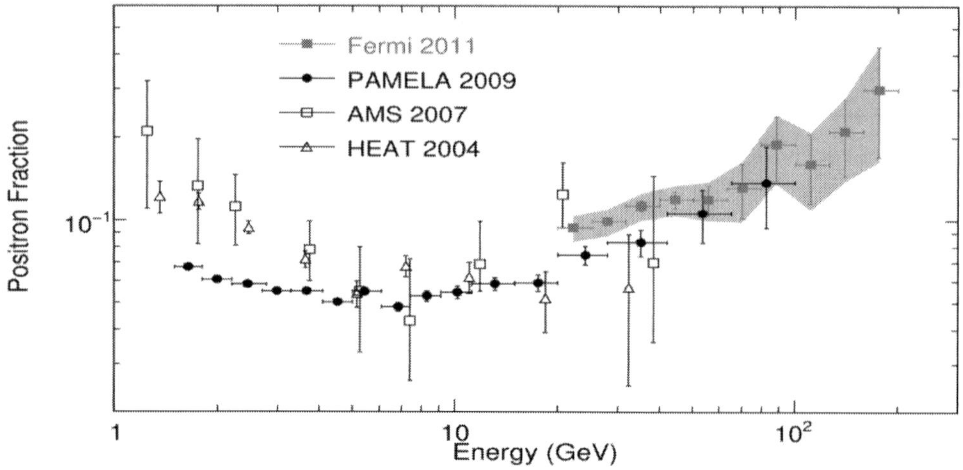

Figure 5. Positron fraction measured by the Fermi LAT and by other experiments [16, 17, 15]. The Fermi statistical uncertainty is shown with error bars and the total (statistical plus systematic uncertainty) is shown as a shaded band.

3.4. The Cosmic Ray Electron spectrum

Recently the experimental information available on the Cosmic Ray Electron (CRE) spectrum has been dramatically expanded with a high precision measurement of the electron spectrum from 7 GeV to 1 TeV [13], [14]. The spectrum shows no prominent spectral features and it is significantly harder than that inferred from several previous experiments.

More recently we provided a further, and stronger, evidence of the positron anomaly by providing direct measurement of the absolute e^+ and e^- spectra, and of their fraction, between 20 and 200 GeV using the Earth magnetic field (see figure 4). A steady rising of the positron fraction was observed up to that energy in agreement with that found by PAMELA. In the same energy range, the e^- spectrum was fitted with a power-law with index $\gamma(e^-) = -3.19 \pm 0.07$ which is in agreement with what recently measured by PAMELA between 1 and 625 GeV [15]. Most importantly, Fermi-LAT measured, for the first time, the e^+ spectrum in the 20 - 200 GeV energy interval (see figure 5) . The e^+ spectrum is fitted by a power-law with index $\gamma(e^+) = -2.77 \pm 0.14$. These measurements seems to rule out the standard scenario in which the bulk of electrons reaching the Earth in the GeV - TeV energy range are originated by Supernova Remnants (SNRs) and only a small fraction of secondary positrons and electrons comes from the interaction of CR nuclei with the interstellar medium (ISM). An additional electron + positron component peaked at ~ 1 TeV seems necessary for a consistent description of all the available data sets. The temptation to claim the discovery of dark matter from detection of electrons from annihilation of dark matter particles is strong but there are competing astrophysical sources, such as pulsars, that can give a strong flux of primary positrons and electrons (see [18] and references therein). At energies between 100 GeV and 1 TeV the electron flux reaching the Earth may be the sum of an almost homogeneous and isotropic component produced by Galactic supernova remnants and the local contribution of a few pulsars with the latter expected to contribute more and more significantly as the energy increases. If a single pulsar give the dominant contribution to the extra component a large anisotropy and a small bumpiness should be expected; if several pulsars contribute the opposite scenario is expected.

So far no positive detection of CRE anisotropy was reported by the Fermi-LAT collaboration, but some stringent upper limits were published [19] the pulsar scenario is still compatible with these upper limits.

Forthcoming experiments like AMS-02 and CALET are expected to reduce drastically the uncertainties on the propagation parameters by providing more accurate measurements of the spectra of the nuclear components of CR. Fermi-LAT and those experiments are also expected to provide more accurate measurements of the CRE spectrum and anisotropy looking for features which may give a clue of the nature of the extra component.

4. Conclusions

Fermi turned three years in orbit on June, 2011, and it is definitely living up to its expectations in terms of scientific results delivered to the com-munity. The mission is planned to continue at least two more years (likely more) with many remaining opportunities for discoveries.

5. Acknowledgments

The Fermi LAT Collaboration acknowledges support from a number of agencies and institutes for both development and the operation of the LAT as well as scientific data analysis. These include NASA and DOE in the United States, CEA/Irfu and IN2P3/CNRS in France, ASI and INFN in Italy, MEXT, KEK, and JAXA in Japan, and the K. A. Wallenberg Foun- dation, the Swedish Research Council and the National Space Board in Sweden. Additional support from INAF in Italy and CNES in France for science analysis during the operations phase is also gratefully acknowledged.

References

[1] Meegan, C.*et al.*, 662, 469 (2003)
[2] W.B.Atwood *et al.* [Fermi Coll.] ApJ 697 No 2 (2009 June 1) 1071-1102 [arXiv:0902.1089]
[3] A.Abdo *et al.* [Fermi Coll.] submitted to Astrophys. J. Suppl. Series [arXiv:1108.1435]
[4] A.Abdo *et al.*[Fermi Coll.] Astrophys. J. Suppl. Series 188, 405 (2010) [arXiv:1002.2280]
[5] A. Morselli *et al.*, Nucl.Phys. 113B (2002) 213
[6] A.Cesarini, F.Fucito, A.Lionetto, A.Morselli, P. Ullio, Astropart. Phys. 21 (2004) 267 [astro-ph/0305075]
[7] E. Baltz *et al.* , JCAP07 (2008) 013 [arXiv:0806.2911]
[8] V. Vitale and A. Morselli for the Fermi/LAT Collaboration, 2009 Fermi Symposium [arXiv:0912.3828]
[9] A. Morselli, B.Cañadas, V.Vitale, Il Nuovo Cimento Vol. 34 C, N. 3 (2011) [arXiv:1012.2292]
[10] A.Abdo *et al.* [Fermi Coll.] , Physical Review Letters 104, 091302 (2010) [arXiv:1001.4836].
[11] A.Abdo *et al.* [Fermi Coll.] ApJ 712 (2010) 147-158 [arXiv:1001.4531]
[12] M.Ackermann *et al.* [Fermi Coll.] 2011, PRL accepted [arXiv:1108.3546]
[13] A.A.Abdo*et al.* [Fermi Coll.], PRL 102, 181101 (2009) [arXiv:0905.0025]
[14] M.Ackermann *et al.* [Fermi Coll.], PRD 82, 092004 (2010) [arXiv:1008.3999]
[15] O.Adriani. *et al.* [PAMELA Coll.], *Physical Review Letters* **106**, p. 201101(May 2011).
[16] M. A. DuVernois *et al.*, [HEAT Coll.) ApJ 559, 296 (2001)
[17] M. Aguilar *et al.*, [AMS Coll.] Physics Reports 366, 331 (2002)
[18] D. Grasso, S. Profumo, A. W. Strong, L. Baldini, R. Bellazzini, E. D. Bloom, J. Bregeon, G. di Bernardo, D. Gaggero, N. Giglietto, T. Kamae, L. Latronico, F. Longo, M. N. Mazziotta, A. A. Moiseev, A. Morselli, J. F. Ormes, M. Pesce-Rollins, M. Pohl, M. Razzano, C. Sgro, G. Spandre and T. E. Stephens, *Astroparticle Physics* (2009) **32**, 140.
[19] M.Ackermann *et al.* [Fermi Coll.], Physical Review D 82, 092003 (2010) [arXiv:1008.5119]

Journal of Physics: Conference Series **384** (2012) 012003 doi:10.1088/1742-6596/384/1/012003

Some considerations on formalism for the spin-dependent WIMP-nucleus elastic scattering

Mirco Cannoni

Departamento de Física Aplicada, Facultad de Ciencias Experimentales, Universidad de Huelva, 21071 Huelva, Spain

E-mail: mirco.cannoni@dfa.uhu.es

Abstract. We review the formalism for spin-dependent WIMP-nucleus elastic scattering discussing the simplification obtained using the normalized spin structure functions and some applications.

1. Introduction

During the past year some of the numerous direct detection experiments have reported excess of events that can be indication of the interaction in the detectors of the weakly interacting massive particles (WIMP) that should form the halo of the galaxy.

In particular, the CoGENT collaboration [1] reported evidence for the annual modulation in the event rate thus confirming the long-standing evidence claimed in the last ten years by the DAMA collaboration [2]. The CRESST collaboration [3] reported an unexplained excess of events in their detector. On the other hand, the best fit areas of the three experiments in the plane $(m_\chi, \sigma^{\rm SI})$ –WIMP's mass, spin-independent (SI) WIMP–nucleon cross section– are not consistent between themselves and other experiments, CDMS [4], XENON100 [5], have reported upper limits that exclude their favoured parameter space. At present the experimental evidences are thus unclear and controversial.

Other experiments with light nuclei, for example COUPP [6], SIMPLE [7], are more sensitive to the spin-dependent (SD) scattering.

Here we review the simplification of the SD formalism obtained using the normalized spin structure functions and discuss its application in various aspects of the SD scattering [8, 9].

2. Simplifying the structure functions

The differential WIMP-nucleus cross section, as function of the recoil energy of the nucleus $E_R = q^2/2m_A$ being q the modulus of the momentum transfer, is given by

$$\frac{d\sigma_A^{\rm SD}}{dE_R} = \frac{m_A}{2\mu_A^2 v^2}\sigma_A(0)\Phi^{\rm SD}(E_R). \tag{1}$$

m_A is the mass of the nucleus with mass number A, μ_A the neutralino-nucleus reduced mass and v the relative velocity. The total cross section at $q = 0$ reads

$$\sigma_A^{\rm SD}(0) = \frac{\mu_A^2}{\pi}\, 4\, \frac{J+1}{J}\, |a_p\langle \mathbf{S}_p\rangle + a_n\langle \mathbf{S}_n\rangle|^2. \tag{2}$$

$\langle \mathbf{S}_{p,n} \rangle$ are the spin matrix elements of the proton and neutron groups calculated in a nuclear state with maximal projection of the ground state angular momentum, $\langle \mathbf{S}_{p,n} \rangle \equiv \langle J, M_J = J | S^z_{p,n} | J, M_J = J \rangle$. The function $\Phi^{\rm SD}(E_R)$ contains the spin structure functions that depend on the momentum transfer and is normalized as $\Phi^{\rm SD}(0) = 1$.

In the standard formalism introduced by Engel in Refs. [10, 11] we have

$$\Phi_E^{SD}(E_R) = \frac{S(E_R)}{S(0)}, \tag{3}$$

$$S(E_R) = a_0^2 S_{00}(E_R) + a_0 a_1 S_{01}(E_R) + a_1^2 S_{11}(E_R), \tag{4}$$

$$S(0) = \frac{2J+1}{\pi}\frac{J+1}{J}(a_p\langle \mathbf{S}_p\rangle + a_n\langle \mathbf{S}_n\rangle)^2 \tag{5}$$

Here $a_{0,1}$ are the neutralino-proton/neutron scattering amplitudes in the isospin representation and are related to the proton-neutron representation by $a_{p,n} = (a_0 \pm a_1)/2$. At $q = 0$ $S_{ij}(0) \neq 1$, the functions are not normalized to one and the function S_{01} for some nuclei can be negative. Furthermore $\Phi_E^{SD}(E_R)$ depends not only from nuclear physics but also on particle physics. This situation is different from the spin-independent cross section where $\Phi_E^{SI}(q) = F^2(q^2)$ is the nuclear form-factor.

An alternative formalism was introduced by Vergados and collaborators [12]. In this framework we can write

$$\Phi_V^{SD}(E_R) = \frac{\mathcal{F}(E_R)}{\mathcal{F}(0)}, \tag{6}$$

$$\mathcal{F}(E_R) = a_0^2 F_{00}(E_R) + 2a_0 a_1 F_{01}(E_R) + a_1^2 F_{11}(E_R), \tag{7}$$

$$\mathcal{F}(0) = a_0^2 + 2a_0 a_1 + a_1^2. \tag{8}$$

In this case $F_{ij}(0) = 1$ by construction.

In Refs. [8, 9] it is shown that the two formalisms are equivalent and connected by

$$F_{ij}(E_R) = \frac{S_{ij}(E_R)}{S_{ij}(0)}, \tag{9}$$

thus if the S_{ij} are known also the F_{ij} are known and vice versa.

In Fig. 1, upper panels, from left to right, we show the functions $F_{ij}(E_R)$ for one light nucleus, ^{19}F, one medium-heavy, ^{73}Ge, and one heavy nucleus, ^{127}I, all of them employed in numerous experiments. In the abscissas we use the dimensionless variable $y = (qb/2)^2$ where $b = 1\,\text{fm}\,A^{1/6}$ is the oscillator size parameter appearing in shell-model calculations using harmonic oscillator wave functions. The functional form of S_{ij} and F_{ij} is typically a polynomial times an exponential in y. The function F_{ij} for ^{19}F are taken from Ref. [12], the functions for ^{73}Ge are from Ref. [13], for ^{127}I from Ref. [14] (set calculated with the Bonn A potential). The functions are practically identical in the recoil energy interval of interest for experiments,

$$F_{00}(E_R) \simeq F_{01}(E_R) \simeq F_{11}(E_R). \tag{10}$$

Thanks to Eq. (10), Eq. (8) reduces to

$$\Phi_V^{SD}(E_R) = F_{11}(E_R). \tag{11}$$

Hence the SD "form factor" is determined by only one SSF. It does not depend on particles physics.

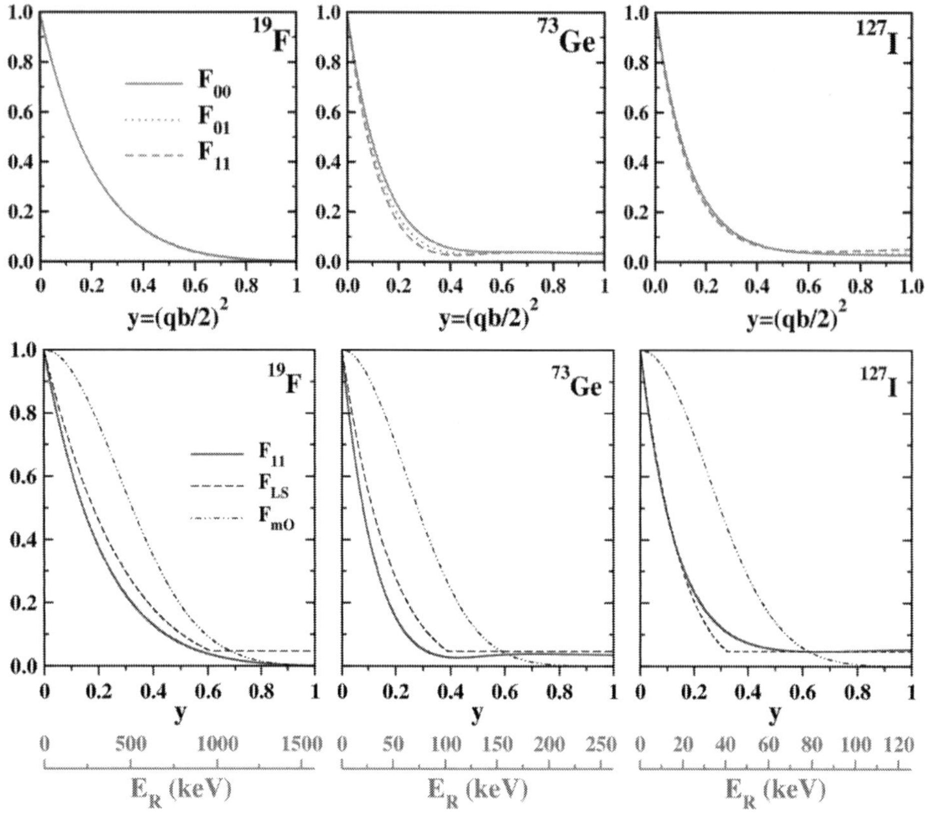

Figure 1. Top panels, from left to right: the momentum transfer dependent spin structure functions F_{ij} for the nuclei ^{19}F, ^{73}Ge, ^{127}I. Bottom panels, from left to right: blue line the normalized structure function $F_{11} = S_{11}(q)/S_{11}(0)$, for the nuclei ^{19}F, ^{73}Ge and ^{127}I. The dashed line refers to the parametrization of Eq. (12) and the dashed-dotted line to the parametrization of Eq. (13). See text for details.

This simplification of the formalism is largely overlooked in literature. In some cases phenomenological parametrizations are used. One is the parametrization of Lewin and Smith given in [15, 16]

$$F_{\rm LS}(qr_n) = \begin{cases} \left(\frac{\sin(qr_n)}{qr_n}\right)^2 & qr_n < 2.55,\ qr_n > 4.5, \\ 0.047 & 2.55 \leq qr_n \leq 4.5, \end{cases} \tag{12}$$

with the nuclear radius $r_n \simeq 1.0A^{1/3}$ fm, or Gaussian parametrizations like the one used in the code micrOMEGAs [17] for the case of nuclei for which the S_{ij} are not available:

$$F_{\rm mO} = \frac{S_{ij}(q)}{S_{ij}(0)} = \exp\left(-\frac{q^2 R_A^2}{4}\right), \tag{13}$$

where $R_A = 1.7A^{1/3} - 0.28 - 0.78(A^{1/3} - 3.8 + [(A^{1/3} - 3.8)^2 + 0.2]^{1/2})$. A different Gaussian parametrization is given for example in Ref. [18]. Figure 1, bottom panels, shows the normalized SSF F_{11}, $F_{\rm LS}$ and $F_{\rm mO}$. The approximation furnished by $F_{\rm LS}$ is reasonable both at low recoil energies and at higher energies in the region of the plateau, especially for the heavy nucleus, while the approximation furnished by $F_{\rm mO}$ is much worse in all the cases.

Journal of Physics: Conference Series **384** (2012) 012003 doi:10.1088/1742-6596/384/1/012003

3. The problem of setting model independent upper limits

In supersymmetric models where the WIMP is the lightest neutralino the SI proton and neutron cross sections are to a very good approximation equal and the WIMP-nucleus cross section is determined only by one WIMP-nucleon cross section $\sigma^{\rm SI}$. Setting constrains on the SD couplings is complicated by the fact that the two elementary cross sections, WIMP-proton and WIMP-neutron, can differ up to 30-40% and in principle both can contribute.

The origin of the problems is the fact that in the standrd formula for the neutralino-nucleus cross section, Eqs. (6-8) the particle physics degrees of freedom $a_{0,1}$ are not factorized from the momentum dependent spin structure functions S_{ij}, thus when setting an upper limit one is forced to fix the neutralino "composition" by the ratio of the couplings a_0/a_1 or equivalently a_p/a_n. A way to avoid this problem was discussed in Ref. [19]. These assumptions and the method were anyway criticized in in Refs. [21, 22]. With the formalism of Section 2 we can gain some new insight.

The total cross section for a single nucleon is $\sigma^{SD}_{p,n} = 3\frac{\mu_p^2}{\pi}|a_{p,n}|^2$, thus

$$\sigma^{\rm SD}_A(0) = \left(\frac{\mu_A}{\mu_p}\right)^2 \frac{4}{3}\frac{J+1}{J}\left(\langle \mathbf{S}_p\rangle\sqrt{\sigma^{\rm SD}_p} + \varrho\langle \mathbf{S}_n\rangle\sqrt{\sigma^{\rm SD}_n}\right)^2. \tag{14}$$

In general both the SD WIMP-nucleon scattering amplitudes a_p and a_n and the spin matrix elements can have opposite sign, hence $\varrho = \pm 1$ is the relative sign between $|a_p\langle \mathbf{S}_p\rangle|$ and $|a_n\langle \mathbf{S}_n\rangle|$. The case of more general phases is treated in Ref. [20]. We introduce the factors

$$\phi_A = \frac{\rho_0 v_0}{m_\chi m_A}, \quad (\mathcal{C}^{p,n}_A)^2 = \left(\frac{\mu_A}{\mu_p}\right)^2 \frac{4}{3}\frac{J+1}{J}\langle \mathbf{S}_{p,n}\rangle, \quad t^{\rm SD}_A = \int \frac{dE_R}{\epsilon_0} F_{11}(E_R)\int \frac{dv}{v} f(v), \tag{15}$$

with ρ_0 the local dark matter density, v_0 the circular velocity of the Sun and $t^{\rm SD}_A$ a suitably normalized dimensionless integral over the velocity distribution and the recoil energy, see for details Ref. [9]. The expression for the total event rate thus becomes

$$R^{\rm SD} = \phi_A\left(\mathcal{C}^p_A\sqrt{\sigma^{\rm SD}_p} \pm \mathcal{C}^n_A\sqrt{\sigma^{\rm SD}_n}\right)^2 t^{\rm SD}_A. \tag{16}$$

If an experiment with exposure $\mathcal{E}_A$ set an upper limit at some confidence level on the number of events, N^{UL}, then this is converted in an upper limit on the cross section requiring $R_A \times \mathcal{E}_A < N^{UL}$,

$$\left(\mathcal{C}^p_A\sqrt{\sigma^{\rm SD}_p} \pm \mathcal{C}^n_A\sqrt{\sigma^{\rm SD}_n}\right)^2 < \frac{N^{UL}}{\phi_A t^{\rm SD}_A \mathcal{E}_A}. \tag{17}$$

The right-hand side of (17) is by definition the experimental upper limit on the neutralino-nucleus SD cross section. As in Ref. [19], we call it $\sigma^{\rm lim}_A$ and introduce the rescaled limts per nucleon $\sigma^{\rm lim(A)}_{p,n}$

$$\sigma^{\rm lim}_A \equiv \frac{N^{UL}}{\phi_A t^{\rm SD}_A \mathcal{E}_A}, \tag{18}$$

$$\sigma^{\rm lim(A)}_{p,n} \equiv \frac{\sigma^{\rm lim}_A}{(\mathcal{C}^{p,n}_A)^2}. \tag{19}$$

Dividing both members of (17) by (18) and using the quantities (19) we arrive at

$$\left(\frac{\sqrt{\sigma^{\rm SD}_p}}{\sqrt{\sigma^{\rm lim(A)}_p}} \pm \frac{\sqrt{\sigma^{\rm SD}_n}}{\sqrt{\sigma^{\rm lim(A)}_n}}\right)^2 < 1, \tag{20}$$

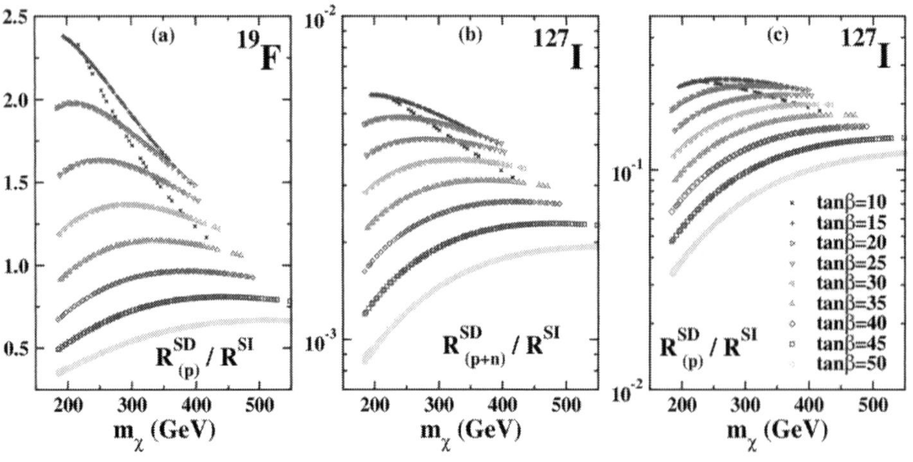

Figure 2. Ratio of the spin-dependent total event rate over the spin-independent rate $R^{\mathrm{SD}}/R^{\mathrm{SI}}$ varying $\tan\beta$ in the stau co-annihilation region of the CMSSM with $A_0 = 0$ and $\mu > 0$. In panel (a) for ^{19}F. In panel (b), the ratio is plotted for ^{127}Itaking into account both the protons and neutrons contribution in the spin-dependent rate; in panel (c) only the proton contribution is included.

that is exactly Eq. (13) of Ref. [19] in the case of the allowed region in (σ_p, σ_n) plane.

In Ref. [19] the nucleon cross section limits in Eq. (19) are defined as basic quantities that then are combined to give Eq. (20). To do this it is necessary to assume that for a given nucleus it is possible to set separately limits on the SD-proton and SD-neutron cross sections even in the case that one contribution is clearly sub-dominant. In reality our derivation shows that such hypothesis are unnecessary and that the full justification of Eq. (20) only relies on the fact that as the spin-dependent form factor one can take the F_{11} instead of Eq. (8). This function, included in the factor t^{SD}, also provide the correct momentum dependent behaviour of the upper limit $\sigma^{\mathrm{lim(A)}}$.

3.1. The meaning of the limits on the nucleon cross sections

At this stage the "upper limits" on the single proton or neutron cross sections, Eq. (19), must be considered only as useful quantities introduced to write Eq. (17) in the compact form (20). They become the actual experimental upper limits if, for the nucleus from which these are determined and in a specific WIMP model, one can prove that the protons contribution is dominant over the neutrons contribution or vice-versa (given the dominance of the SD rate over the SI rate).

To appreciate this point let us consider the following framework: a light nucleus, ^{19}F, (employed in COUPP, PICASSO, SIMPLE for example) a heavy nucleus, ^{127}I, (COUPP, DAMA, KIMS) and the stau co-annihilation region of the constrained minimal supersymmetric standard model [9].

The first nuclear shell-model calculation for ^{19}F [23] found $\langle \mathbf{S}_p \rangle = 0.441$, $\langle \mathbf{S}_n \rangle = -0.109$. The successive calculation of Ref. [12] using a more realistic interaction, found $\langle \mathbf{S}_p \rangle = 0.4751$ and $\langle \mathbf{S}_n \rangle = -0.0087$. The protons contribution is thus similar but the neutrons contribution is clearly negligible. In the case of ^{127}I, although proton favouring, the neutrons group contribution to the nuclear spin is of the same order of magnitude. If the neutralino couplings to the proton and neutron are similar, the neutrons contribution to the nuclear spin must be considered.

We show the ratio $R^{\mathrm{SD}}/R^{\mathrm{SI}}$ for fluorine in Fig. 2(a). The SD rate is bigger by a factor up more than 2 at low and medium $\tan\beta$ but it is smaller than the SI rate at large $\tan\beta$; in any

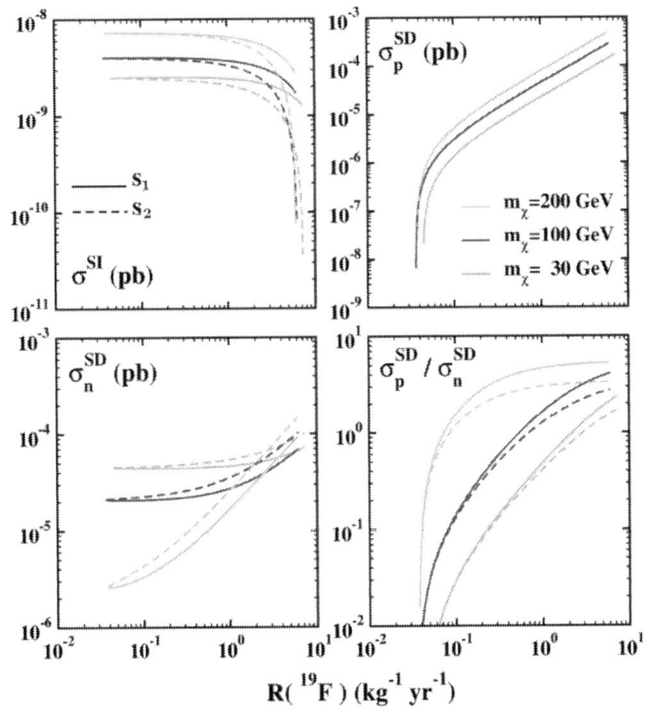

Figure 3. The solutions of the system Eq. (21) for the nuclei ^{127}I, ^{73}Ge, ^{19}F as a function of $R(^{19}\mathrm{F})$: s_+, solid lines, and s_-, dashed lines. We fix $R(^{127}\mathrm{I}) = R(^{73}\mathrm{Ge}) = 1$ kg^{-1} yr^{-1} and three values of the neutralino mass.

case the two rates are always of the same order of magnitude. The SI rate cannot be completely neglected at high $\tan\beta$ and for lower $\tan\beta$, neglecting it, one underestimates the total rate.

Fig. 2(c) shows the ratio of $R^{\mathrm{SD}}/R^{\mathrm{SI}}$ in ^{127}Ionly considering the proton contribution: due to the A^2 proportionality, the SI rate always dominate by a factor from 4 to 25. On the other hand in Fig. 2(b) both are included: given that $a_p < 0$ and $a_n > 0$, a cancellation in the SD rate is expected because the products $a_p\langle\mathbf{S}_p\rangle$ and $a_n\langle\mathbf{S}_n\rangle$ are of the same order and have opposite sign. This makes the SD rate from 2 to 3 orders of magnitude smaller than the SI.

In the case of the specific WIMP model examined, hence, iodine can only constrain the SI interaction, while the exclusion plots in the (m_χ, σ_p^{SD}) from fluorine are inaccurate In this case one has to draw an exclusion plot in the (σ_p^{SD}, σ^{SI}) plane for each fixed mass, the so-called mixed coupling approach [24].

4. Extracting the cross sections from total events rate

The simplified formalism allows to analytically determine the three elementary cross sections and connect the solutions to the relative sign between the proton and the neutron spin scattering amplitudes once the measurements of total event rate from three appropriate targets become available.

Consider again the three nuclei already discussed. Neglecting the neutron contribution in ^{19}F and the proton contribution in ^{127}I, and keeping both for ^{127}I, we can write a system of three equations using the expression for the total rate in the following form:

$$\begin{cases} \sigma^{SI} + \mathcal{R}_{\rm I}\left(\Omega_p^{\rm I}\sqrt{\sigma_p^{SD}} + \varrho\Omega_n^{\rm I}\sqrt{\sigma_n^{SD}}\right)^2 - \mathcal{S}_{\rm I} = 0 \\ \sigma^{SI} + \mathcal{R}_F(\Omega_p^{\rm F})^2\sigma_p^{SD} - \mathcal{S}_{\rm F} = 0 \\ \sigma^{SI} + \mathcal{R}_{\rm Ge}(\Omega_n^{\rm Ge})^2\sigma_n^{SD} - \mathcal{S}_{\rm Ge} = 0. \end{cases} \tag{21}$$

We refer the reader to Ref. [8] for meaning and the definitions of the various factors in these expressions. Here it is sufficient to note that system is non-linear and that there exist two sets of solutions, indicated as s_1 and s_2, each set being associated with one sign of ϱ. We find numerically that, for the chosen nuclei s_1 (s_2) is the set of solutions associated to $\varrho = -1$ ($\varrho = +1$). As discussed in Section 3, in the stau coannihilation region, a_p is negative and a_n is positive, the fundamental cross sections are thus provided by the set s_1. An example is given in Fig. 3 where we fix $R(^{127}\text{I}) = R(^{73}\text{Ge}) = 1$ kg^{-1} yr^{-1} and three values of the neutralino mass and plot the extracted cross sections as a function of the rate in fluorine.

Acknowledgments

The author acknowledges J. D. Vergados and M. E. Gomez for discussions and collaboration. Support by MultiDark, grant CSD2009-00064 of the Spanish MICINN project Consolider-Ingenio 2010 Programme and partial support from MICINN project FPA2009-10773, MICINN-INFN(PG21)FPA2009-10773 and from Junta de Andalucia under grant P07FQM02962 are acknowledged.

References

[1] Aalseth C E *et al.* (CoGeNT Collaboration) 2011 *Phys. Rev. Lett.* **07** 141301 (*Preprint* 1106.0650)
[2] Bernabei R *et al.* (DAMA/LIBRA Collaboration) 2010 *Eur. Phys. J.* C **67** 39 (*Preprint* 1002.1028)
[3] Angloher G *et al.* (CRESST II Collaboration) 2011 Results from 730 kg days of the CRESST-II Dark Matter Search (*Preprint* 1109.0702)
[4] Ahmed Z *et al.* (CDMS-II Collaboration) 2011 *Phys. Rev. Lett.* **106** 131302 (*Preprint* 1011.2482)
[5] Aprile E *et al.* (XENON100 Collaboration) 2011 *Phys. Rev. Lett.* **107** 131302 (*Preprint* 1104.2549)
[6] Behnke E *et al.* (COUPP Collaboration) 2011 *Phys. Rev. Lett.* **106** 021303 (*Preprint* 1008.3518)
[7] Felizardo M *et al.* (SIMPLE Collaboration) Final analysis and results of the phase II SIMPLE dark matter search (*Preprint* 1106.3014)
[8] Cannoni M, Vergados J D and Gomez M E 2011 *Phys. Rev.* D **83** 075010 (*Preprint* 1011.6108)
[9] Cannoni M 2011 *Phys. Rev.* D **84** 095017 (*Preprint* 1108.4337)
[10] Engel J 1991 *Phys. Lett.* B **264** 114
[11] Engel J, Pittel S and Vogel P 1992 *Int. J. Mod. Phys.* E **1** 1
[12] Divari P C, Kosmas T S, Vergados J D and Skouras L D 2000 *Phys. Rev.* C **61** 054612
[13] Dimitrov V I, Engel J and Pittel S 1995 *Phys. Rev.* D **51**, 291 (*Preprint* hep-ph/9408246)
[14] Ressell M T and Dean D J 1997 *Phys. Rev.* C **56** 535 (*Preprint* hep-ph/9702290)
[15] Smith P F *et al.* 1996 *Phys. Lett.* B **379** 299
[16] Lewin J D and Smith P F 1996 *Astropart. Phys.* **6** 87
[17] Belanger G, Boudjema F, Pukhov A and Semenov A 2009 *Comput. Phys. Commun.* **180** 747 (*Preprint* 0803.2360)
[18] Gondolo P 1996 Phenomenological introduction to direct dark matter detection *Preprint* hep-ph/9605290
[19] Tovey D R, Gaitskell R J, Gondolo P, Ramachers Y A and Roszkowski L 2000 *Phys. Lett.* B **488** 17 (*Preprint* hep-ph/0005041)
[20] Vergados J D 2010 *Nucl. Phys.* B **829** 383 (*Preprint* 0907.3587)
[21] Bednyakov V A 2004 *Phys. Atom. Nucl.* **67** 1931 (*Preprint* hep-ph/0310041); Bednyakov V A and Klapdor-Kleingrothaus H V 2009 *Phys. Part. Nucl.* **40** 583 (*Preprint* 0806.3917)
[22] Bednyakov V A and Simkovic F 2005 *Phys. Part. Nucl.* **36** 131 (*Preprint* hep-ph/0406218); 2006 *ibid.* **37** S106 (*Preprint* hep-ph/0608097)
[23] Pacheco A F and Strottman D 1989 *Phys. Rev.* D **40** 2131
[24] Bernabei R *et al.* 2001 *Phys. Lett.* B **509** 197

7th International Workshop on the Dark Side of The Universe (DSU 2011) IOP Publishing
Journal of Physics: Conference Series **384** (2012) 012004 doi:10.1088/1742-6596/384/1/012004

Fermi-LAT prospects for the detection of $\mu\nu$SSM gravitino Dark Matter

Carlos Muñoz

Departamento de Física Teórica and Instituto de Física Teórica UAM/CSIC, Universidad Autónoma de Madrid, Cantoblanco, E-28049, Madrid, Spain

E-mail: carlos.munnoz@uam.es

Abstract. The $\mu\nu$SSM is a supersymmetric model that has been proposed to solve the problems generated by other supersymmetric extensions of the standard model of particle physics. Given that R-parity is broken in the $\mu\nu$SSM, the gravitino is a natural candidate for decaying dark matter since its lifetime becomes much longer than the age of the Universe. In this model, gravitino dark matter could be detectable through the emission of a monochromatic gamma ray in a two-body decay. We study the prospects of the Fermi-LAT telescope to detect such monochromatic lines in 5 years of observations of the most massive nearby extragalactic objects. The dark matter halo around the Virgo galaxy cluster is selected as a reference case, since it is associated to a particularly high signal-to-noise ratio and is located in a region scarcely affected by the astrophysical diffuse emission from the galactic plane. The simulation of both signal and background gamma-ray events is carried out with the Fermi Science Tools, and the dark matter distribution around Virgo is taken from a N-body simulation of the nearby extragalactic Universe, with constrained initial conditions provided by the CLUES project. We find that a gravitino with a mass range of 0.6–2 GeV, and with a lifetime range of about 3×10^{27}–2×10^{28} s would be detectable by the Fermi-LAT with a signal-to-noise ratio larger than 3. We also obtain that gravitino masses larger than about 4 GeV are already excluded in the $\mu\nu$SSM by Fermi-LAT data of the galactic halo.

1. Introduction

Evidence indicating the presence of dark matter (DM) can be obtained at very different scales: from cosmological ones through the analysis of the angular anisotropies in the cosmic microwave background radiation [1], down to galactic scales considering lensing and galaxy dynamics studies. However, all these evidences are not able to provide us with complete information on the nature of DM, beyond the fact that it has to be mainly non-baryonic. Since, within the standard model of particle physics there are no viable non-baryonic candidates, the existence of DM represents one of the most compelling evidences for physics beyond the standard model [2].

The "μ from ν" Supersymmetric Standard Model ($\mu\nu$SSM) was proposed in the literature to solve the so called μ-problem and explain the origin of neutrino masses by simply introducing right-handed neutrinos ν [3, 4]. It is an interesting model that can be tested at the LHC and, as a consequence, its phenomenology has been analized in detail recently [5, 6, 7]. In the $\mu\nu$SSM, R-parity is broken and therefore the lightest supersymmetric particle (LSP) decays. Thus, neutralinos [2] or sneutrinos [8], having very short lifetimes, are no longer viable candidates for the DM of the Universe. Nevertheless, if the role of the LSP is played by the gravitino, $\Psi_{3/2}$, its decay is suppressed both by the feebleness of the gravitational interaction and by the small

R-parity violating coupling. As a consequence, its lifetime can be much longer than the age of Universe and the $\mu\nu$SSM gravitino can represent a good DM candidate [6, 7].

Since the gravitino decays producing a monochromatic photon with an energy equal to half of the gravitino mass, its presence can, in principle, be inferred indirectly from the data of the Fermi gamma-ray space telescope [6, 7]. Fermi was launched on June 11th, 2008, and its main instrument, the Large Area Telescope (LAT), covers an energy range from roughly 20 MeV to 300 GeV with an angular resolution of ~ 0.15 degrees at 68% containment above 10 GeV [9].

In the DM halo of the Milky Way the gamma-ray flux coming from DM decay is maximized in the direction of the galactic center where the DM density is larger. This region, however, should be considered with particular care since the gamma-ray emission due to the presence of conventional astrophysical sources is not fully understood [10].

The expected diffuse gamma-ray emission from DM decay in the mid-latitude range ($10^\circ \leq |b| \leq 20^\circ$) was computed for a Navarro-Frenk-White (NFW) profile [11] in Ref. [6] and compared with the 5-month measurement reported by Fermi-LAT [12]. The non-observation of sharp monochromatic lines in the gamma-ray spectrum permitted to draw bounds on the parameter space of the $\mu\nu$SSM gravitino. In particular, values of gravitino mass $m_{3/2}$ larger than about 10 GeV were excluded, as well as lifetimes $\tau_{3/2}$ smaller than about 3 to 5×10^{27} s. Notice that because of this upper bound on $m_{3/2}$, three body decay modes of the gravitino [13, 14] are not relevant, and therefore we do not considered them in this work.

It is worth noticing here that in Ref. [15] ([16]) the Fermi-LAT collaboration presented constraints on monochromatic emission using 11 (23) months of data for $|b| > 10^\circ$ *plus* a $20^\circ \times 20^\circ$ square around the galactic center. However, the derived limits only refer to the emission above 30 (7) GeV, covering, in the context of the $\mu\nu$SSM, gravitinos with masses larger than 60 (14) GeV, leaving our region of interest unconstrained. On the other hand, in work [17], two-years Fermi-LAT data for $|b| \geq 10^\circ$ have been used to constrain the DM gamma-ray line flux in the energy range between 1 and 300 GeV. Lower bounds on $\tau_{3/2}$ of about 5×10^{28} s were obtained in our region of interest below 10 GeV. Recently, these bounds together with those obtained in [18] by analyzing the data from EGRET were used in [19] to constrain the parameter space of gravitino dark matter in the bilinear R-parity violating model.

We know from X-ray and lensing studies [20, 21] that the main mass component of structures like galaxy clusters is DM. Such objects are, therefore, promising targets for DM searches. The main goal of this talk based on work [7] is to explore the prospects for detecting $\mu\nu$SSM gravitino DM in galaxy clusters using the prediction for 5 years of operation of the Fermi-LAT telescope, and taking into account the above mentioned bounds. Although in a very recent work [22] based on three years of Fermi-LAT gamma-ray data, the flux coming from nearby clusters was analyzed, the relevant bounds obtained on the lifetime correspond to gravitino masses larger than 250 GeV.

In our analysis we will use the following strategy. The DM density field of the nearby extragalactic Universe is described using the maps provided in Ref. [23], and based on a constrained N-body simulation provided by the CLUES project [24]. This density is then taken as an input for the Fermi observation simulation tool to predict the photon signal. We will use the most recent version of the public Fermi Science Tools [25] to describe the performance of the telescope and to simulate both the DM signal and the astrophysical background. From the analysis of the simulated photon maps we finally compute our prospects of detection for the $\mu\nu$SSM gravitino.

The paper is organized as follows. In Section 2, the flux of gamma rays from $\mu\nu$SSM gravitino decay is discussed. In Section 3, the cosmological simulation used to infer the DM distribution in the Local Universe is briefly described, while Section 4 is devoted to summarize the basic of the simulation technique implemented in the Fermi Science Tools. Finally, the prospects for DM detection are derived in Section 5, and the conclusions are left for Section 6.

2. Gamma-rays from gravitino decay in the $\mu\nu$SSM

In the supergravity Lagrangian an interaction term is predicted between the gravitino, the field strength for the photon, and the photino. Since, due to the breaking of R-parity, the photino and the left-handed neutrinos are mixed, the gravitino will be able to decay through the interaction term into a photon and a neutrino [26]. The gravitino lifetime $\tau_{3/2}$ results to be:

$$\tau_{3/2} \simeq 3.8 \times 10^{27}\text{s} \left(\frac{|U_{\tilde{\gamma}\nu}|^2}{10^{-16}}\right)^{-1} \left(\frac{m_{3/2}}{10\text{ GeV}}\right)^{-3}, \tag{1}$$

where $|U_{\tilde{\gamma}\nu}|^2$ is the photino content of the neutrino, and is constrained to be $|U_{\tilde{\gamma}\nu}|^2 \sim 10^{-16} - 10^{-12}$ in the $\mu\nu$SSM, in order to reproduce neutrino masses [6]. As a consequence, the gravitino will be very long lived. Additionally, adjusting the reheating temperature one can reproduce the correct relic density for each possible value of the gravitino mass (see [6] and references therein).

The detection of DM in several R-parity breaking scenarios has been studied in the literature [26, 27, 6] considering the case of gravitinos emitting gamma-rays when decaying in *i*) the smooth galactic halo, and *ii*) extragalactic regions at cosmological distances. As mentioned in the Introduction, in this talk we will analyze the detection of DM in the $\mu\nu$SSM considering the contribution of gamma-rays coming from *iii*) nearby extragalactic structures [7].

In *i*), the gamma-ray signal is an anisotropic sharp line and the flux is given by

$$\frac{d\Phi}{dE}(E) = \frac{\delta(E - \frac{m_{3/2}}{2})}{4\pi\tau_{3/2}m_{3/2}} \int_{\text{los}} \rho_{halo}(\vec{l})d\vec{l}\,, \tag{2}$$

where the halo DM density is integrated along the line of sight l, and we will use a NFW density profile for the Milky Way halo compatible with the latest observational constraints as modeled in [28]. Let us remark, nevertheless, that since the region that we will study below does not include the inner galaxy, any density profile will give rise to similar results.

On the other hand, in *ii*), the photons produced by gravitinos decaying at cosmological distances are red-shifted during their journey to the observer, and we obtain the isotropic extragalactic flux applying the analysis of Refs. [26, 29] to the $\mu\nu$SSM. As can be seen e.g. in Figs. 3 and 4 of Ref. [6], the sharp line produced by the galactic halo dominates over this extragalactic signal.

Finally, for nearby extragalactic structures *iii*), the gamma-ray signal is a monochromatic line similarly to *i*), and Eq. (2) can also be used for the computation of the flux. Actually the contribution from the smooth galactic DM halo is practically isotropic in the region around a particular nearby extragalactic structure (at least at high latitudes) and less important than the contribution from the object itself. Moreover at high latitudes, the galactic foreground is smaller than near the galactic center, so that objects with a lower DM-induced gamma-ray flux can potentially be associated to larger prospects of detection than the region near the galactic center. Thus, the study of the extragalactic density field is something worth carrying out. In our work this is taken from the results of Ref. [23], where an N-body simulation with constrained initial condition was used. We will review the most important features of this simulation in the next Section.

3. CLUES simulations

CLUES (Constrained Local UniversE Simulations) [24] N-body simulations aim at describing the formation and evolution of DM halos in a way to reproduce, as precise as possible, our Local Universe. To this goal, constrained initial conditions are set up using the information from radial and peculiar velocities of galaxies from astrophysical catalogs, together with the determination of the masses of the galaxy clusters detected in X-rays [30, 31].

In Ref. [23], a CLUES simulation was used to compute projected Cartesian grid all-sky maps of the DM distribution in the Local Universe. In particular, the Box160CR simulation was used, containing 1024^3 particles in a box with a side of 160 h^{-1} [23, 30]. The characteristics of the most massive clusters such as Virgo, Coma and Perseus, together with the Great Attractor, are well reproduced compared to the real objects, apart from a typical mismatch around 5 Mpc h^{-1} in their position. The all-sky maps are available at Ref. [32] both in the case of an annihilating and a decaying DM particle. Here we use the map corresponding to the case of decaying DM. The gamma-ray flux can be derived from the values in the map simply multiplying them by the particle physics factor shown outside the integral in Eq. (2). This result is used as input for the Fermi-LAT observation simulations, which we will describe in the next section.

4. Simulations with the Fermi Science Tools

For this work, the simulation of gamma-ray events was carried out with the `gtobssim` routine, part of the Fermi Science Tools package v9r23p1. Its output is a list of mock gamma-ray events with corresponding spatial direction, arrival time and energy, distributed according to an input source model. Our source model accounts for the gamma-ray signal from $\mu\nu$SSM gravitino decay as described in previous sections, and for the galactic and extragalactic background diffuse emission. In particular, the galactic background emission is mainly correlated with structures in the Milky Way since it arises from the interaction of high-energy cosmic rays with the interstellar medium and the interstellar radiation field. The far extragalactic background, on the contrary, is supposed to be almost isotropic. Its value is based on the modelization of the galactic component, on detected Fermi-LAT sources, and on the solar gamma-ray emission. We used the so-called RING model [33] as recommended by the Fermi-LAT collaboration, which is obtained as a fit to the real Fermi-LAT data.

The simulation of gamma-ray events through `gtobssim` is based on in-flight Instrument Response Functions (IRFs), accounting for the telescope effective area, energy dispersion and point-spread function (PSF). Two IRFs publicly available are called P6_V3_DIFFUSE and P6_V3_DATACLEAN. They both profit from the improvement in the knowledge of the telescope performances after the first two years of data taking [34]. The main difference between the two IRFs is the fact that DATACLEAN event selection perform most stringent cuts than DIFFUSE on the interpretation of an event as a real photon. As will be discussed below, in this work we present results from P6_V3_DIFFUSE, but the sensitivity of our results on the choice of the IRFs have been checked without finding any significant effect.

5. $\mu\nu$SSM gravitino dark matter: prospects of detection with Fermi-LAT

Let us finally study the prospects for $\mu\nu$SSM gravitino DM detection, taking into account the contributions discussed in Sect. 2. First, we simulate the all-sky map of 5-years Fermi-LAT observations, separately for DM gamma-ray events and background events. For DM, as an example we show in Fig. 1 from [7] the case of a gravitino with mass $m_{3/2}$ = 8 GeV and lifetime $\tau_{3/2} = 2.5 \times 10^{26}$ s. Gamma-rays in the energy range between 3.4 GeV and 4.6 GeV are simulated. This energy range corresponds to an interval of $\pm 2\ \Delta E$ around the position of the line (4 GeV), where ΔE=0.3 GeV is the energy resolution (at 68% containment normal incidence) of the telescope at that energy.

Since we want to determine the Fermi-LAT capability of detecting local extragalactic DM structures, the gamma-ray emission from DM is what we will refer to as *signal*. For each direction of the sky ψ, the number of signal (S) and background (B) photons are determined integrating over a 3×3 degrees region centered in ψ. The S/N ratio is then defined as $S/\sqrt{S+B}$. From the map we can infer which extragalactic structure is the best target to derive our prospects

Journal of Physics: Conference Series **384** (2012) 012004 doi:10.1088/1742-6596/384/1/012004

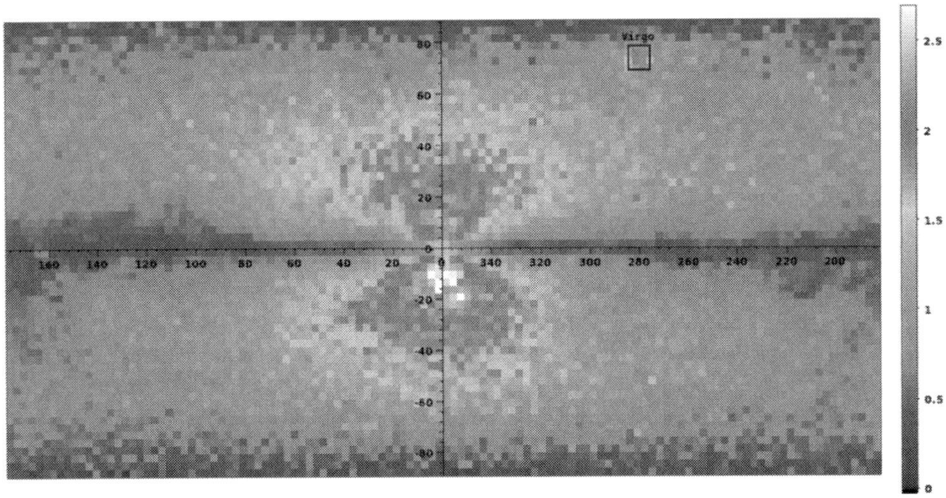

Figure 1. S/N all-sky map of gamma-ray emission from gravitino DM decay simulated using the Fermi Science Tools. DM events are from the decay of a $\mu\nu$SSM gravitino with $m_{3/2} = 8$ GeV and $\tau_{3/2} = 2.5 \times 10^{26}$ s. The signal comes from all three DM components: galactic halo, and both, local (from CLUES simulation) and far (isotropic) extragalactic DM density. The black square indicates the position of the Virgo cluster. Pixels in the map have an angular dimension of 3×3 degrees.

for detecting gravitino DM. Let us remark first that we are not simulating the contribution of the point sources already detected by Fermi-LAT and present in the 2 year catalog [35]. Since this contribution represents a source of uncertainty, the S/N ratio should not be considered as a good estimator of the prospects for detecting DM in the regions where the contamination of the point sources is dominant. Also the discover in the Fermi-LAT data of lobes structures [36], extending up to 50 degrees above and below the galactic plane introduce a new source of background not included in this work. The inner galaxy region is characterized by a quite large S/N ratio (around 3). However, contamination from point sources as well as the poorly-understood Fermi lobes is known to be large at low galactic latitudes, significantly affecting the reliability of our predictions in this region. It is worth noticing here that, as mentioned in the Introduction, in [15, 16] the region $|b| \geq 10°$ *plus* a $20° \times 20°$ square around the galactic center was used as target. Let us remark that these works follow a different kind of analysis, searching spectral deviations from a power-law behavior. We note that in such a large region of the sky the energy spectra used as background can be well described as a power law. Therefore, these works have no problems concerning background uncertainties.

Thus we neglect the zone with large S/N close to the galactic center and plane, and focus only on the case of extragalactic sources. Among these, the object with the largest S/N ratio ($S/N = 2.1$) is the Virgo cluster. In our analysis we only select this cluster, indicated with a black square in Fig.1. Although in [23] filamentary regions of the cosmic web were pointed out as good targets for DM decay searches, in this analysis, where we are using a different particle physics model (and energy range), we do not find in principle a significant S/N ratio in those regions.

As described in Ref. [23], the projected map is generated with Virgo being fixed in its real observed position because it is the best constrained object in the simulation, and therefore it is certainly the best object to consider for our purposes. Besides, there are two point sources detected near Virgo, M87 and 2E1228+1437, with an integrated flux of $(3.3 \pm 0.8) \times 10^{-10}$

ph/cm^{2}s and $(1.6 \pm 0.6) \times 10^{-10}$ ph/cm^{2}s between 3 and 10 GeV, respectively [35]. In a 5×5 degree region around Virgo and free of the two point sources, the total gamma-ray flux from DM decay is 2.5×10^{-9} ph/cm^{2}s (for an example of a 8 GeV gravitino with a lifetime of 5×10^{27}s), one order of magnitude larger than the contribution of the point sources. Smaller regions like 1×1 or 3×3 degrees are more affected by the emission due to the point sources. On the other hand, going to larger regions like 7×7 degrees, the signal contribution is not significantly increasing and therefore the S/N ratio decreases. Thus, hereafter we select a 5×5 degree region around Virgo as the best target to obtain our predictions.

With the purpose of scanning the most interesting portion of the $\mu\nu$SSM parameter space, we re-simulate the gamma-ray events from the region of 5×5 degrees around Virgo changing the value of the gravitino mass. We run 17 different simulations of this region, each one with a different value for the gravitino mass, ranging from 0.6 to 10 GeV, for a given decay lifetime. The lower bound of 0.6 GeV on the gravitino mass is chosen because it corresponds to a line energy of about 0.3 GeV, where the PSF (point spread function) of the Fermi LAT becomes larger than our region of interest. The energy interval covered by each simulation is $[(m_{3/2}/2 - \Delta E), (m_{3/2}/2 + \Delta E)]$, where the energy resolution (at 68% containment normal incidence) ΔE is computed at the position of the line. Using the results of those simulations, we determine the values of lifetimes corresponding to a $S/N = 5$ ($S/N = 3$) . These are plotted as blue (green) dots in Fig. 2 from [7], as a function of $m_{3/2}$. The blue (green) region indicates points with $S/N \geq 5$ (3). This is the main result of our work.

Let us remark that the errors in the figure are obtained propagating the statistical errors on the number of signal and background events (assuming Poissonian statistics). Regarding possible systematic errors, we note here that the points in Fig. 2 are obtained with the P6_V3_DIFFUSE IRF but the simulations were repeated using both, P6_V3_DATACLEAN and P7CLEAN_V6, and the results are found to be compatible. The use of P6_V3_DIFFUSE allows us to estimate the systematic error on our limits to be between 5 and 20%.

As mentioned in the Introduction, in Ref. [6] the area below the red dot-dashed line was disfavored by Fermi-LAT data of the diffuse gamma-ray galactic emission in the mid-latitude range $10° \leq |b| \leq 20°$. In addition, in Ref. [17], the area below the black dots was also disfavored. From a likelihood analysis focused on the region $|b| \geq 10°$, lower bounds on $\tau_{3/2}$ of about 5×10^{28} s were obtained in our region of interest below 10 GeV. On the other hand, the area below the yellow dashed line is disfavored by the bounds obtained in [18] by analyzing the data from EGRET in the galactic center region. In particular, we used the upper limits on the gamma-ray line fluxes obtained in that work to constrain the $\mu\nu$SSM gravitino lifetime. Finally, points in the magenta shaded region are excluded by gamma-ray observations from the galactic center obtained with the SPI spectrometer on INTEGRAL satellite, and the isotropic diffuse photon background as determined from SPI, COMPTEL and EGRET data [37].

On the other hand, the black dashed lines correspond to the predictions of the $\mu\nu$SSM for several representative values of the R-parity mixing parameter. As mentioned in Sect. 2, this is constrained to be $|U_{\tilde{\gamma}\nu}|^2 \sim 10^{-16} - 10^{-12}$ in the $\mu\nu$SSM [6], in order to reproduce the correct neutrino masses. As a consequence, any acceptable point must be in the area between the left and right black dashed lines. Let us remark, however, that these bounds are very conservative, as discussed in [6], and in fact the results of a scan of the low-energy parameter space of the $\mu\nu$SSM implied that the range $10^{-15} \leq |U_{\tilde{\gamma}\nu}|^2 \leq 5 \times 10^{-14}$ is specially favored. The corresponding lines are also shown in the figure.

The combination of the constraints associated to red dot-dashed and black dashed lines, implies already that values of the gravitino mass larger than about 10 GeV are excluded, as well as lifetimes smaller than about 3 to 5×10^{27} s [6]. Actually, in the region of gravitino masses between 0.6 and about 1.5 GeV, lifetimes smaller than about 7 to 3 $\times 10^{27}$ s, respectively, are excluded because of the constraints associated to the yellow dashed line. When constraints

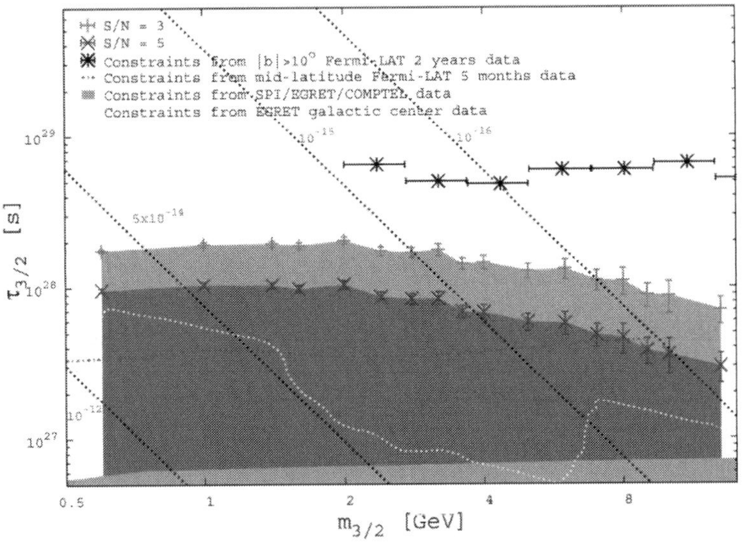

Figure 2. Constraints on lifetime versus mass for gravitino DM in the $\mu\nu$SSM. Blue (green) points indicate values of $\tau_{3/2}$ and $m_{3/2}$ of the $\mu\nu$SSM gravitino corresponding to a detection of gamma-rays with a $S/N = 5$ (3) in the 5×5 degree region centered on the position of the Virgo cluster, for a 5 years simulation using the Fermi Science Tools. The blue (green) region indicates points with S/N larger than 5 (3). The red dot-dashed line indicates the lower limit on $\tau_{3/2}$ obtained from the Fermi-LAT measurements of the mid-latitude gamma-ray diffuse emission after 5 months [6]. The yellow dashed line indicates the lower limit on $\tau_{3/2}$ obtained from the EGRET measurements of the galactic center gamma-ray emission. The black dots show the lower limit on $\tau_{3/2}$ obtained in the adopted energy bands [17], from the Fermi-LAT measurements of the $|b| \geq 10°$ gamma-ray diffuse emission after 2 years. The black dashed lines correspond to the predictions of the $\mu\nu$SSM [6] for several representative values $|U_{\tilde{\gamma}\nu}|^2 = 10^{-16}$, 10^{-15}, 5×10^{-14}, 10^{-12} (see Eq. (1)). The magenta shaded region is excluded by gamma-ray observations such as SPI, COMPTEL and EGRET [37].

associated to black dots are also imposed, it turns out that the gravitino mass has to be smaller than about 4 GeV, and lifetimes have to be larger than about 6×10^{28} s for gravitino masses between 2 and 4 GeV. Thus, the combination of these results with the one obtained above for detection of DM from Virgo in 5 years of Fermi-LAT observations, leaves us with the blue and green areas above the yellow dashed and red dot-dashed lines, and gravitino mass smaller than 2 GeV, as those with good prospects for DM detection. Summarizing, we find that a gravitino DM with a mass range of 0.6–2 GeV, and with a lifetime range of about 3×10^{27}–2×10^{28} s would be detectable by the Fermi-LAT with a signal-to-noise ratio larger than 3. If no gamma-ray lines are detected in 5 years, these regions of the parameter space of the $\mu\nu$SSM would be excluded.

6. Conclusions and outlook

In this talk based on work [7], we have obtained the regions of the parameter space $(m_{3/2}, \tau_{3/2})$ of the $\mu\nu$SSM with the best prospects for the detection of a gamma-ray monochromatic line from the decay of gravitino DM (see Fig. 2). Summarizing, we find that a gravitino DM with a mass range of 0.6–2 GeV, and with a lifetime range of about 3×10^{27}–2×10^{28} s would be detectable by the Fermi-LAT with a signal-to-noise ratio larger than 3. We also obtain that gravitino masses larger than about 4 GeV are now disfavored in the $\mu\nu$SSM by Fermi-LAT data of the galactic halo.

In the analysis we have assumed 5 years of observation of the Virgo cluster by the Fermi-LAT space telescope. This cluster was selected as our optimal target due to its particularly high S/N ratio. Of course, a more precise determination of the Fermi-LAT possibilities of detecting gamma-ray lines towards Virgo would require the simulation of both, M87 and 2E1228+1437. Also a more sophisticated analysis pipeline than the computation of the S/N ratio, possibly through the determination of a test-statistics (TS) likelihood in order to derive the lower value of $\tau_{3/2}$ for which the signature of a line would be detectable with respect to the background.

Let us remark that the simulation of the gamma-ray flux was carried out with the use of the `gtobssim` routine from the Fermi Science Tools, whereas the DM distribution around the cluster has been modeled following the results of Ref. [23] based on a constrained N-body simulation from the CLUES project [24]. With the present work we have also confirmed the potential of using extragalactic massive structures as optimal targets for decaying DM detection. For such a goal the maps of the local extragalactic DM distribution produced in Ref. [23] represent a unique, ready-to-use tool.

We conclude that there are good prospects for Fermi to detect monochromatic lines from gravitino decay in the energy range spanning from hundreds MeV to few GeV. That is also the energy range where it is more difficult to extract information from the data, due to imperfect parametrization of the background as a simple power law. Nevertheless, our results in Fig. 2 can be considered as an additional motivation to extend the Fermi-LAT analysis on lines to energies below 2 GeV.

Acknowledgments

This work was supported by the Spanish MINECO's Consolider-Ingenio 2010 Programme under grant MultiDark CSD2009-00064. The work of C. Muñoz was supported in part by MINECO under grants FPA2009-08958 and FPA2009-09017, by the Comunidad de Madrid under grant HEPHACOS S2009/ESP-1473, and by the European Union under the Marie Curie-ITN program PITN-GA-2009-237920.

References

[1] E. Komatsu *et al.* [WMAP Collaboration], Astrophys. J. Suppl. **192** (2011) 18 [arXiv:1001.4538 [astro-ph.CO]].
[2] For a review, see e.g.: C. Muñoz, Int. J. Mod. Phys. **A19** (2004) 3093 [arXiv:hep-ph/0309346].
[3] D. E. López-Fogliani and C. Muñoz, Phys. Rev. Lett. **97** (2006) 041801 [arXiv:hep-ph/0508297].
[4] For reviews, see: C. Muñoz, AIP Conf. Proc. **1200** (2010) 413 [arXiv:0909.5140 [hep-ph]]; D. E. López-Fogliani, arXiv:1004.0884 [hep-ph].
[5] N. Escudero, D. E. López-Fogliani, C. Muñoz and R. R. de Austri, JHEP **12** (2008) 099 [arXiv:0810.1507 [hep-ph]]; P. Ghosh and S. Roy, JHEP **04** (2009) 069 [arXiv:0812.0084 [hep-ph]]; A. Bartl, M. Hirsch, A. Vicente, S. Liebler and W. Porod, JHEP **05** (2009) 120 [arXiv:0903.3596 [hep-ph]]; J. Fidalgo, D. E. López-Fogliani, C. Muñoz and R. Ruiz de Austri, JHEP **08** (2009) 105 [arXiv:0904.3112 [hep-ph]]; P. Ghosh, P. Dey, B. Mukhopadhyaya and S. Roy, JHEP **05** (2010) 087 [arXiv:1002.2705 [hep-ph]]; D.J.H. Chung and A. Long, Phys. Rev. **D81** (2010) 123531 [arXiv:1004.0942[hep-ph]]; P. Bandyopadhyay, P. Ghosh and S. Roy, Phys. Rev. **D84** (2011) 115022 [arXiv:1012.5762[hep-ph]]; S. Liebler and W. Porod, Nucl. Phys. **B855** (2012) 774 [arXiv:1106.2921[hep-ph]]; J. Fidalgo, D. E. López-Fogliani, C. Muñoz and R. Ruiz de Austri, JHEP **10** (2011) 020 [arXiv:1107.4614[hep-ph]].
[6] K. Y. Choi, D. E. López-Fogliani, C. Muñoz and R. R. de Austri, JCAP **03** (2010) 028 [arXiv:0906.3681 [hep-ph]].
[7] G.A. Gómez-Vargas, M. Fornasa, F. Zandanel, A.J. Cuesta, C. Muñoz, F. Prada and G. Yepes, JCAP **02** (2012) 001 [arXiv:1110.3305 [astro-ph.HE]].
[8] See e.g.: D.G. Cerdeño, C. Muñoz and O. Seto, Phys. Rev. **D79** (2009) 023510 [arXiv:0807.3029 [hep-ph]], and references therein.
[9] W. B. Atwood *et al.* [Fermi LAT Collaboration], Astrophys. J. **697** (2009) 1071 [arXiv:0902.1089]; http://fgst.slac.stanford.edu/ .

Journal of Physics: Conference Series **384** (2012) 012004 doi:10.1088/1742-6596/384/1/012004

[10] A. Lapi, A. Paggi, A. Cavaliere, A. Lionetto, A. Morselli and V. Vitale, arXiv:0912.1766 [astro-ph.HE]; V. Vitale, A. Morselli and f. t. F. Collaboration, arXiv:0912.3828 [astro-ph.HE].
[11] J.F. Navarro, C.S. Frenk and S.D.M. White, Astrophys. J. **462** (1996) 563.
[12] A. A. Abdo *et al.* [Fermi LAT Collaboration], Phys. Rev. Lett. **103** (2009) 251101 [arXiv:0912.0973 [astro-ph.HE]].
[13] K. Y. Choi, D. Restrepo, C. E. Yaguna and O. Zapata, JCAP **10** (2010) 033 [arXiv:1007.1728 [hep-ph]].
[14] M.A. Diaz, S. Garcia Saenz, B. Koch, Phys. Rev. **D84** (2011) 055007 [arXiv:1106.0308 [hep-ph]].
[15] A. A. Abdo *et al.* [Fermi LAT Collaboration], Phys. Rev. Lett. **104** (2010) 091302 [arXiv:1001.4836 [astro-ph.HE]].
[16] E. Bloom by the Fermi-LAT Collaboration, talk at Aspen Winter Workshop 'Indirect and Direct Detection of Dark Matter', Feb. 6-8, 2011, http://www.slac.stanford.edu/exp/glast/aspen11
[17] G. Vertongen and C. Weniger, JCAP **05** (2011) 027. [arXiv:1101.2610 [hep-ph]].
[18] A.R. Pullen, R.-R. Chary and M. Kamionkowski, Phys. Rev. **D76** (2007) 063006, Erratum-ibid. **D83** (2011) 029904 [arXiv:1109.0512 [hep-ph]]
[19] D. Restrepo, M. Taoso, J.W.F. Valle and O. Zapata, arXiv:1109.0512 [hep-ph]
[20] M. Markevitch *et al.*, Astrophys. J. **567** (2002) L27 [arXiv:astro-ph/0110468].
[21] D. Clowe, M. Bradac, A. H. Gonzalez, M. Markevitch, S. W.Randall, C. Jones and D. Zaritsky, Astrophys. J. **648** (2006) L109 [arXiv:astro-ph/0608407].
[22] X. Huang, G. Vertongen and C. Weniger, arXiv:1110.1529 [hep-ph].
[23] A. J. Cuesta *et al.*, Astrophys. J. **726** (2011) L6 [arXiv:1007.3469 [astro-ph.HE]].
[24] http://www.clues-project.org
[25] http://fermi.gsfc.nasa.gov/ssc/data/analysis
[26] F. Takayama and M. Yamaguchi, Phys. Lett. B **485** (2000) 388 [arXiv:hep-ph/0005214].
[27] W. Buchmuller, L. Covi, K. Hamaguchi, A. Ibarra and T. Yanagida, JHEP **03** (2007) 037 [arXiv:hep-ph/0702184]; G. Bertone, W. Buchmuller, L. Covi and A. Ibarra, JCAP **11** (2007) 003 [arXiv:0709.2299 [astro-ph]]; A. Ibarra and D. Tran, Phys. Rev. Lett. **100** (2008) 061301 [arXiv:0709.4593 [astro-ph]]; K. Ishiwata, S. Matsumoto and T. Moroi, Phys. Rev. **D78** (2008) 063505 [arXiv:0805.1133 [hep-ph]]. W. Buchmuller, A. Ibarra, T. Shindou, F. Takayama and D. Tran, JCAP **09** (2009) 021 [arXiv:0906.1187 [hep-ph]].
[28] F. Prada, A. Klypin, J. Flix Molina, M. Martinez, E. Simonneau, Phys. Rev. Lett. **93** (2004) 241301 [arxiv:astro-ph/0401512].
[29] J. M. Overduin and P. S. Wesson, Phys. Rept. **402** (2004) 267 [arXiv:astro-ph/0407207].
[30] S. Gottloeber, Y. Hoffman and G. Yepes, arXiv:1005.2687 [astro-ph.CO].
[31] A. Klypin, Y. Hoffman, A. Kravtsov and S. Gottloeber, Astrophys. J. **596** (2003) 19 [arXiv:astro-ph/0107104].
[32] http://www.clues-project.org/articles/darkmattermaps.html
[33] http://fermi.gsfc.nasa.gov/ssc/data/access/lat/BackgroundModels.html
[34] R. Rando and f. t. F. Collaboration, arXiv:0907.0626 [astro-ph.IM].
[35] T. L. Collaboration, arXiv:1108.1435 [astro-ph.HE].
[36] M. Su, T. R. Slatyer and D. P. Finkbeiner, Astrophys. J. **724** (2010) 1044 [arXiv:1005.5480 [astro-ph.HE]].
[37] H. Yuksel and M. D. Kistler, Phys. Rev. **D78** (2008) 023502 [arXiv:0711.2906 [astro-ph]].

DAMA/LIBRA results and perspectives

R. Bernabei[1,2], P. Belli[1,2], F. Cappella[3,4], V. Caracciolo[5], R. Cerulli[5], C.J. Dai[6], A. d'Angelo[3,4], A. Di Marco[1,2], H.L. He[6], A. Incicchitti[3,4], X.H. Ma[6], F. Montecchia[1,2,7], X.D. Sheng[6], R.G. Wang[6], Z.P. Ye[6,8]

[1] Dip. di Fisica, Università di Roma "Tor Vergata", I-00133 Rome, Italy
[2] INFN, sez. Roma "Tor Vergata", I-00133 Rome, Italy
[3] Dip. di Fisica, Università di Roma "La Sapienza", I-00185 Rome, Italy
[4] INFN, sez. Roma, I-00185 Rome, Italy
[5] Laboratori Nazionali del Gran Sasso, I.N.F.N., Assergi, Italy
[6] IHEP, Chinese Academy, P.O. Box 918/3, Beijing 100039, China
[7] Lab. Sperim. Policentrico di Ingegneria Medica, Università di Roma "Tor Vergata", I-00133 Rome, Italy
[8] University of Jing Gangshan, Jiangxi, China

E-mail: riccardo.cerulli@lngs.infn.it

Abstract.
The DAMA/LIBRA experiment is running at LNGS; it has a sensitive mass of about 250 kg highly radiopure NaI(Tl) and it is mainly devoted to the investigation of Dark Matter (DM) particles in the Galactic halo by exploiting the model independent DM annual modulation signature. The present DAMA/LIBRA experiment and the former DAMA/NaI one (the first generation experiment having an exposed mass of about 100 kg) have released so far results corresponding to a total exposure of 1.17 ton $\times$ yr, collected over 13 annual cycles. They provide a model independent evidence of the presence of DM particles in the galactic halo at 8.9 σ C.L.. Some obtained results are shortly summarized and future perspectives mentioned.

1. Introduction

The DAMA project is based on the development and use of low background scintillators. The main experimental set-ups are: i) DAMA/NaI ($\simeq$ 100 kg of highly radiopure NaI(Tl)) that took data for 7 annual cycles and completed its data taking on July 2002 [1, 2, 3, 4, 5, 6, 7, 8, 9, 10, 11, 12]; ii) DAMA/LXe, $\simeq$ 6.5 kg liquid Kr-free Xenon enriched either in ^{129}Xe or in ^{136}Xe [13]; iii) DAMA/R&D, a facility dedicated to tests on prototypes and to perform experiments developing and using various kinds of low background crystal scintillators in order to investigate various rare processes [14]; iv) DAMA/Ge, where sample measurements are carried out and where dedicated measurements on rare events are performed [15]; v) the second generation DAMA/LIBRA set-up, $\simeq$ 250 kg highly radiopure NaI(Tl)) [16, 17, 18, 19] mainly devoted to the investigation of the presence of Dark Matter (DM) particles in the galactic halo. Profiting of the low background features of these set-ups, many rare processes have been studied and competitive results have been obtained.

DAMA/LIBRA is the main apparatus, it is investigating the presence of DM particles in the galactic halo by exploiting the model independent DM annual modulation signature originally suggested in the mid 80's[20]. As a consequence of the annual revolution of the Earth around

Journal of Physics: Conference Series **384** (2012) 012005 doi:10.1088/1742-6596/384/1/012005

the Sun, moving in the Galaxy with respect to the Local Standard of Rest towards the star Vega near the constellation of Hercules, our planet should be crossed by a larger flux of Dark Matter particles around $\sim$2 June (when the Earth orbital velocity has the same versus of the solar system with respect to the Galaxy) and by a smaller one around $\sim$2 December (when the two velocities are opposite). Thus, this signature depends on the composition of the Earth and Sun velocitites and it is not correlated with seasons. This DM annual modulation signature is very distinctive since the effect induced by DM particles must simultaneously satisfy all the following requirements: (1) the rate must contain a component modulated according to a cosine function; (2) with one year period; (3) with a phase that peaks roughly around $\sim$ 2nd June; (4) modulation must be present only in a well-defined low energy range, where DM particles can induce signals; (5) it must be present only in those events where just a single detector, among all the available ones in the used set-up, actually "fires" (*single-hit* events), since the probability that DM particles experience multiple interactions is negligible; (6) the modulation amplitude in the region of maximal sensitivity has to be less about 7% in case of usually adopted halo distributions, but it may be significantly larger in case of some particular scenarios such as e.g. those in refs. [21, 22]. Only systematic effects or side reactions able to simultaneously fulfil all the mentioned requirements and able to account for the whole observed modulation amplitude might mimic this DM signature; no one has been found or suggested. At present status of technology it is the only model independent signature which can effectively be exploited by direct Dark Matter investigation.

The DAMA/LIBRA data released so far correspond to six annual cycles for an exposure of 0.87 ton$\times$yr [17, 18]. Considering these data together with those previously collected by DAMA/NaI over 7 annual cycles (0.29 ton$\times$yr), the total exposure collected over 13 annual cycles is 1.17 ton$\times$yr; this is orders of magnitude larger than the exposures typically collected in the field.

2. DAMA/LIBRA results

The sensitive part of the DAMA/LIBRA set-up is made of 25 highly radiopure NaI(Tl) crystal scintillators placed in a 5-rows by 5-columns matrix; each crystal is coupled to two low background photomultipliers working in coincidence at single photoelectron level. The detectors are placed inside a sealed copper box flushed with HP nitrogen and surrounded by a low background and massive shield made of Cu/Pb/Cd-foils/polyethylene/paraffin; moreover, about 1 m concrete (made from the Gran Sasso rock material) almost fully surrounds (mostly outside the barrack) this passive shield, acting as a further neutron moderator. The installation has a 3-levels sealing system which excludes the detectors from environmental air and the whole installation is air-conditioned. The detectors' responses range from 5.5 to 7.5 photoelectrons/keV. Energy calibrations with X-rays/γ sources are regularly carried out down to few keV in the same conditions as the production runs. The software energy threshold is 2 keV. The DAMA/NaI set up and its performances are described in ref.[1, 3, 4, 5], while for the DAMA/LIBRA set-up and its performances see refs. [16, 18].

Several analyses on the model-independent DM annual modulation signature have been performed [17, 18]. Fig. 1 shows the time behaviour of the experimental residual rates of the *single-hit* events collected by DAMA/NaI and by DAMA/LIBRA in the (2–6) keV energy interval [17, 18]. The superimposed curve is the cosinusoidal function: $A\cos\omega(t-t_0)$ with a period $T=\frac{2\pi}{\omega}=1$ yr, with a phase $t_0=152.5$ day (June 2nd), and modulation amplitude, A, obtained by best fit over the 13 annual cycles. The hypothesis of absence of modulation in the data can be discarded [17, 18] and, when the period and the phase are released in the fit, values well compatible with those expected for a DM particle induced effect are obtained [18]. In particular, in the cumulative (2–6) keV energy interval one gets: $A=(0.0116\pm0.0013)$ cpd/kg/keV, $T=(0.999\pm0.002)$ yr and $t_0=(146\pm7)$ day. Thus, the analysis of the *single-hit*

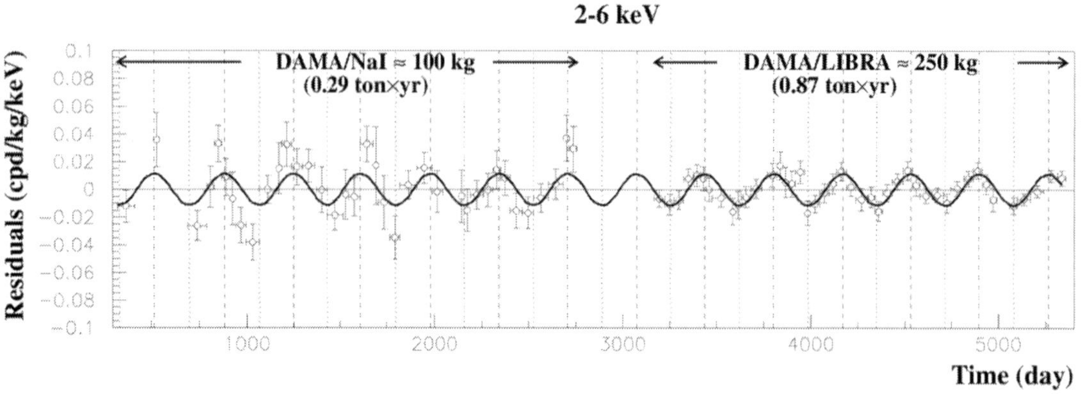

Figure 1. Experimental model-independent residual rate of the *single-hit* scintillation events, measured by DAMA/NaI over seven and by DAMA/LIBRA over six annual cycles in the (2 – 6) keV energy interval as a function of the time [4, 5, 17, 18]. The zero of the time scale is January 1st of the first year of data taking. The experimental points present the errors as vertical bars and the associated time bin width as horizontal bars. The superimposed curve is $A\cos\omega(t-t_0)$ with period $T = \frac{2\pi}{\omega} = 1$ yr, phase $t_0 = 152.5$ day (June 2nd) and modulation amplitude, A, equal to the central value obtained by best fit over the whole data. The dashed vertical lines correspond to the maximum expected for the DM signal (June 2nd), while the dotted vertical lines correspond to the minimum. See Refs. [17, 18] and text.

residual rate favours the presence of a modulated cosine-like behaviour with proper features at 8.9 σ C.L.[18].

The same data of Fig. 1 have also been investigated by a Fourier analysis, obtaining a clear peak corresponding to a period of 1 year [18]; this analysis in other energy regions shows only aliasing peaks instead. Moreover, in order to verify absence of annual modulation in other energy regions and, thus, to also verify the absence of any significant background modulation, the energy distribution in energy regions not of interest for DM detection has also been investigated. This has allowed the exclusion of a background modulation in the whole energy spectrum at a level much lower than the effect found in the lowest energy region for the *single-hit* events [18]. A further relevant investigation has been done by applying the same hardware and software procedures, used to acquire and to analyse the *single-hit* residual rate, to the *multiple-hits* events in which more than one detector "fires". In fact, since the probability that a DM particle interacts in more than one detector is negligible, a DM signal can be present just in the *single-hit* residual rate. Thus, this allows the study of the background behaviour in the same energy interval of the observed positive effect. The result of the analysis is reported in Fig. 2 where it is shown the residual rate of the *single-hit* events measured over the six DAMA/LIBRA annual cycles, as collected in a single annual cycle, together with the residual rates of the *multiple-hits* events, in the same (2–6) keV energy interval. A clear modulation is present in the *single-hit* events, while the fitted modulation amplitudes for the *multiple-hits* residual rate are well compatible with zero [18]. Similar results were previously obtained also for the DAMA/NaI case [5]. Thus, again evidence of annual modulation with proper features, as required by the DM annual modulation signature, is present in the *single-hit* residuals (events class to which the DM particle induced events belong), while it is absent in the *multiple-hits* residual rate (event

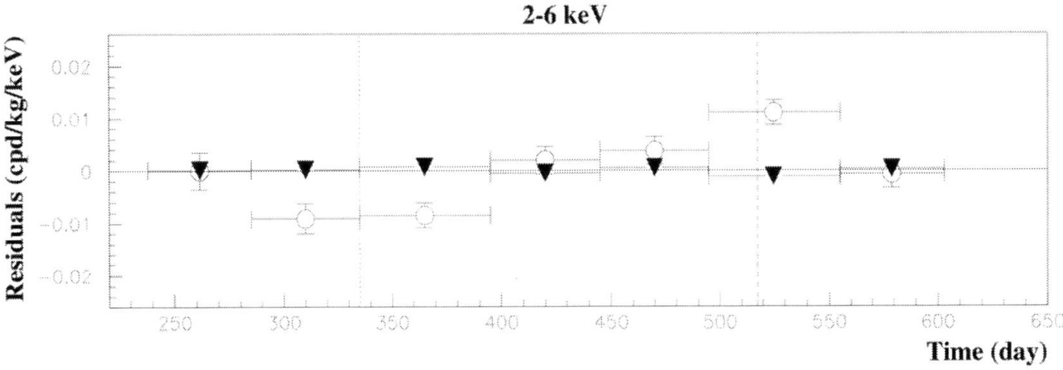

Figure 2. Experimental residual rates over the six DAMA/LIBRA annual cycles for *single-hit* events (open circles) (class of events to which DM events belong) and for *multiple-hit* events (filled triangles) (class of events to which DM events do not belong). They have been obtained by considering for each class of events the data as collected in a single annual cycle and by using in both cases the same identical hardware and the same identical software procedures. The initial time of the figure is taken on August 7th. The experimental points present the errors as vertical bars and the associated time bin width as horizontal bars [17, 18]. See text.

class to which only background events belong). Since the same identical hardware and the same identical software procedures have been used to analyse the two classes of events, the obtained result offers an additional strong support for the presence of a DM particle component in the galactic halo further excluding any side effect either from hardware or from software procedures or from background. The annual modulation present at low energy has also been analyzed by depicting the differential modulation amplitudes, S_m, as a function of the energy; the S_m is the modulation amplitude obtained by maximum likelihood method over the data, considering $T = 1$ yr and $t_0 = 152.5$ day. The S_m values are reported as function of the energy in Fig. 3. It can be inferred that a positive signal is present in the (2–6) keV energy interval, while S_m values compatible with zero are present just above; in particular, the S_m values in the (6–20) keV energy interval have random fluctuations around zero with χ^2 equal to 27.5 for 28 degrees of freedom. It has been also verified that the measured modulation amplitudes are statistically well distributed in all the crystals, in all the annual cycles and energy bins; these and other discussions can be found in ref. [18]. Other interesting results have been obtained by analysing the data introducing a sine-like modulation in the maximum likelihood procedure. In this case alternatively the signal can be written as: $S_{0,k} + S_{m,k} \cos\omega(t - t_0) + Z_{m,k} \sin\omega(t - t_0) = S_{0,k} + Y_{m,k} \cos\omega(t - t^*)$, where $S_{0,k}$, $S_{m,k}$ and $Z_{m,k}$ are the constant part, the cosine-like and sine-like modulation amplitudes of the signal in k-th energy interval. Obviously, for signals induced by DM particles one would expect: i) $Z_{m,k} \sim 0$ (because of the orthogonality between the cosine and the sine functions); ii) $S_{m,k} \simeq Y_{m,k}$; iii) $t^* \simeq t_0 = 152.5$ day. In fact, these conditions hold for most of the dark halo models; however, it is worth noting that slight differences in the phase could be expected in case of possible contributions from non-thermalized DM components, such as e.g. the SagDEG stream [7] and the caustics [23]. The 2σ contours in the plane (S_m, Z_m) for the (2–6) keV and (6–14) keV energy intervals and those in the plane (Y_m, t^*) are reported in Fig. 4 [18]. The best fit values for the (2–6) keV energy interval are (1σ errors): $S_m = (0.0111 \pm 0.0013)$ cpd/kg/keV; $Z_m = -(0.0004 \pm 0.0014)$ cpd/kg/keV; $Y_m = (0.0111 \pm 0.0013)$ cpd/kg/keV; $t^* = (150.5 \pm 7.0)$

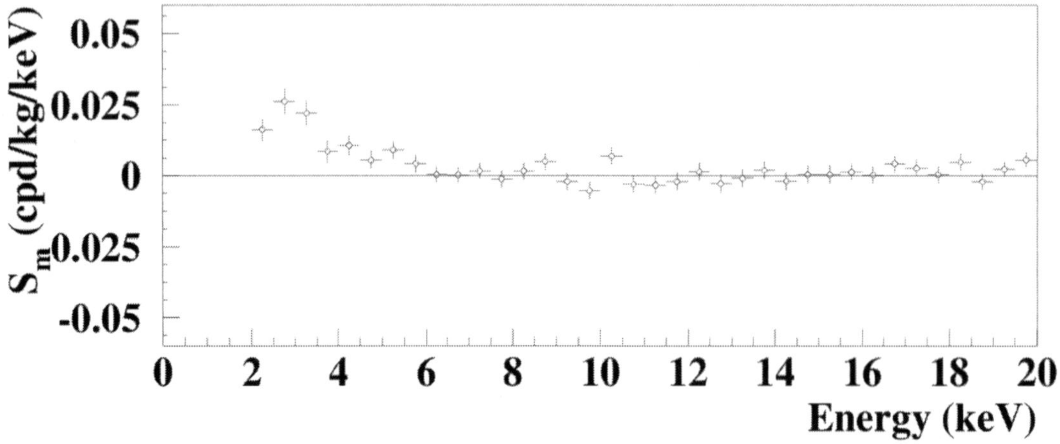

Figure 3. Energy distribution of the modulation amplitudes S_m for the total cumulative exposure 1.17 ton×yr. The energy bin is 0.5 keV. A clear modulation is present in the lowest energy region, while S_m values compatible with zero are present just above. In fact, the S_m values in the (6–20) keV energy interval have random fluctuations around zero with χ^2 equal to 27.5 for 28 degrees of freedom. See Refs. [17, 18].

day; while for the (6–14) keV energy interval are: $S_m = -(0.0001 \pm 0.0008)$ cpd/kg/keV; $Z_m = (0.0002 \pm 0.0005)$ cpd/kg/keV; $Y_m = -(0.0001 \pm 0.0008)$ cpd/kg/keV and t^* obviously not determined. These results confirm those achieved by other kinds of analyses. In particular, a modulation amplitude is present in the *single-hit* events of the lower energy intervals and the period and the phase agree with those expected for DM signals [18].

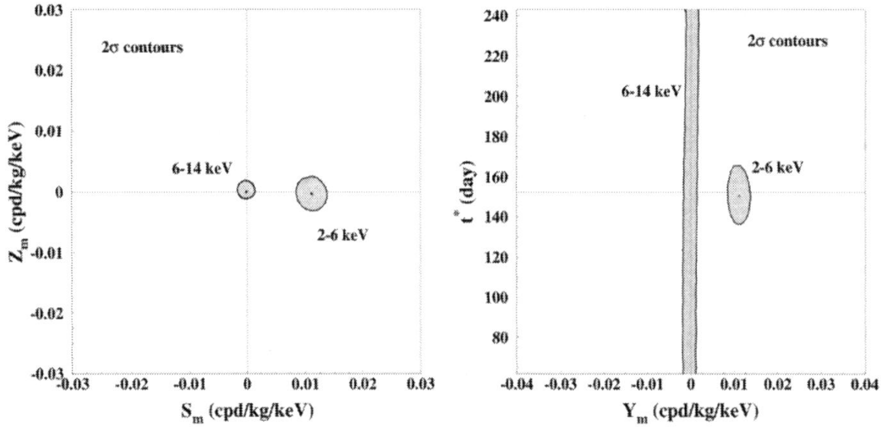

Figure 4. 2σ contours in the plane (S_m, Z_m) (*left*) and in the plane (Y_m, t^*) (*right*) for the (2–6) keV and (6–14) keV energy intervals. The contours have been obtained by the maximum likelihood method, considering the cumulative exposure of 1.17 ton × yr. See text.

Both the data of DAMA/LIBRA and of DAMA/NaI fulfil all the requirements of the DM annual modulation signature.

Careful investigations on absence of any significant systematics or side reaction have been quantitatively carried out (see e.g. ref. [4, 5, 16, 17, 18, 24], and refs therein). No systematics or side reactions able to mimic the signature (that is, able to account for the measured modulation amplitude and simultaneously satisfy all the requirements of the signature) has been found or suggested by anyone over more than a decade.

3. About interpretation

A rich variety of theoretical patterns are present in literature. The scenarios are stimulating and intriguing to interpret in model dependent way the results coming from different experiments on dark matter investigation.

Anyhow, it is worth noting that DAMA has firstly a model independent result due to the exploitation of a DM signature with specific peculiarities.

About the interpretation of this model independent result in terms of a particular astrophysical, nuclear and particle physics scenario and in order to consider a comparative overview with measurements by other experiments, it is important to refresh that there is neither an unique reference theoretical model of interpretation, nor a single set of assumptions for parameters in astrophysical, nuclear and particle Physics. In addition often comparisons are not performed in a fully consistent way. Thus every model dependent analysis chooses a model framework by fixing many parameters and assumptions, but many uncertainties are present. Uncertainties are present on the models about the nature of the candidate particle, on the interaction coupling, on the form factors for each target material, on spin factors, on the scaling laws, on the nuclear Physics framework, on the dark halo model and related parameters, on possible galactic streams and so on. Uncertainties can also exist in some cases on some experimental aspects as: marginal exposures, poorly known detector response, extrapolated energy threshold, extrapolated energy scale, extrapolated energy resolution, definition of a fiducial volume, possible non-uniformity of some detector response, not regular and/or suitable calibration procedures, on quenching factors, on stability of the operating conditions, etc.. In addition, in some cases, there are significant uncertainties on subtraction/rejection procedures when applied and on the stability of all the selection windows and related quantities. All these aspects can affect results and comparisons at various extent, precluding the comparisons from an universal value.

The obtained DAMA model independent evidence is compatible with a wide set of scenarios regarding the nature of the DM candidate and related astrophysical, nuclear and particle Physics. For examples some given scenarios and parameters are discussed e.g. in Refs. [2, 4, 5, 6, 7, 8, 9, 10, 11] and in Appendix A of Ref. [17]. Further large literature is available on the topics [25, 26]; other possibilities are open and we just recall the recent paper [26] on the case of a light-mass Dark Matter candidate particle interacting with the detector nuclei by coherent elastic process; comparison with recent possible positive hint [27] is also given.

It is worth noting that no other experiment exists, whose result can be directly compared in a model-independent way with those by DAMA/NaI and DAMA/LIBRA. Moreover, concerning those activities claiming model dependent exclusion under some largely arbitrary assumptions (see for example discussions in [4, 5, 17, 28, 29]) and generally using marginal exposures, it is worth noting that often important critical points exist in some of their experimental aspects (as discussed before and in many papers); in addition existing experimental and theoretical uncertainties are not considered.

Similar considerations hold for the indirect detection searches, in fact also in this case no direct model-independent comparison can be performed between the results obtained in direct and indirect activities, since it does not exist a biunivocal correspondence between the observables in the two kinds of experiments. Moreover, these searches are restricted to some DM candidates and scenarios and their results are strongly model dependent. Anyhow, measurements published up to now are not in conflict with the effect observed by DAMA experiments.

4. Upgrades and perspectives

A first upgrade of the DAMA/LIBRA set-up was performed in September 2008. A further and more important upgrade has been performed in the end of 2010 when all the PMTs have been replaced with new ones with higher quantum efficiency; this will allow the lowering of the software energy threshold and, hence, the improvement of the performance and of the sensitivity of the experiment. A deeper corollary information on the nature of the DM candidate particle(s) and on the various related astrophysical, nuclear and particle Physics scenarios will be possible.

Since January 2011 the DAMA/LIBRA experiment is again in data taking in the new configuration. Further improvements are foreseen with new preamplifiers and trigger modules realised to further implement the low energy studies. In the future DAMA/LIBRA will also continue its study on several other rare processes [19] as also the former DAMA/NaI apparatus did [12]. Further developments are in progress.

5. Conclusion

The positive model independent evidence for the presence of DM particles in the galactic halo is now supported by DAMA/NaI and DAMA/LIBRA at 8.9σ C.L.; the cumulative exposure is 1.17 ton$\times$yr, collected over 13 annual cycles. Further corollary analyses in some of the many possible scenarios for DM candidates, interactions, halo models, nuclear/atomic properties, etc.are in progress as well as analyses/data taking to investigate other rare processes. The last upgrade in fall 2010 was successfully concluded and further improvements are planned. The strictly quality control allows DAMA/LIBRA to be still the highest radiopure set-up in the field with the largest exposed sensitive mass, the full control of running conditions, the largest duty-cycle and an exposure orders of magnitude larger than any other activity in the field.

References

[1] R. Bernabei et al. *Il Nuovo Cim.* A **112**, 545 (1999).
[2] R. Bernabei et al. *Phys. Lett.* B **389**, 757 (1996); *Phys. Lett.* B **424**, 195 (1998); *Phys. Lett.* B **450**, 448 (1999); *Phys. Rev.* D **61**, 023512 (2000); *Phys. Lett.* B **480**, 23 (2000); *Phys. Lett.* B **509**, 197 (2001); *Eur. Phys. J.* C **23**, 61 (2002); *Phys. Rev.* D **66**, 043503 (2002).
[3] R. Bernabei et al. *Eur. Phys. J.* C **18**, 283 (2000).
[4] R. Bernabei et al. *La Rivista del Nuovo Cimento* **26**, n.1, 1 (2003).
[5] R. Bernabei et al. *Int. J. Mod. Phys.* D **13**, 2127 (2004).
[6] R. Bernabei et al. *Int. J. Mod. Phys.* A **21**, 1445 (2006).
[7] R. Bernabei et al. *Eur. Phys. J.* C **47**, 263 (2006).
[8] R. Bernabei et al. *Int. J. Mod. Phys.* A **22**, 3155 (2007).
[9] R. Bernabei et al. *Eur. Phys. J.* C **53**, 205 (2008).
[10] R. Bernabei et al. *Phys. Rev.* D **77**, 023506 (2008).
[11] R. Bernabei et al. *Mod. Phys. Lett.* A **23**, 2125 (2008).
[12] See in the publication list in: http://people.roma2.infn.it/dama
[13] P. Belli et al. *Astropart. Phys.* **5**, 217 (1996); P. Belli et al. *Nuovo Cim.* C **19**, 537 (1996); P. Belli et al. *Phys. Lett.* B **B387**, 222 (1996); *Phys. Lett.* B **389**, 783 (err.) (1996); R. Bernabei et al. *Phys. Lett.* B **436**, 379 (1998); P. Belli et al. *Phys. Lett.* B **465**, 315 (1999); P. Belli et al. *Phys. Rev.* D **61**, 117301(2000); R. Bernabei et al. *New J. of Phys.* **2**, 15.1(2000); R. Bernabei et al. *Phys. Lett.* B **493**, 12 (2000); R. Bernabei et al. *Nucl. Instr. & Meth* A **482**, 728 (2000); R. Bernabei et al. *Eur. Phys. J. direct* C **11**, 1 (2001); R. Bernabei et al. *Phys. Lett.* B **527**, 182 (2002); R. Bernabei et al. *Phys. Lett.* B **546**, 23 (2002); R. Bernabei

et al. *Beyond the Desert 2003* (Berlin: Springer) p. 365 (2003); R. Bernabei et al. *Eur. Phys. J.* A **27**, s01 35 (2006).

[14] Bernabei R et al. *Astropart. Phys.* **7**, 73 (1997); R. Bernabei et al. *Nuovo Cim.* A **110**, 189 (1997); P. Belli et al. *Astropart. Phys.* **10**, 115 (1999); P. Belli et al. *Nucl. Phys.* B **563**, 97 (1999); R. Bernabei et al. *Nucl. Phys.* A **705**, 29 (2002); P. Belli et al. *Nucl. Instr. & Meth* A **498**, 352 (2003); R. Cerulli et al. *Nucl. Instr. & Meth* A **525**, 535 (2004); R. Bernabei et al. *Nucl. Instr. & Meth* A **555** 270 (2005); R. Bernabei et al. *Ukr. J. Phys.* **51**, 1037 (2006); P. Belli et al. *Nucl. Phys.* A **789**, 15 (2007); P. Belli et al. *Phys. Rev.* C **76**, 064603 (2007); P. Belli et al. *Phys. Lett.* B**658**, 193 (2008); P. Belli et al. *Eur. Phys. J.* A **36**, 167 (2008); P. Belli et al. *Nucl. Phys.* A**826**, 256 (2009); P. Belli et al. *Nucl. Instr. & Meth* A **615**, 301 (2010); P. Belli et al. *Nucl. Instr. & Meth* A **626-627**, 31 (2011); P. Belli et al. *J. Phys. G: Nucl. Part. Phys.* **38**, 015103 (2011).

[15] P. Belli et al. *Nucl. Instr. & Meth.* A **572**, 734 (2007); P. Belli et al. *Nucl. Phys.* A **806**, 388 (2008); P. Belli et al. *Nucl. Phys.* A **824**, 101 (2009); P. Belli et al. *Proceed. of the Int. Conf. NPAE 2008* (ed. INR-Kiev, Kiev), p. 473 (2009); P. Belli et al. *Eur. Phys. J.* A **42**, 171 (2009); P. Belli et al. *Nucl. Phys.* A **846**, 143 (2010); P. Belli et al. *Nucl. Phys.* A **859**, 126 (2011); P. Belli et al. *Phys. Rev.* C **83**, 034603 (2011); P. Belli et al. *Eur. Phys. J.* A **47**, 91 (2011).

[16] R. Bernabei et al. *Nucl. Instr. & Meth.* A **592**, 297 (2008).

[17] R. Bernabei et al. *Eur. Phys. J.* C **56**, 333 (2008).

[18] R. Bernabei et al. *Eur. Phys. J.* C **67**, 39 (2010).

[19] R. Bernabei et al. *Eur. Phys. J.* **C62**, 327–332 (2009).

[20] Drukier K A et al. *Phys. Rev.* D **33**, 3495 (1986); Freese K et al. *Phys. Rev.* D **37**, 3388 (1988).

[21] D. Smith and N. Weiner *Phys. Rev.* D **64**, 043502 (2001); D. Tucker-Smith and N. Weiner *Phys. Rev.* D **72**, 063509 (2005).

[22] K. Freese et al. *Phys. Rev.* D **71**, 043516 (2005); *Phys. Rev. Lett.* **92**, 111301 (2004).

[23] F.S. Ling, P. Sikivie and S. Wick *Phys. Rev.* D **70**, 123503 (2004).

[24] R. Bernabei et al. *AIP Conf. Proceed.* **1223**, 50 (2010) (arXiv:0912.0660); *J. Phys.: Conf. Ser.* **203**, 012040 (2010)(arXiv:0912.4200); http://taup2009.lngs.infn.it/slides/jul3/nozzoli.pdf, talk given by F. Nozzoli; *Can. J. Phys.* **89**, 11 (2011); *SIF Atti Conf.* **103**, (2011) (arXiv:1007.0595); pre-print ROM2F/2011/12, TIPP2011 Conf., Chicago, USA (2011).

[25] A.Bottino et al. arXiv:1112.5666; A. Bottino et al. *Phys. Rev.* D **81**, 107302 (2010); N. Fornengo et al. *Phys. Rev.* D **83**, 15001 (2011); A.L. Fitzpatrick et al. arXiv:1003.0014 (2010); D. Hooper et al. arXiv:1007.1005v2 (2010); D. G. Cerdeno and O. Seto *JCAP* **0908**, 032 (2009); D. G. Cerdeno, C. Munoz and O. Seto *Phys. Rev.* D **79**, 023510 (2009); D. G. Cerdeno et al. *JCAP* **0706**, 008 (2007); J.F. Gunion et al. arXiv:1009.2555 (2009); A.V. Belikov et al. arXiv:1009.0549 (2010); C. Arina and N. Fornengo *JHEP* **11**, 029 (2007); G. Belanger et al, arXiv:1105.4878; S. Chang et al. Phys. Rev. D **79**, 043513 (2009); S. Chang et al. arXiv:1007.2688; R. Foot arXiv:1001.0096, arXiv:1106.2688, *Phys. Rev.* D **82**, 095001 (2010); Y. Mambrini *JCAP* **1107**, 009 (2011), *JCAP* **1009**, 022 (2010); Y. Bai and P.J. Fox *JHEP* **0911**, 052 (2009); J. Alwall et al. arXiv:1002.3366; M. Yu Khlopov et al. arXiv:1003.1144; S. Andreas et al. *Phys. Rev.* D **82**, 043522 (2010); M.S. Boucenna, S. Profumo arXiv:1106.3368 P.W. Graham et al. Phys. Rev. D **82**, 063512 (2010); B. Batell, M. Pospelov and A. Ritz Phys. Rev. D **79**, 115019 (2009); E. Del Nobile et al. Phys. Rev. D **84**, 027301 (2011); J. Kopp et al. *JCAP* **1002**, 014 (2010); V. Barger et al. Phys. Rev. D **82**, 035019 (2010); S. Chang et al. *JCAP* **1008**, 018 (2010); J.L. Feng et al. *Phys. Lett.* B **703**, 124 (2011); M.T. Frandsen et al. Phys. Rev. D **84**, 041301 (2011); Y.G. Kim and S. Shin *JHEP* **0905**, 036 (2009); S. Shin arXiv:1011.6377; M.R. Buckley arXiv:1106.3583 N. Fornengo et al. arXiv:1108.4661 P. Gondolo et al. arXiv:1106.0885 E. Kuflik et al. arXiv:1003.0682; C. Arina et al. arXiv:1105.5121; M.R. Buckley et al. 1011.1499.

[26] P. Belli et al. *Phys. Rev.* D **84**, 055014 (2011).

[27] C.E. Aalseth et al. arXiv:1002.4703; arXiv:1106.0650

[28] R. Bernabei et al. "Liquid Noble gases for Dark Matter searches: a synoptic survey", Exorma Ed., Roma, ISBN 978-88-95688-12-1, pp. 1–53 (arXiv:0806.0011v2) (2009).

[29] J.I. Collar and D.N. McKinsey arXiv:1005.0838; arXiv:1005.3723; J.I. Collar arXiv:1006.2031; arXiv:1010.5187; arXiv:1103.3481; arXiv:1106.0653; arXiv:1106.3559.

Analyzing direct dark matter detection data with unrejected background events by the AMIDAS website

Chung-Lin Shan

Institute of Physics, Academia Sinica
No. 128, Sec. 2, Academia Road, Nankang, Taipei 11529, Taiwan, R.O.C.

E-mail: `clshan@phys.sinica.edu.tw`

Abstract. In this talk I have presented the data analysis results of extracting properties of halo WIMPs: the mass and the (ratios between the) spin-independent and spin-dependent couplings/cross sections on nucleons by the **AMIDAS** website by taking into account possible *unrejected* background events in the analyzed data sets. Although *non-standard* astronomical setup has been used to generate *pseudodata* sets for our analyses, it has been found that, *without* prior information/assumption about the local density and velocity distribution of halo Dark Matter, these WIMP properties have been reconstructed with $\sim 2\%$ to $\lesssim 30\%$ deviations from the input values.

1. Introduction

In order to extract properties of halo WIMPs (Weakly Interacting Massive Particles) by using data from direct Dark Matter detection experiments as model-independently as possible, we have developed a series of data analysis method for reconstructing the one-dimensional WIMP velocity distribution [1] as well as determining the WIMP mass [2], the spin-independent (SI) WIMP coupling on nucleons [3] and the ratios between different WIMP couplings/cross sections [4]. Moreover, in collaboration with the DAMNED (DArk Matter Network Exclusion Diagram) Dark Matter online tool [5], part of the ILIAS Project [6], the "**AMIDAS**" (A Model-Independent Data Analysis System) website for online simulation/data analysis has also been established [7, 8, 9].

In this article, in order to demonstrate the usefulness and powerfulness as well as the model-independence of the **AMIDAS** package for direct Dark Matter detection experiments, I will analyze *blindly* some *pseudodata* sets generated theoretically for different detector materials and present the reconstructed WIMP properties. This means that I will simply upload these data sets onto the **AMIDAS** website and follow the instructions to reconstruct different WIMP properties *without* using any information about the input setup used for generating the analyzed pseudodata. For cases in which some information about WIMPs (e.g., the mass m_χ) and/or Galactic halo (e.g., the local Dark Matter density ρ_0) is required, I will naively use the commonly used/favorite values for the data analyses. Moreover, due to the time limit on preparing the presentation for the CYNGUS 2011 workshop, in Ref. [10] we considered only the case of *null* background events. In this article, we have taken into account a small fraction of possible *unrejected* background events in the analyzed data sets [11, 12, 13, 14].

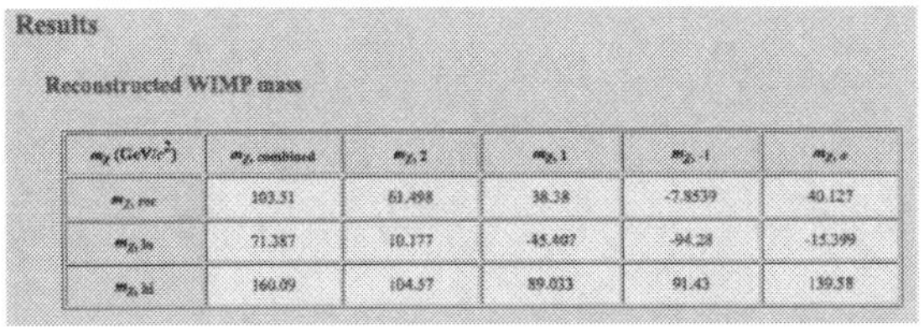

Results

Reconstructed WIMP mass

m_χ (GeV/c^2)	$m_{\chi,\rm combined}$	$m_{\chi,2}$	$m_{\chi,1}$	$m_{\chi,-1}$	$m_{\chi,\sigma}$
$m_{\chi,\rm rec}$	103.31	61.498	38.38	-7.8539	40.127
$m_{\chi,\rm lo}$	71.387	10.177	-45.407	-94.28	-15.399
$m_{\chi,\rm hi}$	160.09	104.57	89.033	91.43	139.58

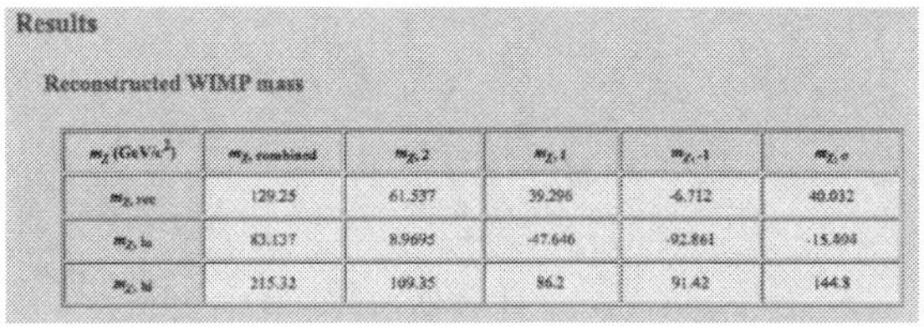

Results

Reconstructed WIMP mass

m_χ (GeV/c^2)	$m_{\chi,\rm combined}$	$m_{\chi,2}$	$m_{\chi,1}$	$m_{\chi,-1}$	$m_{\chi,\sigma}$
$m_{\chi,\rm rec}$	129.25	61.537	39.296	-6.712	40.032
$m_{\chi,\rm lo}$	83.137	8.9695	-47.646	-92.861	-15.894
$m_{\chi,\rm hi}$	215.32	169.35	86.2	91.42	144.8

Figure 1. The reconstructed WIMP mass m_χ by means of the method introduced in Ref. [2] with a target combination of ^{28}Si + ^{76}Ge nuclei. Two forms of the elastic nuclear form factor given in Eqs. (1) and (2) have been used in the left and right frames, respectively.

After that I show the blindly reconstructed properties of halo WIMPs in Sec. 2, in Sec. 3 I will reveal the input setup and the background spectrum used for generating the analyzed data and compare the reconstructed results to them. Finally, I conclude in Sec. 4.

2. Reconstructed WIMP properties

In this section, I present the reconstructed WIMP properties analyzed by the `AMIDAS` website. While in each uploaded file there are *exactly* 50 data sets, in each data set there are *on average* 50 recorded events (i.e., 50 measured recoil energies); the exact number of total events is Poisson distributed. For simplicity, the experimental minimal and maximal cut–off energies have been set as 0 and 100 keV for all data sets.

In order to check the effect of using a "wrong" elastic nuclear form factor, two forms have been considered for the SI WIMP–nucleus cross section in our analyses. One is the simple exponential form:

$$F^2_{\rm ex}(Q) = e^{-Q/Q_0} \,. \tag{1}$$

Here Q is the recoil energy transferred from the incident WIMP to the target nucleus, Q_0 is the nuclear coherence energy given by $Q_0 = 1.5/m_{\rm N} R_0^2$, where $R_0 = \left[0.3 + 0.91\,(m_{\rm N}/{\rm GeV})^{1/3}\right]$ fm is the radius of the nucleus and $m_{\rm N}$ is the mass of the target nucleus. Meanwhile, we used also a more realistic analytic form for the elastic nuclear form factor:

$$F^2_{\rm SI}(Q) = \left[\frac{3 j_1(qR_1)}{qR_1}\right]^2 e^{-(qs)^2} \,. \tag{2}$$

Here $j_1(x)$ is a spherical Bessel function, $q = \sqrt{2m_{\rm N}Q}$ is the transferred 3-momentum, for the effective nuclear radius we use

$$R_1 = \sqrt{R_A^2 - 5s^2} \tag{3}$$

with $R_A \simeq 1.2\,A^{1/3}$ fm and a nuclear skin thickness $s \simeq 1$ fm. For the SD WIMP–nucleus cross section, we only used the "thin–shell" nuclear form factor:

$$F^2_{\rm TS}(Q) = \begin{cases} j_0^2(qR_1)\,, & \text{for } qR_1 \le 2.55 \text{ or } qR_1 \ge 4.5\,, \\ {\rm const.} \simeq 0.047\,, & \text{for } 2.55 \le qR_1 \le 4.5\,. \end{cases} \tag{4}$$

2.1. WIMP mass m_χ

As one of the most important properties of halo WIMPs as well as the basic information for reconstructing other quantities in our model–independent analysis methods, I consider at first the determination of the WIMP mass m_χ by means of the method introduced in Ref. [2].

In Figs. 1 I show the reconstructed WIMP masses and the lower and upper bounds of their 1σ statistical uncertainties. The usual target combination of ^{28}Si + ^{76}Ge nuclei has been used

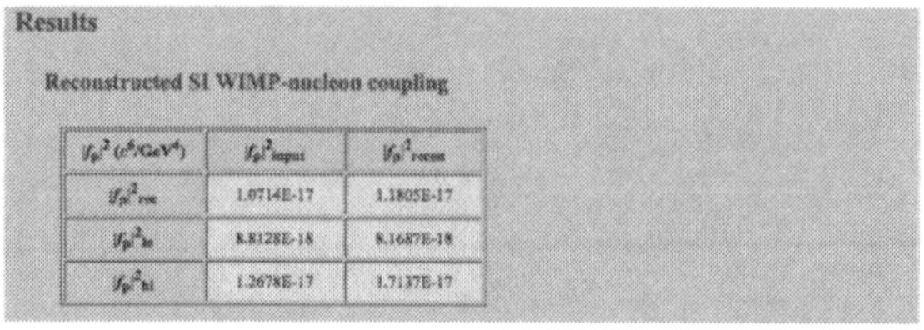

Results

Reconstructed SI WIMP-nucleon coupling

$\lvert f_p\rvert^2$ (c^4/GeV^4)	$\lvert f_p\rvert^2_{input}$	$\lvert f_p\rvert^2_{recon}$
$\lvert f_p\rvert^2_{rec}$	1.0714E-17	1.1805E-17
$\lvert f_p\rvert^2_{lo}$	8.8128E-18	8.1687E-18
$\lvert f_p\rvert^2_{hi}$	1.2678E-17	1.7137E-17

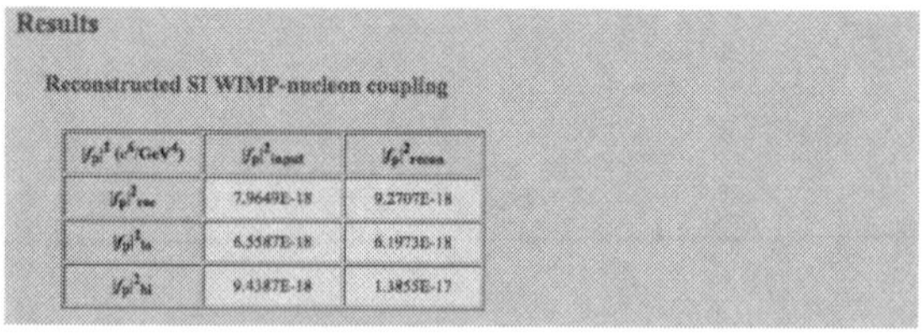

Results

Reconstructed SI WIMP-nucleon coupling

$\lvert f_p\rvert^2$ (c^4/GeV^4)	$\lvert f_p\rvert^2_{input}$	$\lvert f_p\rvert^2_{recon}$
$\lvert f_p\rvert^2_{rec}$	7.9649E-18	9.2707E-18
$\lvert f_p\rvert^2_{lo}$	6.5587E-18	6.1973E-18
$\lvert f_p\rvert^2_{hi}$	9.4387E-18	1.3855E-17

Figure 2. The reconstructed *squared* SI WIMP–nucleon coupling $|f_{\rm p}|^2$ by means of the method introduced in Ref. [3] with a ^{76}Ge target. The commonly used value of the local Dark Matter density $\rho_0 = 0.3\ {\rm GeV/cm^3}$ and a larger value of $\rho_0 = 0.4\ {\rm GeV/cm^3}$ as well as the elastic nuclear form factors given in Eqs. (1) and (2) have been used for estimating $|f_{\rm p}|^2$ in the left and right frames, respectively.

for this reconstruction, whereas two analytic forms of the elastic nuclear form factor given in Eqs. (1) and (2) have been used for determining m_χ in the left and right frames, respectively. While $m_{\chi,n}$ with $n = -1, 1, 2$ and $m_{\chi,\sigma}$ have been estimated by Eqs. (34) and (40) of Ref. [2], respectively, $m_{\chi,{\rm combined}}$ has been estimated by the χ^2–fitting defined in Eq. (51) of Ref. [2], which combines the estimators for $m_{\chi,n}$ and $m_{\chi,\sigma}$ with each other. The reconstructed WIMP mass $m_{\chi,{\rm combined}}$ as well as $m_{\chi,n}$ and $m_{\chi,\sigma}$ shown here have been corrected by the iterative $Q_{\rm max}$–matching procedure described in Ref. [2].

It can be found here that, although all single estimators ($m_{\chi,n}$ with $n = -1, 1, 2$ and $m_{\chi,\sigma}$) give generally a (relatively lighter) WIMP mass of ~ 60 GeV or even lighter and a 1σ upper bound of ~ 140 GeV, the *mean* values of the combined (in principle, more reliable) results (second columns) of the reconstructed WIMP mass give $m_\chi \sim 115$ GeV with a *rough* 1σ upper (lower) bound of ~ 190 (80) GeV, or, equivalently,

$$m_\chi \simeq 115^{+75}_{-35}\ {\rm GeV}\,. \tag{5}$$

Moreover, the combined results with two different form factors show not only a large overlap between ~ 85 GeV and ~ 180 GeV, but also a good coincidence: comparing to the $\sim^{+75}_{-35}$ GeV 1σ statistical uncertainty and the $\sim^{+45}_{-30}$ GeV overlap, the difference between two median values is $\lesssim 30$ GeV! This indicates that, for the first approximation of giving/constraining the most plausible range of the WIMP mass with pretty few total events, the uncertainty on the nuclear form factor could be safely neglected.

2.2. *Spin–independent WIMP–nucleon coupling $|f_{\rm p}|^2$*

Following the WIMP mass determination, I consider now the reconstruction of the SI WIMP coupling on nucleons $|f_{\rm p}|^2$ with a ^{76}Ge target [3] [1].

In Figs. 2 I show the reconstructed squared SI WIMP-nucleon couplings and the lower and upper bounds of their 1σ statistical uncertainties estimated by Eqs. (17) and (18) of Ref. [3] with an *assumed* (100±10 GeV, labeled with the subscript "input") and the reconstructed (from Sec. 2.1, labeled with "recon") WIMP masses. The commonly used value of the local Dark Matter density $\rho_0 = 0.3\ {\rm GeV/cm^3}$ and a larger value of $\rho_0 = 0.4\ {\rm GeV/cm^3}$ [15, 16, 17] as well as the elastic nuclear form factors given in Eqs. (1) and (2) have been used for estimating $|f_{\rm p}|^2$ in the left and right frames, respectively.

Among these results, the *mean* value and the *overlap* of two most plausible results (estimated

[1] Remind that the theoretical prediction by most supersymmetric models that the SI scaler WIMP couplings on protons and on neutrons are (approximately) equal: $f_{\rm p} \simeq f_{\rm n}$ has been adopted in the AMIDAS package.

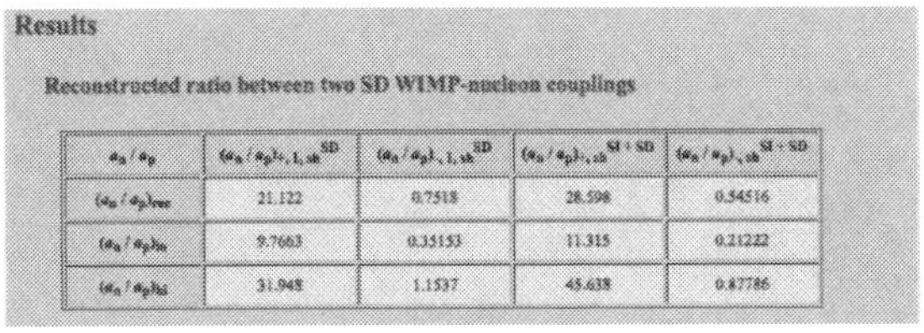

Results

Reconstructed ratio between two SD WIMP-nucleon couplings

a_n/a_p	$(a_n/a_p)_{+,1,sh}^{SD}$	$(a_n/a_p)_{-,1,sh}^{SD}$	$(a_n/a_p)_{+,sh}^{SI+SD}$	$(a_n/a_p)_{-,sh}^{SI+SD}$
$(a_n/a_p)_{rec}$	21.122	0.7518	28.598	0.54516
$(a_n/a_p)_{lo}$	9.7663	0.35153	11.315	0.21222
$(a_n/a_p)_{hi}$	31.948	1.1537	45.638	0.87786

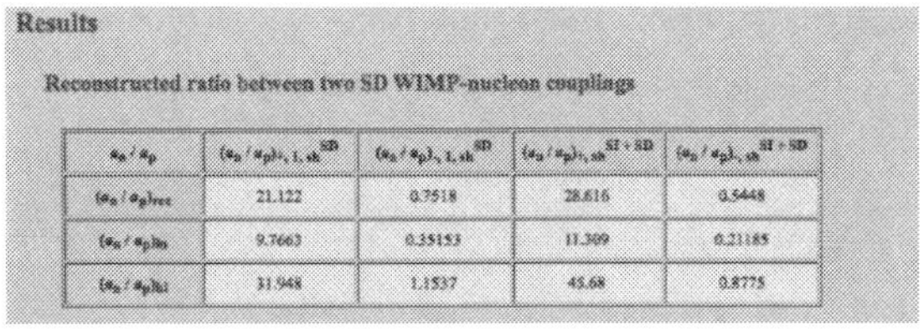

Results

Reconstructed ratio between two SD WIMP-nucleon couplings

a_n/a_p	$(a_n/a_p)_{+,1,sh}^{SD}$	$(a_n/a_p)_{-,1,sh}^{SD}$	$(a_n/a_p)_{+,sh}^{SI+SD}$	$(a_n/a_p)_{-,sh}^{SI+SD}$
$(a_n/a_p)_{rec}$	21.122	0.7518	28.616	0.5448
$(a_n/a_p)_{lo}$	9.7663	0.35153	11.309	0.21185
$(a_n/a_p)_{hi}$	31.948	1.1537	45.68	0.8775

Figure 3. The reconstructed ratio between two SD WIMP–nucleon couplings, $a_{\rm n}/a_{\rm p}$, by means of two methods introduced in Ref. [4]. As usual, the elastic nuclear form factors given in Eqs. (1) and (2) have been used for determining $a_{\rm n}/a_{\rm p}$ in the left and right frames, respectively.

with the reconstructed WIMP mass) give a *rough* 1σ range of

$$|f_{\rm p}|^2 \simeq 1.05^{+0.33}_{-0.24} \times 10^{-17}\ {\rm GeV}^{-4}, \tag{6}$$

or, equivalently,

$$|f_{\rm p}| \simeq 3.24^{+0.51}_{-0.37} \times 10^{-9}\ {\rm GeV}^{-2}. \tag{7}$$

Since the reconstructed WIMP mass given in Sec. 2.1 is $m_\chi \sim 115$ GeV, one can simply use the proton mass $m_{\rm p}$ to approximate the WIMP–proton reduced mass $m_{\rm r,p}$ and give a reconstructed SI WIMP–nucleon cross section as [10]

$$\sigma_{\chi{\rm p}}^{\rm SI} = \left(\frac{4}{\pi}\right) m_{\rm r,p}^2 |f_{\rm p}|^2 \approx \left(\frac{4}{\pi}\right) m_{\rm p}^2 |f_{\rm p}|^2 \simeq 5.03^{+1.58}_{-1.15} \times 10^{-9}\ {\rm pb}. \tag{8}$$

2.3. Ratio of two spin–dependent WIMP–nucleon couplings $a_{\rm n}/a_{\rm p}$

In Figs. 3 I show the reconstructed $a_{\rm n}/a_{\rm p}$ ratios and the lower and upper bounds of their 1σ statistical uncertainties estimated by Eqs. (2.7) and (2.12) of Ref. [4] with $n = 1$ as well as by Eqs. (3.16) and (3.20) of Ref. [4] at the shifted energy points [1, 4]. A combination of ^{19}F + ^{127}I targets has been used for the reconstruction of $a_{\rm n}/a_{\rm p}$ under the assumption that the SD WIMP–nucleus interaction dominates over the SI one (labeled with the superscript "SD"), whereas a third target of ^{28}Si has been combined with ^{19}F and ^{127}I for the case of the general combination of both SI and SD WIMP interactions (labeled with the superscript "SI + SD").

It can be found that, firstly, the "+ (plus)" solutions of the $a_{\rm n}/a_{\rm p}$ ratios given here are obviously too large to be the reasonable choice for $a_{\rm n}/a_{\rm p}$ and the "− (minus)" solutions should be the correct ones[2]. Secondly, although the reconstructed results under the assumption of the SD dominant WIMP interaction (third columns) is in general *larger* than the (in principle more plausible) results obtained without such a prior assumption (last columns), one could still use the *mean* value and the *overlap* of these two results to give a *rough* 1σ range of

$$\frac{a_{\rm n}}{a_{\rm p}} \simeq 0.65^{+0.23}_{-0.30}. \tag{9}$$

2.4. Ratios of the SD and SI WIMP–nucleon couplings $\sigma_{\chi({\rm p,n})}^{\rm SD}/\sigma_{\chi{\rm p}}^{\rm SI}$

In Figs. 4 I show the reconstructed $\sigma_{\chi({\rm p,n})}^{\rm SD}/\sigma_{\chi{\rm p}}^{\rm SI}$ ratios and the lower and upper bounds of their 1σ statistical uncertainties estimated by Eqs. (3.9), (3.10) and (3.21) of Ref. [4] (with $a_{\rm n}/a_{\rm p}$ estimated by Eq. (3.16) of Ref. [4]) as well as by Eqs. (3.25) and (3.29) of Ref. [4] at the shifted energy points. A combination of data sets of ^{19}F, ^{127}I and ^{28}Si targets (labeled with

[2] Remind that, as discussed in Ref. [4], the correct choice from the "+" and "−" solutions could be decided directly by the values of the group spins of protons and neutrons of the used target nuclei, $\langle S_{\rm (p,n)} \rangle$.

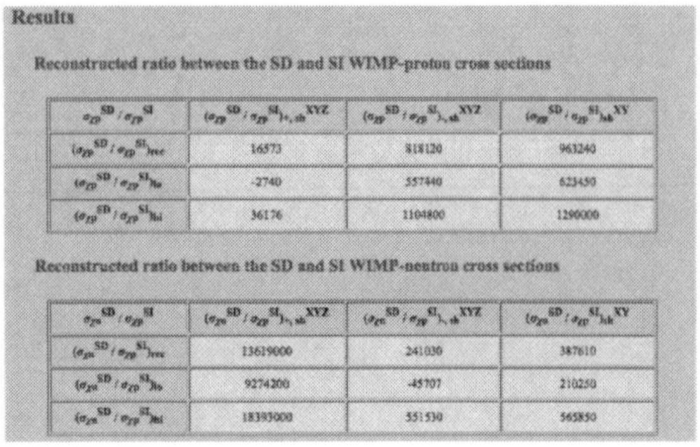

Results

Reconstructed ratio between the SD and SI WIMP-proton cross sections

$\sigma_{\chi p}^{SD}/\sigma_{\chi p}^{SI}$	$(\sigma_{\chi p}^{SD}/\sigma_{\chi p}^{SI})_{+,sh}^{XYZ}$	$(\sigma_{\chi p}^{SD}/\sigma_{\chi p}^{SI})_{-,sh}^{XYZ}$	$(\sigma_{\chi p}^{SD}/\sigma_{\chi p}^{SI})_{sh}^{XY}$
$(\sigma_{\chi p}^{SD}/\sigma_{\chi p}^{SI})_{rec}$	16573	818120	963240
$(\sigma_{\chi p}^{SD}/\sigma_{\chi p}^{SI})_{Na}$	-2740	557440	623450
$(\sigma_{\chi p}^{SD}/\sigma_{\chi p}^{SI})_{Xe}$	36176	1104800	1290000

Reconstructed ratio between the SD and SI WIMP-neutron cross sections

$\sigma_{\chi n}^{SD}/\sigma_{\chi p}^{SI}$	$(\sigma_{\chi n}^{SD}/\sigma_{\chi p}^{SI})_{+,sh}^{XYZ}$	$(\sigma_{\chi n}^{SD}/\sigma_{\chi p}^{SI})_{-,sh}^{XYZ}$	$(\sigma_{\chi n}^{SD}/\sigma_{\chi p}^{SI})_{sh}^{XY}$
$(\sigma_{\chi n}^{SD}/\sigma_{\chi p}^{SI})_{rec}$	13619000	241030	387610
$(\sigma_{\chi n}^{SD}/\sigma_{\chi p}^{SI})_{Na}$	9274200	-45707	210250
$(\sigma_{\chi n}^{SD}/\sigma_{\chi p}^{SI})_{Xe}$	18393000	551530	565850

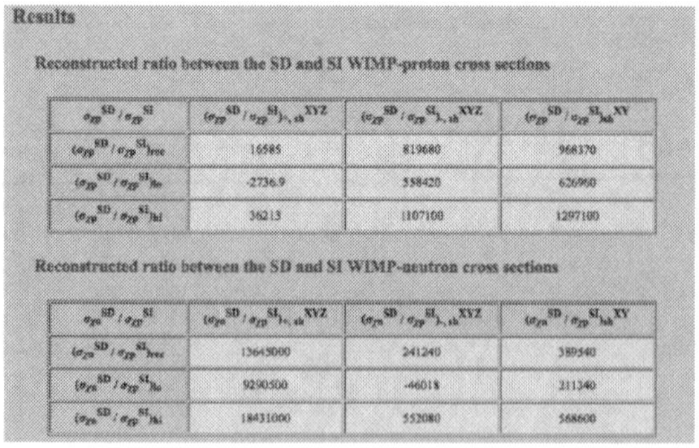

Results

Reconstructed ratio between the SD and SI WIMP-proton cross sections

$\sigma_{\chi p}^{SD}/\sigma_{\chi p}^{SI}$	$(\sigma_{\chi p}^{SD}/\sigma_{\chi p}^{SI})_{+,sh}^{XYZ}$	$(\sigma_{\chi p}^{SD}/\sigma_{\chi p}^{SI})_{-,sh}^{XYZ}$	$(\sigma_{\chi p}^{SD}/\sigma_{\chi p}^{SI})_{sh}^{XY}$
$(\sigma_{\chi p}^{SD}/\sigma_{\chi p}^{SI})_{rec}$	16585	819680	968370
$(\sigma_{\chi p}^{SD}/\sigma_{\chi p}^{SI})_{Na}$	-2736.9	558420	626960
$(\sigma_{\chi p}^{SD}/\sigma_{\chi p}^{SI})_{Xe}$	36213	1107100	1297100

Reconstructed ratio between the SD and SI WIMP-neutron cross sections

$\sigma_{\chi n}^{SD}/\sigma_{\chi p}^{SI}$	$(\sigma_{\chi n}^{SD}/\sigma_{\chi p}^{SI})_{+,sh}^{XYZ}$	$(\sigma_{\chi n}^{SD}/\sigma_{\chi p}^{SI})_{-,sh}^{XYZ}$	$(\sigma_{\chi n}^{SD}/\sigma_{\chi p}^{SI})_{sh}^{XY}$
$(\sigma_{\chi n}^{SD}/\sigma_{\chi p}^{SI})_{rec}$	13645000	241240	389540
$(\sigma_{\chi n}^{SD}/\sigma_{\chi p}^{SI})_{Na}$	9290500	-46018	211340
$(\sigma_{\chi n}^{SD}/\sigma_{\chi p}^{SI})_{Xe}$	18431000	552080	568600

Figure 4. The reconstructed ratios between the SD and SI WIMP–nucleon couplings, $\sigma_{\chi(\mathrm{p,n})}^{\mathrm{SD}}/\sigma_{\chi\mathrm{p}}^{\mathrm{SI}}$, by means of two methods introduced in Ref. [4]. As usual, the elastic nuclear form factors given in Eqs. (1) and (2) have been used for determining $\sigma_{\chi(\mathrm{p,n})}^{\mathrm{SD}}/\sigma_{\chi\mathrm{p}}^{\mathrm{SI}}$ in the left and right frames, respectively.

the superscript "XYZ") and that of data sets of ^{23}Na or ^{131}Xe with the (common) one of ^{76}Ge (labeled with the superscript "XY") have been used and the *mean* value and the *overlap* of these two results give a *rough* 1σ range of

$$\frac{\sigma_{\chi\mathrm{p}}^{\mathrm{SD}}}{\sigma_{\chi\mathrm{p}}^{\mathrm{SI}}} \simeq 8.94_{-2.67}^{+2.13} \times 10^5\,, \qquad \frac{\sigma_{\chi\mathrm{n}}^{\mathrm{SD}}}{\sigma_{\chi\mathrm{p}}^{\mathrm{SI}}} \simeq 3.16_{-1.05}^{+2.36} \times 10^5\,. \tag{10}$$

Then, firstly, from these results one can further obtain that [10][3]

$$\left|\frac{a_\mathrm{n}}{a_\mathrm{p}}\right| \simeq 0.59_{-0.13}^{+0.23}\,. \tag{11}$$

Secondly, combining the results in Eq. (10) with $\sigma_{\chi\mathrm{p}}^{\mathrm{SI}}$ given in Eq. (8), one can obtain that [10]

$$\sigma_{\chi\mathrm{p}}^{\mathrm{SD}} \simeq 4.50_{-1.69}^{+1.77} \times 10^{-3}\ \mathrm{pb}\,, \qquad \sigma_{\chi\mathrm{n}}^{\mathrm{SD}} \simeq 1.59_{-0.64}^{+1.08} \times 10^{-3}\ \mathrm{pb}\,. \tag{12}$$

These results give in turn that [10]

$$|a_\mathrm{p}| = 0.112_{-0.021}^{+0.022}\,, \qquad |a_\mathrm{n}| = 0.067_{-0.013}^{+0.023}\,. \tag{13}$$

On the other hand, one can also use the reconstructed $a_\mathrm{n}/a_\mathrm{p}$ ratio given in Eq. (9) and *one* of the two results given in Eq. (12) to obtain that [10]

$$\sigma_{\chi\mathrm{p}}^{\mathrm{SD}} \simeq 3.76_{-3.81}^{+3.69} \times 10^{-3}\ \mathrm{pb}\,, \qquad \sigma_{\chi\mathrm{n}}^{\mathrm{SD}} \simeq 1.90_{-1.89}^{+1.54} \times 10^{-3}\ \mathrm{pb}\,. \tag{14}$$

These results can also give that

$$|a_\mathrm{p}| \simeq 0.103_{-0.052}^{+0.050}\,, \qquad |a_\mathrm{n}| \simeq 0.073_{-0.036}^{+0.030}\,. \tag{15}$$

It can be found that, not surprisingly, the statistical uncertainties on the reconstructed $\sigma_{\chi(\mathrm{p,n})}^{\mathrm{SD}}$ given in Eq. (14) are ~ 2 or 3 times larger than those given in Eq. (12): Since $\sigma_{\chi(\mathrm{p,n})}^{\mathrm{SD}}/\sigma_{\chi\mathrm{p}}^{\mathrm{SI}}$

[3] Remind that the results given in the second and third columns of the tables in Figs. 4 are reconstructed with the $a_\mathrm{n}/a_\mathrm{p}$ ratio given in the last columns of the tables in Figs. 3.

reconstructed with the F + I + Si combination involve already the reconstructed a_n/a_p ratio given in Eq. (9), the uncertainties on $\sigma_{\chi(p,n)}^{SD}$ given in Eq. (14) are thus *overestimated*. Secondly, the reconstructed a_n/a_p ratio given in Eqs. (9) and (11) and the reconstructed $\sigma_{\chi(p,n)}^{SD}$ given in Eqs. (12) and (14) seem to match to each other pretty well.

The analyses given here show that, firstly, once one can estimate the SI WIMP–nucleon coupling/cross section, $|f_p|$ or $\sigma_{\chi p}^{SI}$, and (one of) the ratios between the SD and SI WIMP–nucleon cross sections, and/or the ratio between two SD WIMP–nucleon couplings, the other couplings/cross sections could in principle be estimated. Secondly, the WIMP couplings/cross sections estimated in different way would be self–cross–checks to each other and the (in)compatibility between the reconstructed results would also help us to check the usefulness of the analyzed data sets offered from different experiments with different detector materials [14].

3. Input setup for generating pseudodata

In Table 1 I give finally the input setup for generating the pseudodata sets used in the analyses demonstrated in the previous section. For comparison, the reconstructed results shown in the previous section are also summarized here.

For generating WIMP signals, we used the commonly used form for the nuclear form factor given in Eq. (2) with another often used analytic form for R_1:

$$R_1 = \sqrt{R_A^2 + \left(\frac{7}{3}\right)\pi^2 r_0^2 - 5s^2} \tag{16}$$

with $R_A \simeq (1.23\,A^{1/3} - 0.6)$ fm, $r_0 \simeq 0.52$ fm, $s \simeq 0.9$ fm. Moreover, the shifted Maxwellian velocity distribution:

$$f_{1,\text{sh}}(v) = \frac{1}{\sqrt{\pi}}\left(\frac{v}{v_e v_0}\right)\left[e^{-(v-v_e)^2/v_0^2} - e^{-(v+v_e)^2/v_0^2}\right] \tag{17}$$

with the Sun's Galactic orbital velocity $v_0 = 230$ km/s has been used; v_e is the *time–dependent* Earth's velocity in the Galactic frame:

$$v_e(t) = v_0\left[1.05 + 0.07\cos\left(\frac{2\pi(t - t_p)}{1\ \text{yr}}\right)\right], \tag{18}$$

where the date on which the Earth's velocity relative to the WIMP halo is maximal has been set as $t_p = 140$ d. In addition, different from our setup used in Ref. [10], the experimental running date has been set as $t_{\text{expt}} = 100$ d and thus $v_e(t_{\text{expt}}) = 254$ km/s, much larger than the usually used values: 200 km/s $\le v_e \le$ 240 km/s. Although these values for the astronomical setup are *non–standard*, we would like to stress that, as shown in the previous section and Table I, such a non–standard halo (model) would not affect the reconstructed results, since for using the **AMIDAS** package and website to analyze (real) data sets, one needs only the form factors for SI and/or SD WIMP–nucleaus cross sections, prior knowledge/assumptions about the WIMP velocity distribution $f_1(v)$ and local density ρ_0 (except the estimation of the SI WIMP–nucleon coupling $|f_p|^2$) are *not required*.

On the other hand, for generating background events, the target–dependent exponential form for the residue background spectrum introduced in Ref. [11]:

$$\left(\frac{dR}{dQ}\right)_{\text{bg,ex}} = \exp\left(-\frac{Q/\text{keV}}{A^{0.6}}\right) \tag{19}$$

Property	Reconstructed value	Input/Estimated value	Remarks
m_χ	115^{+75}_{-35} GeV	130 GeV	
$\sigma_{\chi\mathrm{p}}^{\mathrm{SI}}$	$5.03^{+1.58}_{-1.15} \times 10^{-9}$ pb	4×10^{-9} pb	$f_\mathrm{n} = f_\mathrm{p}$
$\lvert f_\mathrm{p}\rvert^2$	$1.05^{+0.33}_{-0.24} \times 10^{-17}$ GeV^{-4}	9.305×10^{-18} GeV^{-4}	†
$\lvert f_\mathrm{p}\rvert$	$3.24^{+0.51}_{-0.37} \times 10^{-9}$ GeV^{-2}	3.050×10^{-9} GeV^{-2}	†
a_p	$0.112^{+0.022}_{-0.021}$	0.1	
a_n	$0.067^{+0.023}_{-0.013}$	0.07	
$a_\mathrm{n}/a_\mathrm{p}$	$0.65^{+0.23}_{-0.30}$, $0.59^{+0.23}_{-0.13}$	0.7	
$\sigma_{\chi\mathrm{p}}^{\mathrm{SD}}$	$4.50^{+1.77}_{-1.69} \times 10^{-3}$ pb	3.51×10^{-3} pb	†
$\sigma_{\chi\mathrm{n}}^{\mathrm{SD}}$	$1.59^{+1.08}_{-0.64} \times 10^{-3}$ pb	1.72×10^{-3} pb	†
$\sigma_{\chi\mathrm{p}}^{\mathrm{SD}}/\sigma_{\chi\mathrm{p}}^{\mathrm{SI}}$	$8.94^{+2.13}_{-2.67} \times 10^5$	8.77×10^5	†
$\sigma_{\chi\mathrm{n}}^{\mathrm{SD}}/\sigma_{\chi\mathrm{p}}^{\mathrm{SI}}$	$3.16^{+2.36}_{-1.05} \times 10^5$	4.30×10^5	†
$F_{\mathrm{SI}}^2(Q)$		$F_{\mathrm{SI}}^2(Q)$ in Eq. (2)	‡
$F_{\mathrm{SD}}^2(Q)$		$F_{\mathrm{TS}}^2(Q)$ in Eq. (4)	
ρ_0		0.4 GeV/cm^3	
t_p		140 d	
t_{expt}		100 d	‡
v_0		230 km/s	
$v_{\max}$		600 km/s	
$v_\mathrm{e}(t_{\mathrm{expt}})$		253.9 km/s	‡
r_{bg}		0.12	‡

Table 1. The input setup for generating the pseudodata sets used in the analyses demonstrated in this article. The theoretically estimated values and the reconstructed results are also given. †: estimated for 130 GeV m_χ; ‡: different from Ref. [10].

has been used. Here A is the atomic mass number of the target nucleus; the power index of A, 0.6, is an empirical constant, which has been chosen so that the exponential background spectrum is somehow similar to, but still different from the expected recoil spectrum of the target nuclei (see Figs. 5)[4]. Meanwhile, considering possible radioactivity with a characteristic energy, we combined the exponential background spectrum in Eq. (19) with a Gaussian excess:

$$\left(\frac{dR}{dQ}\right)_{\mathrm{bg,Gau}} = \frac{r_{\mathrm{Gau}}}{\sqrt{2\pi}\,(\sigma_{Q,\mathrm{bg}}/\mathrm{keV})} \exp\left[-\frac{(Q - Q_{\mathrm{bg,peak}})^2}{2\sigma_{Q,\mathrm{bg}}^2}\right], \tag{20}$$

[4] Note that, among different possible choices the atomic mass number A has been chosen as the simplest, unique characteristic parameter in the general analytic form (19) for defining the residue background spectrum for *different* target nuclei. However, it does *not* mean that the (superposition of the real) background spectra would depend simply/primarily on A or on the mass of the target nucleus, m_N.

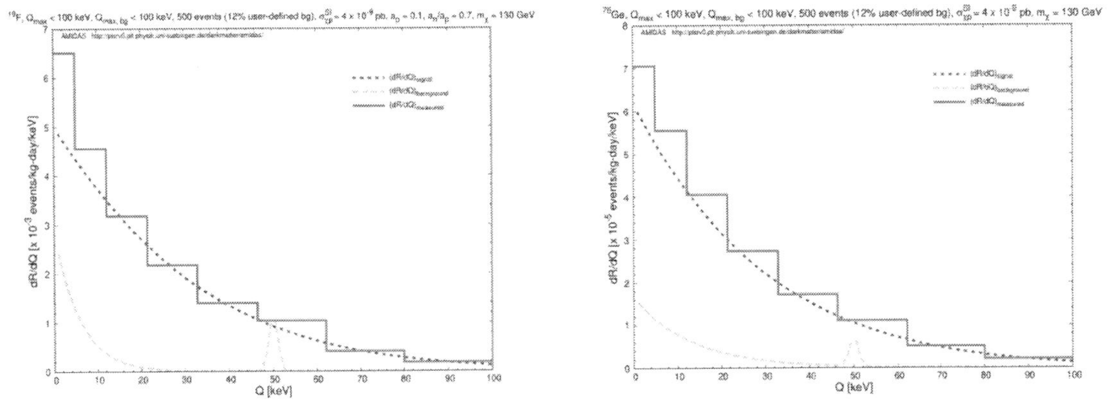

Figure 5. Measured energy spectra (solid red histograms) for a ^{19}F (left) and a ^{76}Ge (right) targets. The dotted blue curves are the elastic WIMP–nucleus scattering spectra, whereas the dashed green curves are the exponential background spectra combined with the Gaussian excesses ($r_{\rm Gau} = 1$, $Q_{\rm bg,peak} = 50$ keV and $\sigma_{Q,\rm bg} = 1$ keV) and normalized to fit to the chosen background ratio $r_{\rm bg} = 12\%$. See the text for further details.

where $Q_{\rm bg,peak}$ and $\sigma_{Q,\rm bg}$ are the characteristic energy and energy dispersion of this background excess, respectively; $r_{\rm Gau}$ is the ratio of this Gaussian excess to the exponential spectrum.

In Figs. 5 I show the measured energy spectra (solid red histograms) for a ^{19}F (left) and a ^{76}Ge (right) targets. The dotted blue curves are the elastic WIMP–nucleus scattering spectra, whereas the dashed green curves are the exponential background spectra given in Eq. (19) combined with the Gaussian excesses given in Eq. (20) ($r_{\rm Gau} = 1$, $Q_{\rm bg,peak} = 50$ keV and $\sigma_{Q,\rm bg} = 1$ keV for both targets), which have been normalized so that the ratios of the areas under the background spectra to those under the (dotted blue) WIMP scattering spectra are equal to the background–signal ratio ($r_{\rm bg} = 12\%$) in the whole data sets. The experimental threshold energies have been assumed to be negligible and the maximal cut-off energies are set as 100 keV. 500 total events on average in one experiment have been simulated.

It can be seen here that[5], firstly, due to the contribution of the unrejected background events in low energy ranges ($Q \lesssim 20$ keV for ^{19}F and $Q \lesssim 40$ keV for ^{76}Ge), the counting rates at the first energy bins (one of the two important quantities required in our model-independent data analyses) have been (strongly) overestimated. Secondly, the Gaussian background excesses around $Q = 50$ keV cause clearly overestimates of the event rates, which would not only contribute (significantly) to the estimates of $I_n = \sum_{\rm all\ events} Q_i^{(n-1)/2}/F^2(Q_i)$ [1], but also cause a larger statistical fluctuation [2].

Nevertheless, our results obtained by analyzing pseudodata sets of $\mathcal{O}(50)$ total events showed that, firstly, the WIMP mass given in Eq. (5) can match the input value very well: the deviations between the input and the reconstructed values with different assumed nuclear form factors are $\lesssim 20\%$ ($\lesssim 30$ GeV). As discussed earlier, this indicates that, for the first approximation of giving/constraining the most plausible range of the WIMP mass with pretty few total events, the uncertainty on the nuclear form factor could be safely neglected.

Secondly, all WIMP–nucleon couplings/cross sections as well as the ratios between them have also been reconstructed with $\sim 2\%$ to $\lesssim 30\%$ deviations from the input/theoretically estimated values. Although the SI WIMP coupling $|f_{\rm p}|$ estimated with the larger (input) local Dark Matter density (right frame of Figs. 2) could be underestimated [3], one can at least give an upper bound

[5] More detailed illustrations and discussions about the effects of residue background events on the measured energy spectrum can be found in Refs. [11, 14].

on $|f_\mathrm{p}|$. Moreover, by combining different methods for estimating different (ratios between the) WIMP couplings/cross sections, one could in principle observe/confirm the (in)compatibility between these results and probably correct the reconstructed values.

4. Summary

In this article I demonstrated the data analysis results of extracted WIMP properties by using theoretically generated pseudodata for different target nuclei, taking into account some unrejected background events mixed in the analyzed data sets. As an extension as well as the complementarity of our earlier theoretical works, I combined reconstructed results of the (ratios between different) WIMP couplings/cross sections on nucleons to estimate each *individual* coupling/cross section. Hopefully, the `AMIDAS` package and website as well as this demonstration can help our experimental colleagues to analyze their *real* direct detection data in the near future and to determine (at least rough ranges of) properties of halo Dark Matter particles.

Acknowledgments
The author appreciates the ILIAS Project and the Physikalisches Institut der Universität Tübingen for kindly providing the opportunity of the collaboration and the technical support of the `AMIDAS` website. The author would also like to thank the friendly hospitality of the Kavli Institute for Theoretical Physics China at the Chinese Academy of Sciences (KITPC) during the DSU workshop and the "Dark Matter and New Physics" program. This work was partially supported by the National Science Council of R.O.C. under contract no. NSC-99-2811-M-006-031 as well as by the National Center of Theoretical Sciences, R.O.C..

References

[1] Drees M and Shan C-L 2007 Reconstructing the velocity distribution of WIMPs from direct dark matter detection data *J. Cosmol. Astropart. Phys.* JCAP06(2007)011 (*Preprint* astro-ph/0703651)
[2] Drees M and Shan C-L 2008 Model-independent determination of the WIMP mass from direct dark matter detection data *J. Cosmol. Astropart. Phys.* JCAP06(2008)012 (*Preprint* 0803.4477)
[3] Shan C-L 2011 Estimating the spin-independent WIMP-nucleon coupling from direct dark matter detection data (*Preprint* 1103.0481)
[4] Shan C-L 2011 Determining ratios of WIMP-nucleon cross sections from direct dark matter detection data *J. Cosmol. Astropart. Phys.* JCAP07(2011)005 (*Preprint* 1103.0482)
[5] `http://pisrv0.pit.physik.uni-tuebingen.de/darkmatter/index1.html`
[6] `http://www-ilias.cea.fr/`
[7] `http://pisrv0.pit.physik.uni-tuebingen.de/darkmatter/amidas/`
[8] Shan C-L 2010 The `AMIDAS` website: an online tool for direct dark matter detection experiments *AIP Conf. Proc.* **1200** 1031 (*Preprint* 0909.1459)
[9] Shan C-L 2009 Uploading user-defined functions onto the `AMIDAS` website (*Preprint* 0910.1971)
[10] Shan C-L 2011 Analyzing direct dark matter detection data on the `AMIDAS` website (*Preprint* 1109.0125)
[11] Chou Y-T and Shan C-L 2010 Effects of residue background events in direct dark matter detection experiments on the determination of the WIMP mass *J. Cosmol. Astropart. Phys.* JCAP08(2010)014 (*Preprint* 1003.5277)
[12] Shan C-L 2010 Effects of residue background events in direct dark matter detection experiments on the reconstruction of the velocity distribution function of halo WIMPs *J. Cosmol. Astropart. Phys.* JCAP06(2010)029 (*Preprint* 1003.5283)
[13] Shan C-L 2011 Effects of residue background events in direct dark matter detection experiments on the estimation of the spin-independent WIMP-nucleon coupling (*Preprint* 1103.4049)
[14] Shan C-L 2011 Effects of residue background events in direct dark matter detection experiments on the determinations of ratios of WIMP-nucleon cross sections (*Preprint* 1104.5305)
[15] Catena R and Ullio P 2010 A novel determination of the local dark matter density *J. Cosmol. Astropart. Phys.* JCAP08(2010)004 (*Preprint* 0907.0018)
[16] Salucci P, Nesti F, Gentile G and Martins C F 2010 The dark matter density at the sun's location *Astron. Astrophys.* **523** A83 (*Preprint* 1003.3101)
[17] Pato M, Agertz O, Bertone G, Moore B and Teyssier R 2010 Systematic uncertainties in the determination of the local dark matter density *Phys. Rev.* D **82** 023531 (*Preprint* 1006.1322)

Recoiling Ion-Channeling in Direct DM Detectors

Graciela B. Gelmini

Department of Physics and Astronomy, University of California Los Angeles (UCLA), 475 Portola Plaza, Los Angeles, CA 90095, USA

E-mail: gelmini@physics.ucla.edu

Abstract. The channeling of the recoiling nucleus in crystalline detectors after a WIMP collision would produce a larger scintillation or ionization signal in direct detection experiments than otherwise expected. I present estimates of channeling fractions obtained using analytic models developed from the 1960's onwards to describe channeling and blocking effects. We find the fractions to be too small to affect the fits to potential WIMP candidates. I also examine the possibility of detecting a daily modulation of the dark matter signal due to channeling.

The potential importance of the channeling effect for direct dark matter detection was first pointed out for NaI(Tl) by Drobyshevski [1] in 2007 and soon after by the DAMA collaboration [2]. Usually only a small fraction $E = QE_R$ of the recoil energy E_R goes to either scintillation or ionization (e.g. $Q_{\rm I} = 0.09$ and $Q_{\rm Na} = 0.30$). Channeled recoiling ions instead give all their energy to electrons, so $Q = 1$ and they produce a larger ionization or scintillation signal. The DAMA paper [2] gave an estimate of the channeling fractions of recoiling Na and I ions as function of the energy in which the fractions grew with decreasing energy to be $\simeq 1$ close to 1 keV (see Fig.4.a). With these channeling fractions, the region of light WIMPs (Weakly Interacting Massive Particle with mass 10 GeV or less) compatible with producing an annual modulation as measured by the DAMA collaboration shifted considerably to lower cross sections, by about one order of magnitude (see e.g. [3] or [4]). Besides channeling being important for fits to WIMP data, it is also important because it could produce a daily modulation of the signal dependent on the orientation of the crystal with respect to the galaxy, which would thus be a background free dark matter signature. In fact, as first pointed out in Ref. [5], Earth's rotation makes the average WIMP velocity change direction with respect to the crystal which produces a daily modulation in the measured recoil energy (equivalent to a modulation of the quenching factors Q), as most recoils successively come in and go out of crystal channels.

My collaborators N. Bozorgnia and P. Gondolo and I, set out more than three years ago to do an analytic calculation of channeling in dark matter detectors, initially to estimate amplitudes of daily modulation due to channeling that could be expected, and realized that the channeling fraction estimates of DAMA did not include "blocking", an effect which makes channeling fractions of recoiling ions much smaller at low energies [6, 7, 8, 9].

Channeling and blocking in crystals refer to the orientation dependence of ion penetration in crystals. "Channeling" occurs when ions propagating in a crystal along symmetry axes and planes suffer a series of small-angle scatterings that maintain them in the open "channels" in between rows (axial channels) or planes (planar channels) of lattice atoms and thus penetrate much further than non-channeled ions and give all their energy to electrons. Channeled ions do not get close to lattice sites, where they would suffer a large-angle scattering which would take

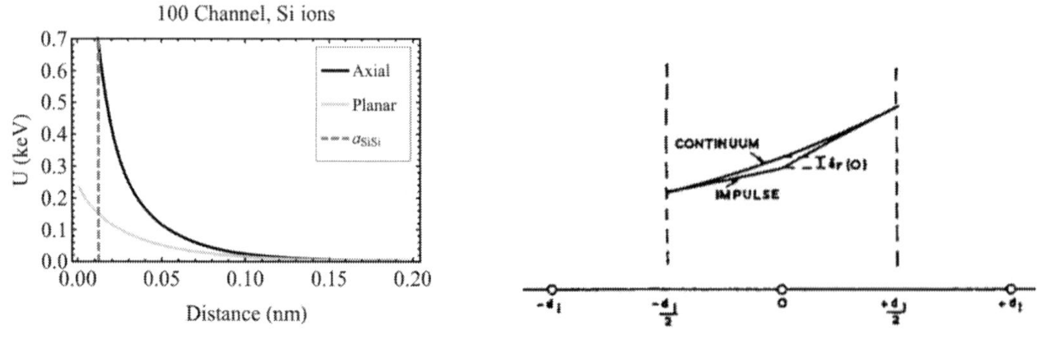

Figure 1. a.(left) Continuum transverse potential U for a string (black) and plane (green) for a Si ion propagating in Si as function of the distance perpendicular to the string or plane [7]. b.(right) The true ion path lies between those predicted by the continuum and impulse approximations, schematically shown in the figure taken from Ref. [13]. The interatomic spacing in the row is here d_1 and two successive half-way planes are shown at $-d_1/2$ and $+d_1/2$.

them out of the channel. "Blocking" is the reduction of the flux of ions originating in lattice sites along symmetry axes and planes due to the shadowing effect of the lattice atoms directly in front of the emitting lattice site (see e.g. the review by D. Gemmell [10] and references therein).

Channeling and blocking effects in crystals are extensively used in crystallography, in measurements of short nuclear lifetimes, in the production of polarized beams etc. However in all these instances the ion energies much larger than the keV to 10's of keV expected in WIMPs collisions. There is one instance in which energies are lower: channeling must be avoided in the implantation of B, P and As atoms in Si crystals to make circuits [11].

There are two main theoretical tools to model channeling and blocking in crystals. One of them consists of large simulations codes based on binary collision models (MARLOWE, UT-MARLOWE, CRYSTAl...) using Monte-Carlo codes with stochastic methods to locate crystal atoms and determine impact parameters and scattering angles (TRIM-SRIM). Some of them commercially available (e.g. from the SILVACO company). The second consists of classical analytic models developed in the 1960's and 70's, in particular by Jens Lindhard in 1965 [12]. These are the models used by the DAMA collaboration and us to estimate channeling fractions.

My collaborators and I used in particular Lindhard's model [12] supplemented by the planar channel models of Morgan and Van Vliet [13] and the 1995-1996 work of G. Hobler [11] on low energy channeling (applied to ion implantation in Si). In these models the rows and planes of lattice atoms are replaced by continuum uniformly charged strings and planes, in which the screened Thomas-Fermi potential is averaged over a direction parallel to a row or plane of lattice atoms to find the transverse continuous potential U, either $U_{\rm axial}(r)$ or $U_{\rm planar}(x)$, which depends only on the distance perpendicular to the string or plane, respectively (see Fig. 1.a).

The continuum model does not imply that the potential energy of an ion moving e.g. near an atomic row is well approximated by the continuum potential U. The actual potential consists of sharp peaks near the atoms and deep valleys in between. The continuum model says that the net deflection due to the succession of impulses from the peaks is identical to the deflection due to a force $-U'$. This is only so if the ion never approaches any individual atom so closely that it suffers a large-angle collision. Channeled ions must not get closer than a critical distance ρ_c (r_c for an axial channel or x_c for a planar channel) to the string or plane. We included just one string or plane per channel in our calculation, which is a good approximation except when ρ_c approaches the half-width of the channel, which happens only at very low energies.

Along each continuum uniformly charged string and plane there is translation invariance, thus the parallel component of the propagating channeled ion velocity is constant. The only change in energy of the propagating ion is due to the change of the perpendicular component of the velocity $\mathrm{v}_{\mathrm{perp}} = \mathrm{v}\sin\phi \simeq \mathrm{v}\phi$, due to the the work of the forces $-U'$, which are only perpendicular too. Thus, for channeled ions the so called "transverse energy" $E_{\mathrm{perp}} = M\mathrm{v}^2_{\mathrm{perp}}/2 + U \simeq E\phi^2 + U$ is conserved (M, $E = M\mathrm{v}^2/2$ are the mass and kinetic energy of the ion).

The conservation of the "transverse energy" implies that for a particular value of E_{perp} (determined by the recoil energy E, initial recoil angle ϕ_i and initial potential and U_i), a channeled ion does not approach the string or plane closer than a minimum distance ρ_{min} (for which $\mathrm{v}_{\mathrm{perp}} = 0$) and that far away from the string or plane, close to the middle of the channel (where the potential is U_{middle}), the ion moves on a trajectory forming an angle ψ with the string or plane, given by $E_{\mathrm{perp}} = E\phi_i^2 + U_i = U(\rho_{\mathrm{min}}) = E\psi^2 + U_{\mathrm{middle}}$. Channeling occurs when $\rho_{\mathrm{min}} > \rho_c$, so $U(\rho_{\mathrm{min}}) < U(\rho_c)$, which amounts to

$$\psi = \sqrt{\left[U(\rho_{\mathrm{min}}) - U_{\mathrm{middle}})\right]/E} \leq \psi_c = \sqrt{\left[U(\rho_c) - U_{\mathrm{middle}})\right]/E}. \tag{1}$$

Here we have ψ_c defined in terms of ρ_c. All the difficulty of this approach resides in calculating the critical distances ρ_c for each channel.

The condition for channeling which defines ρ_c was equated by Lindhard in 1965 to the condition that E_{perp} be conserved. For an axial channel, Lindhard defined "half-way" planes perpendicular to the row of lattice atoms half-way in between two successive atoms, and requiring that E_{perp} be the same at leading order in two successive half-way planes (see Appendix A of Ref. [12]) obtained the axial channeling condition

$$d^2 U''_{\mathrm{axial}}(r)/8E < 1. \tag{2}$$

Here d is the interatomic distance in the row considered, r is the radial distance between the propagating ion and the row, and $U''_{\mathrm{axial}}(r)$ is the second derivative of the axial continuum potential with respect to r. This condition is fulfilled only at distances $r > r_c$, what defines r_c. The same condition for axial channels in Eq. 2 was derived later by Morgan and Van Vliet [13] (with 16 instead of 8) by requiring that the ion position at two successive half-way planes predicted by the continuum and the "impulse" approximations be the same, to first order (see Fig. 1.b). In the impulse approximation the ion suffers an abrupt impulse that changes its momentum only at the closest distance from each lattice atoms. The true situation is in between the impulse and the continuum approximations.

For planar channels, no condition similar to Eq. 2 using the second derivative of $U_{\mathrm{planar}}(x)$ can be derived (thus we believe the planar channeling condition of Ref. [14] not to be correct). Planar channels are more complicated than axial channels. To start with, which lattice atoms on a plane are closest to the projection of the ion path on the plane depends on the direction of the path itself, and these lattice atoms are in general not at regular distances. For planar channels we followed the approach proposed by Morgan and Van Vliet [13], used also by Hobler [11], of defining a "fictitious string" along the projection of the propagating ion path on the plane.

In the following we call $\rho_c(E)$ the critical distance for a static lattice, i.e. one in which vibrations are neglected, obtained from Eq. 2 or the equivalent condition using the "fictitious string" approach for planar channels. As shown in Fig. 2.a, the static critical distance is a decreasing function of E. However, for our purpose we cannot neglect lattice vibrations.

The channeling of an ion depends not only on the initial angle its trajectory makes with a string or plane, but also on its initial position. Only for ions starting their motion close to the middle of the channel the channeling condition is simply $\psi < \psi_c$ However this is not the case in dark matter detectors, since the recoiling ions start their motion at or close to their original lattice sites (and leave those sites empty). For recoiling ions, blocking effects are important. In fact in a perfect static lattice no recoiling ion would be channeled because of what Lindhard

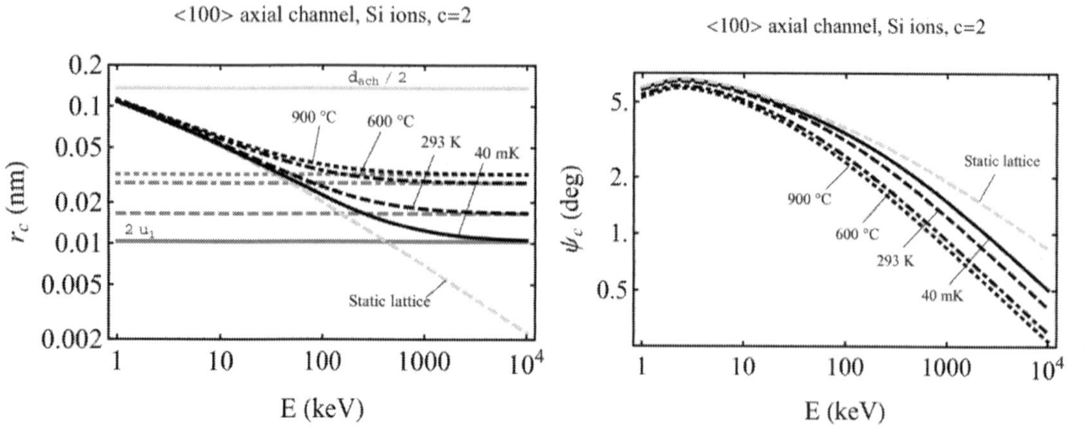

Figure 2. a.(left) Critical channeling distance r_c (Eq. 3 where $\rho_c = r_c$ for an axial channel) and b.(right) critical channeling angles ψ_c (Eq. 1) for a Si ion propagating in the <100> axial channel of a Si crystal, for a static lattice (dashed green) and T corrected (black lines) at different T with $c = 2$ ($2u_1$ shown in red and the half width of the axial channel, $d_{\rm ach}/2$, in solid green) [7].

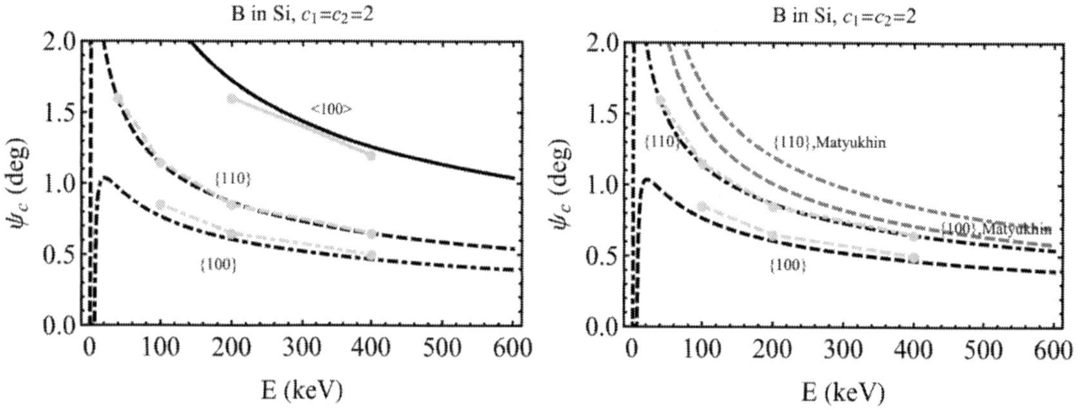

Figure 3. Examples of critical channeling angles ψ_c for B ions propagating in Si at room T. a.(left) Our predicted channeling angles using $c = 2$ (black), are shown here to be very similar to the observed angles given in Ref. [11] (green) for the <100> axial and the {100} and {110} planar channels. b.(right) The corresponding planar critical angles of Ref. [14] (which we believe incorrect) shown in red, are instead larger.

called the "rule of reversibility": due to time reversal symmetry if channeled atoms never go very close to lattice sites, the reversed paths do not happen either. It is due to lattice vibrations that the collision with a WIMP may happen while an atom is displaced with respect to the string or plane where it belongs and, if it is initially far enough into a channel, it may be channeled. The lattice vibrations are strongly temperature (T) dependent. Thus T effects must be taken into account in the calculation of channeling fractions. They can be incorporated through T dependent critical distances [13, 11] (see Fig.2.a)

$$\rho_c(E,T) = \sqrt{\rho_c^2(E) + [c\ u_1(T)]^2}, \tag{3}$$

where $u_1(T)$ is the 1-dimensional amplitude of thermal fluctuations (we used the Debye model)

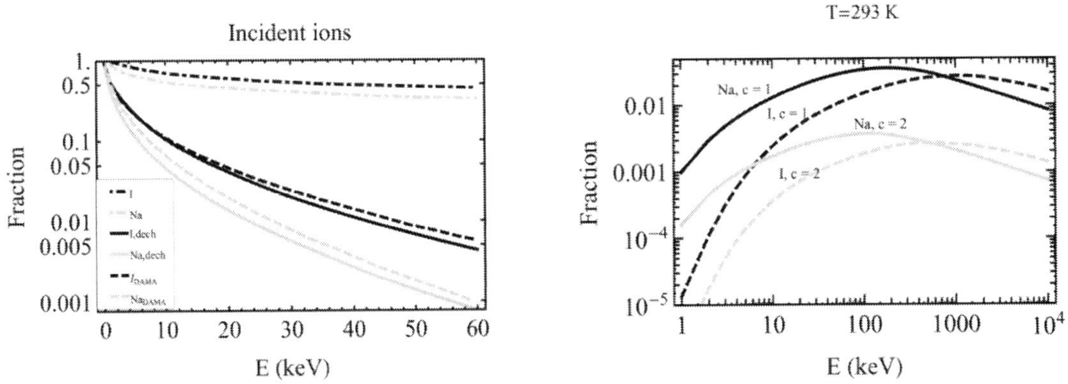

Figure 4. Geometrical channeling fractions of Na and I recoils in NaI(Tl) a.(left) showing the DAMA results [2] (dashed lines) and ours [6] (solid lines) neglecting blocking and b.(right) showing upper bounds to the fractions for the actual situation of ions ejected from lattice sites at room T and with $c =$ (green) and $c = 2$ (black) in Eq. 3 [6].

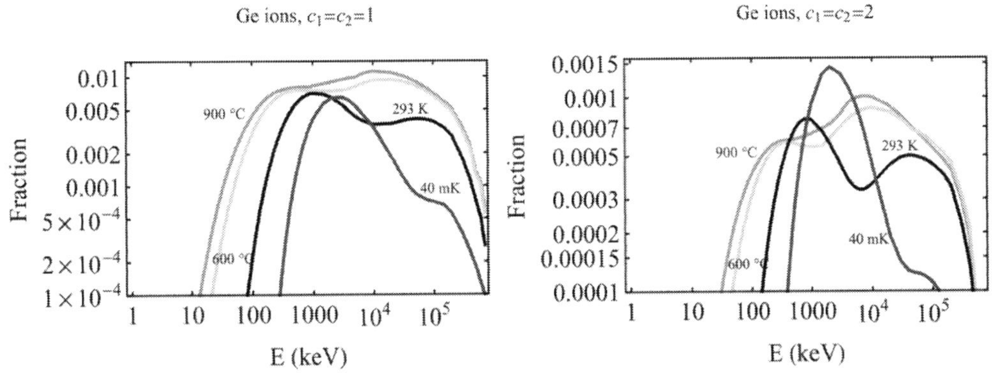

Figure 5. Geometric channeling fractions for Ge ions ejected from lattice sites in a Ge crystal as function of the ion energy for different temperatures T and T effects computed (Eq.3) with a.(left) $c = 1$ and b.(right) $c = 2$.

which increases with T (see the red lines in Fig.2.a), and c is a dimensionless number which (for several crystals and propagating ions, different than the ones we study) was found through data and simulations to be $1 < c < 2$. At large enough energies $\rho_c(E,T) \simeq c\, u_1(T)$ (see Fig.2.a) and thus as T increases the strings are planes become thicker, the channels narrower and ψ_c smaller (see Fig.2.b). Using this formalism with $c = 2$ we could reproduce data on channeling angles of B and P ions in Si measured at room T provided by Hobler [11] (shown in green for B in Figs. 3). Notice in Fig.3.b that the planar angles of Ref. [14] (derived from a planar channeling condition which, as explained above, we believe is not correct) are larger.

The critical angles ψ_c increase as the energy decreases, up to an energy at which they go sharply to zero, signaling that below that particular energy channeling is not possible (see Figs.2.b and 3). In fact, at those very low energies the minimum distance from a string or plane required for channeling should be larger than the half-with of the channel. The critical distances increase as the energy decreases (see e.g. Fig.2.a) and as they approach the half-width of the respective channel, the corresponding critical angle ψ_c goes to zero (see Fig. 2.b).

With our formalism (after including a simple model of dechanneling due to first interactions with Tl impurities) we are able to reproduce the channeling fractions obtained by DAMA, as shown in Fig.4.a [6]. These are calculated using only the channeling condition $\psi < \psi_c$, i.e. as if the recoiling ions would start their motion far away from the row or plane of atoms. In this case, we compute the angle that each recoil direction $\hat{\mathbf{q}}$ forms with each channel, then assigned a value $\chi(E, \hat{\mathbf{q}}) = 1$ to the probability of channeling in the direction $\hat{\mathbf{q}}$ if the angle is smaller than the critical angle for any of the channels and zero otherwise. To obtain the total "geometrical" channeling fraction, i.e. assuming the recoils are isotropically distributed (which in reality they are not - see e.g. Fig.6.b)

$$P_{\text{geometric}}(E) = \frac{1}{4\pi} \int \chi(E, \hat{\mathbf{q}}) d\Omega_q. \tag{4}$$

we average the channeling probability over initial recoil directions. We do it numerically by performing a Riemann sum once the sphere of directions has been divided using the Hierarchical Equal Area isoLatitude Pixelization (HEALPix) [15] (a novel use for HEALPix).

Taking into account that in dark matter detectors the recoiling ions start their motion at or close to their original lattice sites, thus blocking is important, the geometric channeling fractions we obtain are much smaller at low energies (see Fig.4.b). Only if due to lattice vibrations the recoiling ion is initially far enough from its original string or plane it may be channeled. Thus, in our model the probability that an ion is channeled into a particular channel is given by the fraction of nuclei which can be found at a distance larger than a minimum distance $\rho_{i,\text{min}}$ (determined by ρ_c and the initial recoil angle) from the string or plane at the moment of collision. For example, for an axial channel the probability distribution function of the perpendicular distance to the row of the colliding atom due to thermal vibrations can be represented by a Gaussian, $g(\rho) = (r/u_1^2)\exp(-r^2/2u_1^2)$, thus the probability that an ion is channeled is

$$\chi_{\text{axial}}(E, \hat{\mathbf{q}}) = \int_{r_{i,\text{min}}}^{\infty} drg(r) = \exp\left(-r_{i,\text{min}}^2/2u_1^2\right) \tag{5}$$

Notice that any uncertainty in ρ_c is exponentially enhanced in the channeling fraction. Similar equations apply to planar channels and one needs to compute the probability $\chi(E, \hat{\mathbf{q}})$ that an ion with initial energy E is channeled in a given direction $\hat{\mathbf{q}}$ using a recursion of the addition rule in probability theory of the channeling probabilities of all channels [6, 7]. To obtain the total geometrical channeling fraction in Eq. 4, we again average the channeling probability over initial recoil directions numerically using HEALPix

The channeling fractions we obtain are strongly T dependent (see Fig.5). The temperature effects on channeling fractions are complicated. As T increases the probability of finding atoms far from their equilibrium lattice sites increases, which increases the channeling fractions, but the critical distances ρ_c become larger ($\simeq cu_1$ at large enough energies) which decreases the channeling fractions. The fractions are smaller for larger values of c (see Fig.4.b and compare Fig.5.a and Fig.5.b). Also, we do not included any dechanneling effects due to impurities or dopants which would decrease the channeling fractions. Thus we are actually computing upper bounds to the channeling fractions. Using the fractions in Fig. 4.b with $c + 1$ corresponding to the temperature of the DAMA/NaI and the DAMA/LIBRA experiments, T= 293 K, the difference in the WIMP region corresponding to the DAMA annual modulation signal differs only at 7σ is channeling is included or not included [16].

As mentioned above, channeling is a directional effect which depends on the velocity distribution of WIMPs in the dark halo of our Galaxy and could lead to a daily modulation of the signal [5]. The recoil momentum distribution resulting from WIMP collisions is anisotropic, although with a larger dispersion than the WIMP momentum distribution (see Fig.5). Thus Earth's daily rotation changes the direction of the recoil distribution with respect to the crystal axes, changing the amount of recoiling ions that are channeled vs non-channeled.

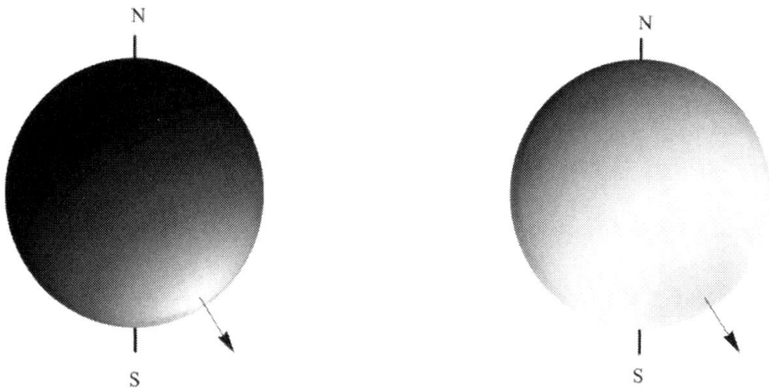

Figure 6. a.(left) Number of 60 GeV WIMPs per solid angle with speed larger than 196.7 km/s (as necessary to produce 10 keV Na recoils) plotted on the sphere of WIMP velocity directions and b.(right) number of 10 keV Na recoils per solid angle plotted on the sphere of recoil directions, for the standard halo model with average WIMPS velocity 288.3 km/s and dispersion 173 km/s. The numbers are normalized to be between 1 (white) and 0 (black). The arrow shows the direction of the average WIMP velocity. Figures from Ref [18].

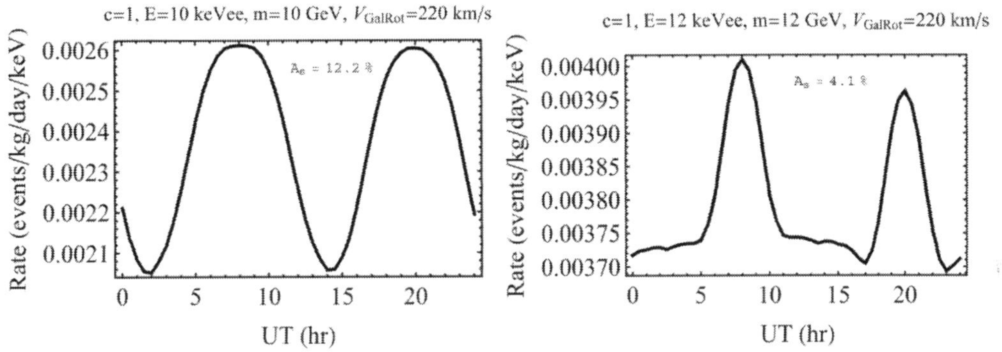

Figure 7. Examples of the daily variations due to channeling of the signal rate in NaI. Rate (in events/kg-day-keVee) shown as function of the Universal Time (UT) during 24 hours for WIMP mass and recoil energy of a.(left)10 GeV and=10 keV, and b.(right) 12 GeV and 12 keV. The parameters used are: 228.4 km/s average WIMP velocity and 300 km/s velocity dispersion, $Q_{\rm Na} = 0.2$, $Q_{\rm I} = 0.09$, $\sigma_p = 2 \times 10^{-40}{\rm cm}^2$, $c = 1$ and $T = 293$ K. Figures from Ref [18].

In order to compute the daily modulation of the interaction rate we need to orient the crystal with respect to the galaxy and convolute the directional recoil rate $dR/dE_R d\Omega_q$, which can be written in terms of the Radon transform [17] of the WIMP velocity distribution, with the probability $p(E, E_R, \hat{\mathbf{q}})dE$ that an energy E is measured when a nucleus recoils in the direction $\hat{\mathbf{q}}$ with initial energy E_R,

$$\frac{dR}{dE} = \int \frac{dR}{dE_R d\Omega_q} p(E, E_R, \hat{\mathbf{q}}) d\Omega_q dE_R, \qquad (6)$$

where

$$p(E, E_R, \hat{\mathbf{q}}) = \chi(E_R, \hat{\mathbf{q}})\delta(E - E_R) + [1 - \chi(E_R, \hat{\mathbf{q}})]\delta(E - QE_R), \tag{7}$$

is normalized so that $\int p(E, E_R, \hat{\mathbf{q}})dE{=}1$. The first term accounts for the channeled events ($E = E_R$) and the second term for the unchanneled events ($E = QE_R$)

After developing the general formalism, in Ref. [18] we examined the possibility of finding a daily modulation due to channeling in the data already collected by the DAMA/NaI and DAMA/LIBRA experiments. We found that the signal daily modulation amplitude, defined as $A_s = [R_{s-\max} - R_{s-\min}]/[R_{s-\max} + R_{s-\min}]$, could be large (of the order of 10% in some instances) in NaI crystals. However, with a simple observability criterium, we find that for this daily modulation to be observable the DAMA total rate should be 1/40 of what it is or the total DAMA exposure should be 40 times larger. The daily modulation due to channeling will be difficult to measure in future experiments too. As an example, one of the best cases we found [18] is that of a solid Ne detector, if the WIMP mass is 5 GeV and its cross section is as large as allowed by direct detection bounds, $\sigma_p = 10^{-39}$ cm^2, then a daily modulation could be detected with a 0.33 ton y exposure, assuming no background.

To conclude, analytic models of channeling and blocking like those we use give good qualitative results but channeling data and or simulations are necessary to get reliable quantitative results (we used some data available for Si only). Moreover, dechanneling and rechanneling effects cannot be easily computed with these models. Montecarlo simulations may be needed to settle these issues (many are used in other applications of channeling).

Acknowledgments
G.G. was supported in part by the US Department of Energy Grant DE-FG03-91ER40662, Task C. She wants to thank the APC (Astroparticles and Cosmology Laboratory), Paris, France, for hospitality while writing this contribution to the DSU2011 conference proceedings.

References

[1] Drobyshevski E M, 2008 *Mod. Phys. Lett. A* **23** 3077 [arXiv:0706.3095 [physics.ins-det]].
[2] Bernabei R *et al.* 2008 *Eur. Phys. J. C* **53**, 205 [arXiv:0710.0288 [astro-ph]].
[3] Savage C, Gelmini G. B., Gondolo P. and Freese K 2009 *JCAP* **0904**, 010 [arXiv:0808.3607 [astro-ph]].
[4] Belli P *et al.* 2011 *Phys. Rev. D* **84**, 055014 [arXiv:1106.4667 [hep-ph]].
[5] Avignone F, Creswick R and Nussinov S, 2008 arXiv:0807.3758 [hep-ph]; Creswick R, Nussinov S and Avignone F 2011 *Astropart. Phys.* **35** [arXiv:1007.0214 [astro-ph.IM]].
[6] Bozorgnia N, Gelmini G B and Gondolo P 2010, *JCAP* **1011**, 019 [arXiv:1006.3110 [astro-ph.CO]];
[7] Bozorgnia N, Gelmini G B and Gondolo P 2010, *JCAP* **1011**, 028 [arXiv:1008.3676 [astro-ph.CO]];
[8] Bozorgnia N, Gelmini G B and Gondolo P 2010, *JCAP* **1011**, 029 [arXiv:1009.3325 [astro-ph.CO]];
[9] Bozorgnia N, Gelmini G B and Gondolo P 2011, *NIM A* **654** 162-69 [arXiv:1011.6006 [astro-ph.CO]]..
[10] Gemmell D. S. 1974 *Rev. Mod. Phys.* **46** 129.
[11] G. Hobler 1996 *Radiation effects and defects in solids* **139** 21; and *NIM B* **115** 323.
[12] Lindhard J, 1965 *Kongel. Dan. Vidensk. Selsk., Mat.-Fys. Medd.* **34** No. 14.
[13] Morgan D V and Van Vliet D, 1968 *Can. J. Phys.* **46**, 503; Morgan D V and Van Vliet, D 1971 *Radiat. Effects and Defects in Solids* **8** 51; Van Vliet D 1973 in *Channeling*, Wiley, London, 37.
[14] Matyukhin S 2008, *Technical Physics* **53** 1578.
[15] K. M. Górski *et al.* 2005 *Ap. J.* **622** 756.
[16] Savage C, Gelmini G, Gondolo P and Freese K, 2010 *Phys. Rev. D* **83** 055002 [arXiv:1006.0972 [astro-ph.CO]].
[17] Gondolo P 2002 *Phys. Rev. D* **66** 103513.
[18] Bozorgnia N, Gelmini G B and Gondolo P, 2011 *Phys. Rev. D* **84**, 023516 [arXiv:1101.2876 [astro-ph.CO]].

An Overview of Dark Matter Experiments at Jefferson Lab

James R. Boyce on behalf of Collaborations: LIPSS, DarkLight, HPS, and APEX

Jefferson Lab, Newport News, VA 23606, USA

boyce@jlab.org

Abstract. Dark Matter research at Jefferson Lab started in 2006 with the **LI**ght **P**seudoscalar and **S**calar **S**earch (LIPSS) collaboration to check the validity of results reported by the PVLAS collaboration. In the intervening years interest in dark matter laboratory experiments has grown at Jefferson Lab. Current research underway or in planning stages probe various mass regions covering 14 orders of magnitude: from 10^{-6} eV to 100 MeV. This presentation will be an overview of our dark matter searches, three of which focus on the hypothesized A' gauge boson.

1. Introduction: Tools of Jefferson Lab: CEBAF and Free-Electron Accelerator (FEL)

The Thomas Jefferson National Accelerator Facility (Jefferson Lab), located in Newport News, VA, uses superconducting RF (SRF) linacs accelerating electrons up to 6 GeV for fixed target experiments in three experiment halls. Details of CEBAF are found elsewhere. [1] An expansion program is currently underway to double the energy to 12 GeV.

The Free-Electron Laser (FEL) facility [2, 3] uses SRF linacs to send a high quality electron beam through a wiggler to generate a photon beam. The photons are captured in an optical cavity that allows a small fraction, < 10%, to be guided with mirrors to experiments conducted in User Laboratories. Figure 1 is a layout of the FEL.

These two accelerator facilities have unique electron and photon beam characteristics for experiments designed to search for Dark Matter (DM) particles in laboratory settings. This overview will focus on two experiments using CEBAF and two using the FEL.

2. LIght Pseudoscalar and Scalar Search (LIPSS) [4]

The LIPSS experiment [5] uses the "Light Shining through a Wall" (LSW) technique [5] whereby an intense, linearly polarized laser beam is passed through a strong magnetic field. See figure 2. A coupling of the photons and magnetic field could produce a Light Neutral Boson, A^0. The A^0 continues in the same direction as the photons - passing through a mirror that deflects the laser beam - and enters into another strong magnetic field where a photon is regenerated from the coupling of the A^0 with the magnetic field. These regenerated photons are then detected by a cooled CCD camera. When the light and/or the A^0 are in the magnetic field, they are in a vacuum chamber that is pumped down to better than 10^{-6} torr. Figure 3 shows the layout of the LSW experiment.

Results of the initial LIPSS experiment have been reported elsewhere. [5, 7] A sample of the results is shown in figure 4 taken from [8] and shows a comparison with other LSW experiments. Such data can be used to set boundaries for scalars, pseudoscalars, Paraphotons, and Millicharged fermions.

Journal of Physics: Conference Series **384** (2012) 012008 doi:10.1088/1742-6596/384/1/012008

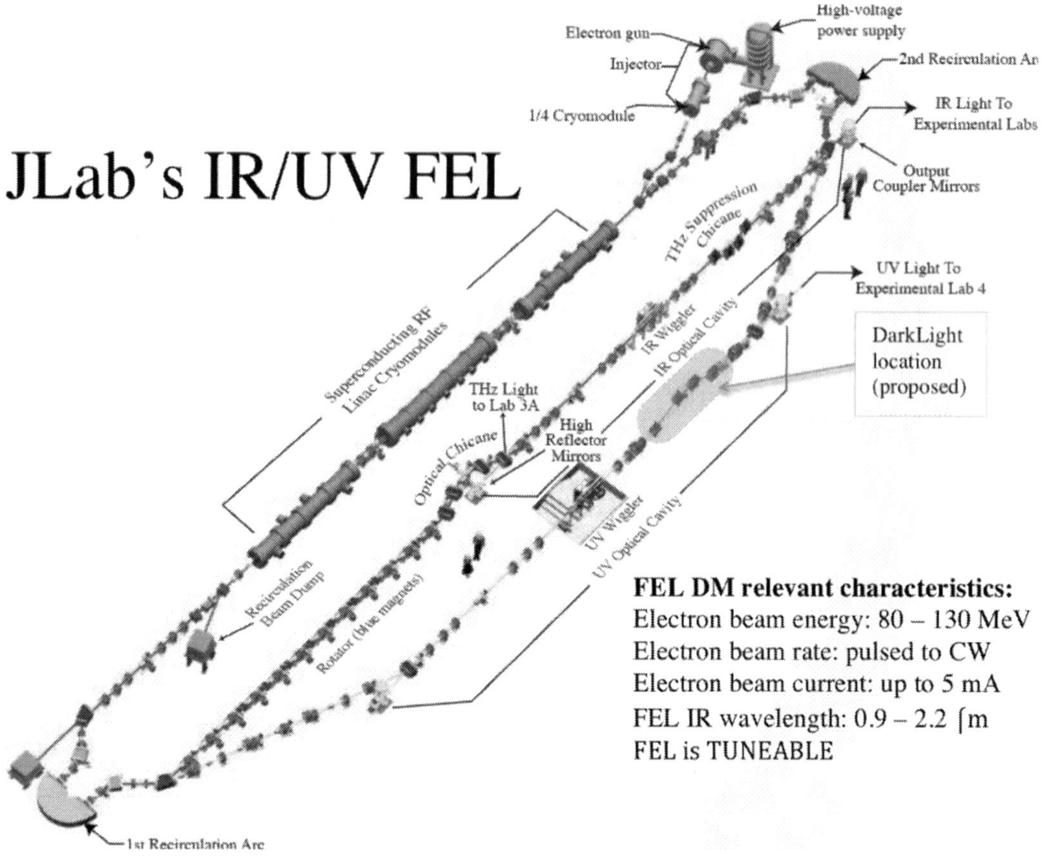

Figure 1. Jefferson Lab's IR/UV FEL Layout. Electrons from the Electron gun (upper right) are injected into the three linac cryomodules, accelerated to maximum energy, travel counter – clockwise around the beam line – either through the IR line or the UV line – and re-injected into the three cryomodules at 180 degrees out of phase, decelerated and deposited in the Recirculating Beam Dump. Laser light - generated when the electrons pass through the wiggler – is captured between two mirrors. A small fraction of the captured light is "out-coupled" from one of the mirrors and is sent to the experimental Labs.

Figure 2. Incident light (g) couples to photons in the magnetic field (B) creating the hypothetical light neutral boson, A^0. The weakly interacting A^0 passes through the optical barrier (the "wall") and into the second magnetic field where photons are regenerated from the interaction of the A^0 with the second magnetic field. The probability of generating γ to A^0 is $P_{\gamma\to A^0}$ where g is coupling constant, B=magnet field, m is mass of A^0, L = mag field length, and ω = photon energy.

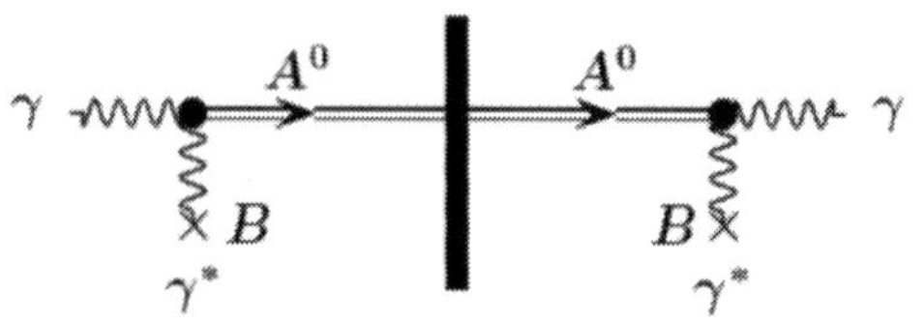

$$P_{\gamma\to A^0} = P_{A^0\to\gamma} = \frac{(gB)^2}{\frac{m^4}{4\omega^2}} \sin^2\left(\frac{m^2 L}{4\omega}\right)$$

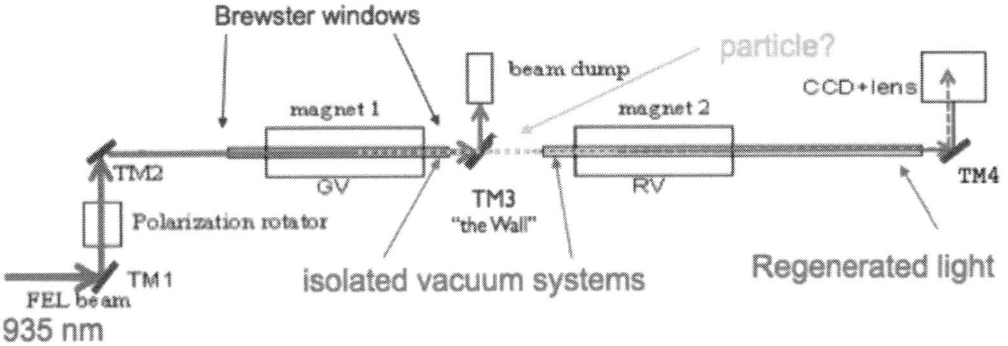

Figure 3. The LIPSS LSW experiment layout. Laser light from the FEL at 935 nm enters from the left, is guided via turning mirrors (TM1 And TM2) into the vacuum chamber GV inside the 1.7 T magnetic field, then exits and is directed by TM3 into a laser beam dump. Any A^0 produced travels through the mirror and enter the second vacuum chamber inside the second 1.7 T magnet. Regenerated photons from the conversion of A^0s are collected by the CCD camera.

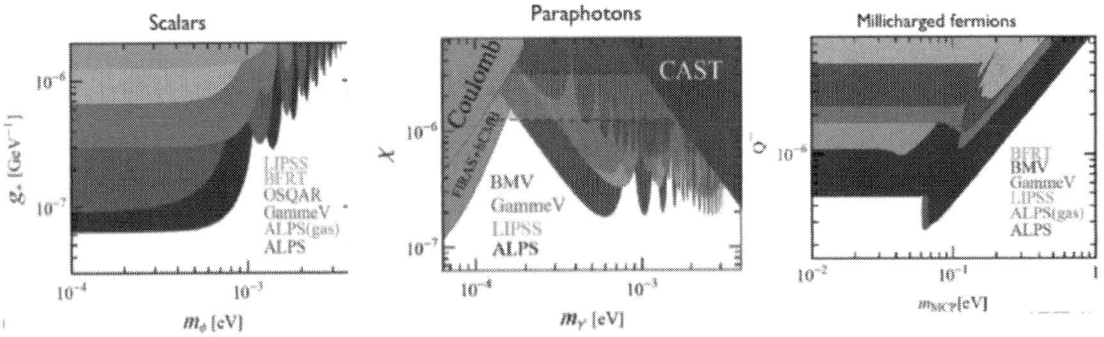

Figure 4. A sample of the results of LSW experiments as compiled in Reference [8].

3. LIPSS-2 Photon-boson kinetic mixing: next steps.

Paraphotons are bosons generated via kinetic mixing. [7,8] The optical cavity of the FEL has ~ ten times more photons than the out-coupled light. A LSW system in line with the optical cavity could search for these Paraphotons. See Figure 6.

Figure 5. Paraphotons can occur via kinetic mixing as described in [5]. The probability of transmission through an LNS setup is P_{trans}, where χ is the coupling constant, ω is the photon energy, and $m_{\gamma'}$ is the Paraphoton mass.

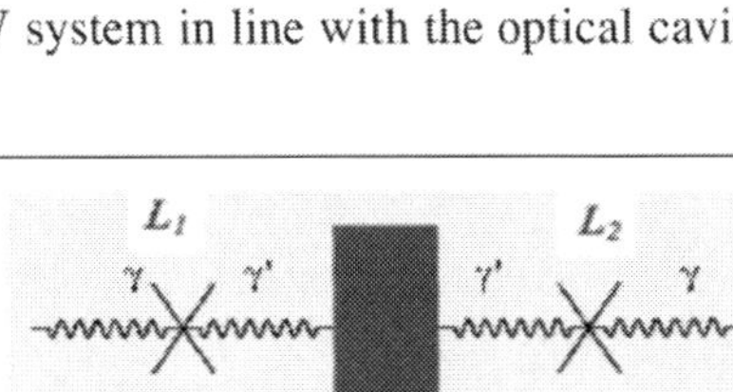

$$P_{trans} = 16\,\chi^4 \left[\sin\left(\frac{\Delta k L_1}{2}\right)\sin\left(\frac{\Delta k L_2}{2}\right)\right]^2$$

where $\Delta k = \omega - \sqrt{\omega^2 - m_{\gamma'}{}^2}$

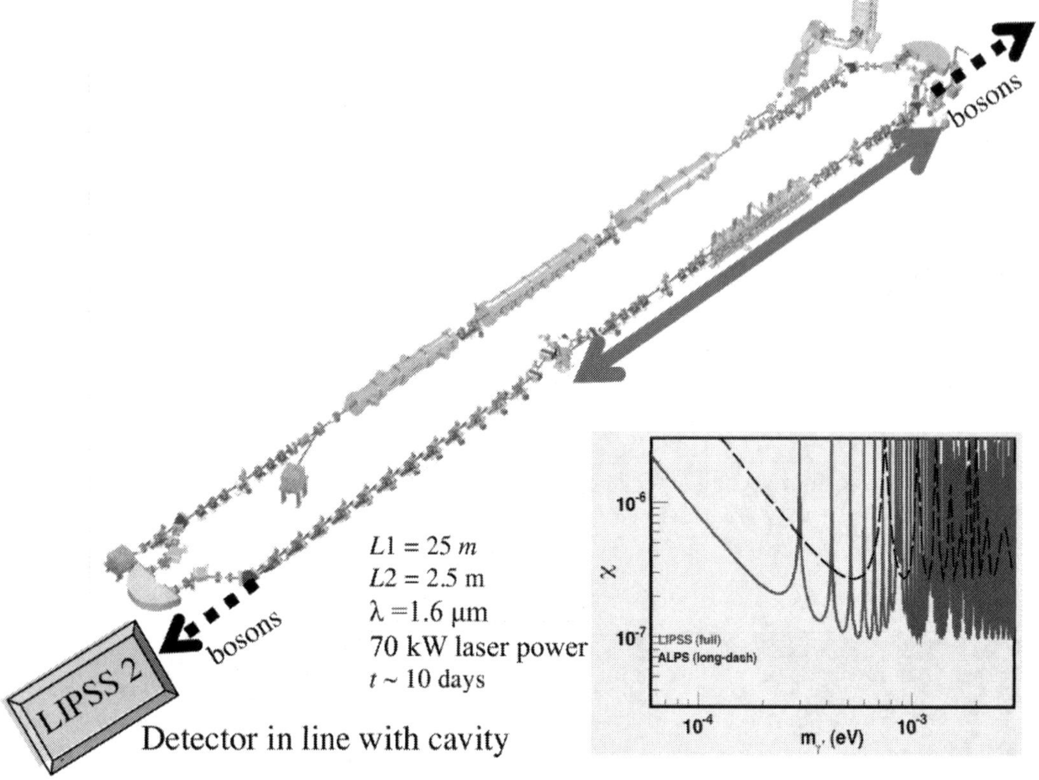

Figure 6. The FEL Optical Cavity contains ~10 times more photons than is available with the "out-coupled" beam. Paraphotons (bosons), generated via kinetic mixing, would exit as shown. A detector system in line with the cavity and with parameters as shown, could extend the explored parameter space, shown in solid blue line, beyond the most recent ALPS results [8], shown in black dashed line. [9]

4. Detecting A Resonance Kinematically with eLectrons Incident on a Gaseous Hydrogen Target (DarkLight). [10]

DarkLight is another proposed search for new light bosons using the Jefferson Lab FEL facility. It uses the high current (~ mA), low energy (~130 MeV) electron beam on a diffuse hydrogen gas target. The experiment is based on predictions described in [11]. Figure 7 depicts the interaction whereby an electron scatters off the proton and a light boson, A', generated near the e^-p scattering, subsequently decays indo an electron positron pair. The detector system being designed will incorporate a solenoid magnet surrounding the interaction region downstream from the UV wiggler (See figure 2.) Evidence for the hypothesized A' would consist of a narrow resonance on a large QED background. Also shown in figure 7 is the region of parameter space DarkLight will probe. The expected Luminosity is about 1 ab^{-1}/month.

Journal of Physics: Conference Series **384** (2012) 012008 doi:10.1088/1742-6596/384/1/012008

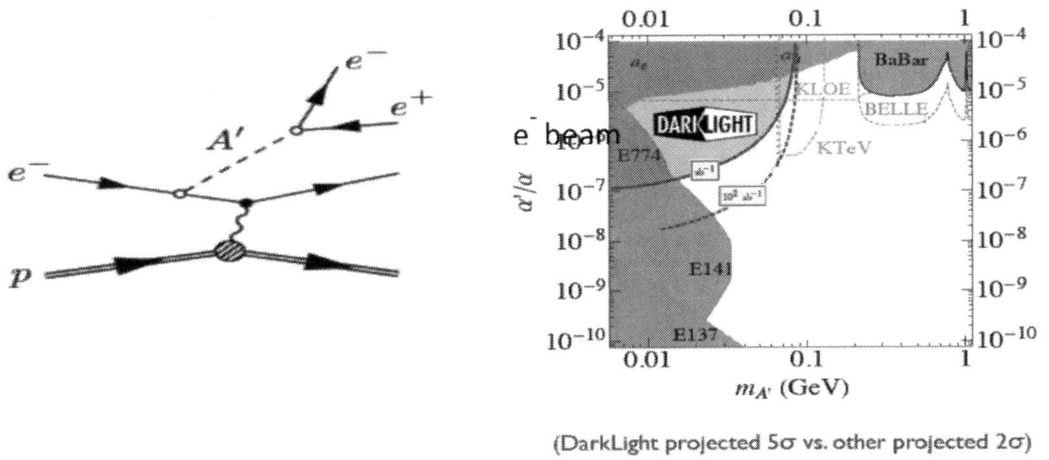

Figure 7. DarkLight reaction and the parameter space explored by DarkLight. The coupling constant, is α'/α, where $\alpha' \equiv \varepsilon^2 \alpha$, $(\alpha = e^2/4\pi)$. Regions in grey, as well as KLOE, Belle, and KTeV, are regions previous experiments have excluded for the generation of A's. The two DarkLight curves are for different run times.

5. Heavy Photon Search (HPS). [12]

The HPS Collaboration has proposed an A' search that would probe two additional regions of parameter space. The HPS would be located downstream from current Hall B equipment. Figures 8 and 9 show the concept and layout. [13]

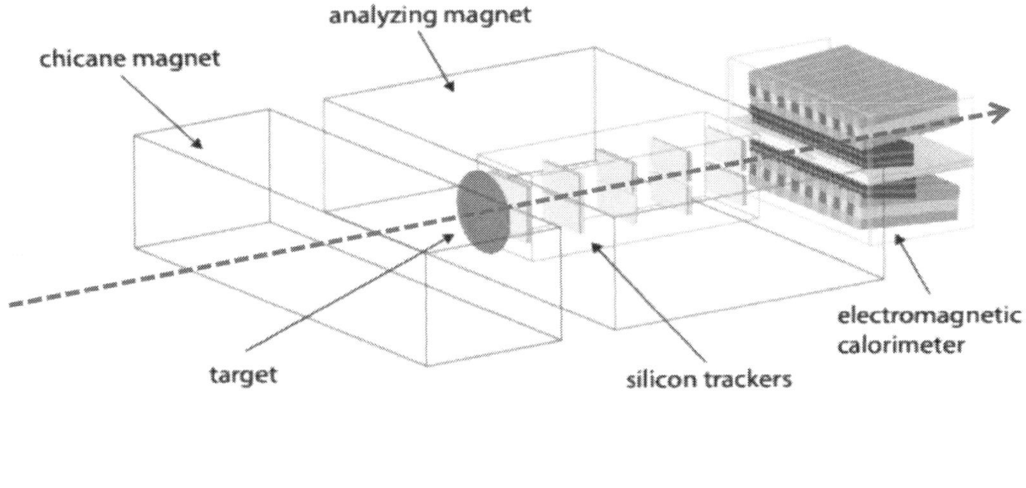

Figure 8. The HPS experiment's target and detection system. The electron beam, traveling left to right, strikes a thin target close to compact Si trackers in a 1T dipole. Downstream of the tracker is a fast, segmented electromagnetic calorimeter for triggering and e- identification. There will also be a μ detector for alternate trigger and μ identification. The detectors are split vertically to avoid the "Dead Zone" occupied by primary beam, bremsstrahlung photons, etc.

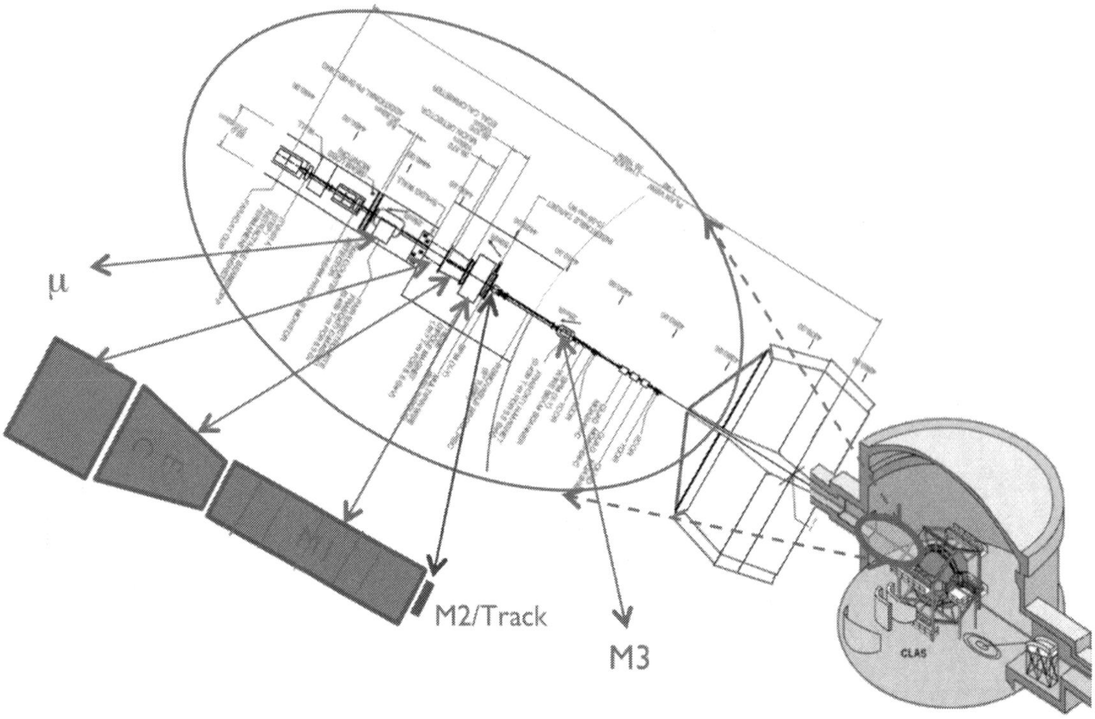

Figure 9. The HPS setup in Hall B. Electrons enter Hall B from the lower right traveling through the CLAS setup and into the HPS setup located between CLAS and the Hall B beam dump. [12, 13].

Figure 10: HPS Reach: Bump Hunt and Vertex Search. These estimates are for a 1 month run @ 400 nA of beam. The bright red is for an e-beam of 5.5 GeV energy and the maroon is for an e- beam of 3.3 GeV. [14]

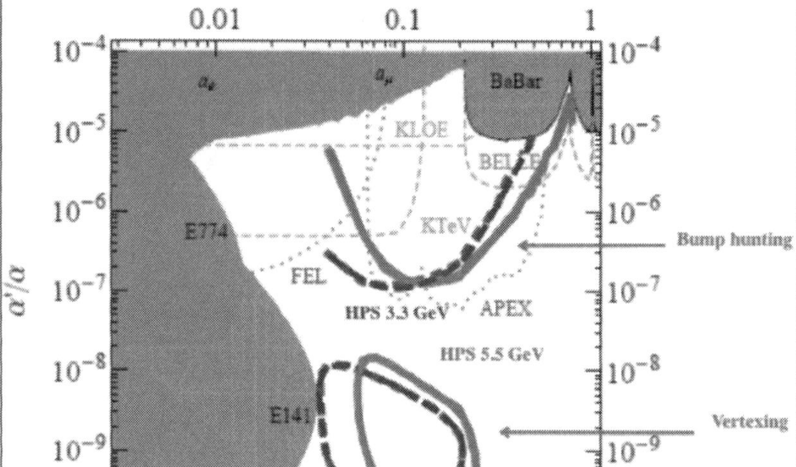

6. A Prime EXperiment (APEX). [15]

The full APEX experiment will be conducted by the APEX Collaboration in Jefferson Lab's Hall A using both spectrometer arms rotated as close to each other as possible. (Shown in figure 11 is a cartoon with the electron arm rotated away from the Hadron arm.) APEX will probe coupling constants $\alpha'/\alpha \gtrsim 10^{-7}$ and A' mass between ~ 50 and 500 MeV.

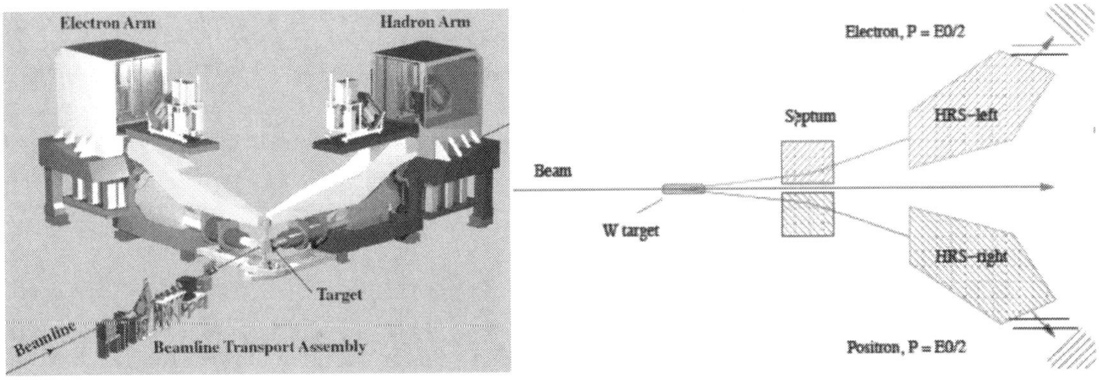

Figure 11. APEX uses the two spectrometers in Hall A. On the left is a graphic of the spectrometers with the electron arm rotated away from the beam line and the Hadron arm. On the right is a sketch of the APEX setup where both spectrometers are ~ +/- 5° from the beam line. A septum magnet is inserted between the W target to provide the separation needed between positrons and electrons for the High Resolution Spectrometers. This arrangement allows for excellent mass resolution, it dramatically suppresses large background from elastic scattered electrons and Moller electrons. The setup also suppresses background from $\pi^0 \to \gamma\, e^+e^-$ and radiated $\gamma \to e^+e^-$.

In July 2010 a Test Run of APEX was conducted to answer specific questions about the feasibility of the full APEX run as proposed by the Collaboration. [16] The Test Run successfully obtained results comparable to results reported in reference [17] and shown in figure 12.

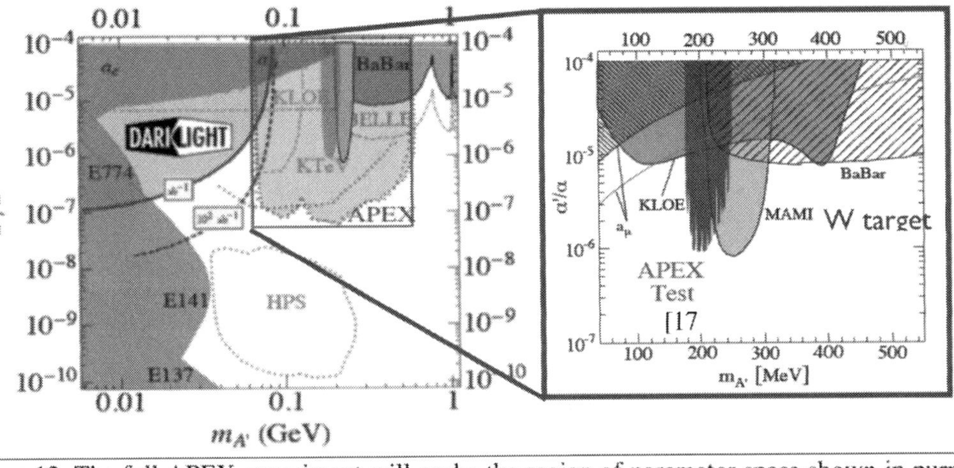

Figure 12. The full APEX experiment will probe the region of parameter space shown in purple on the left. On the right is the expanded region showing the results of the APEX Test Run [16] compared to the results reported by MAMI [17].

7. Conclusions

In recent years, starting with LIPSS, interest in Dark Matter experiments at Jefferson Lab has increased due to the realization that the tools – accelerators and experimental Halls – are ideally suited for Intensity Frontier explorations. This Overview has described several Dark Matter experiments recently conducted or proposed using these facilities. LIPSS successfully conducted and published use of the FEL for probing parameter space for pseudoscalars, scalars, Paraphotons, and millicharged fermions in the mass region from 10^{-6} to 10^{-3} eV. DarkLight proposes to search for a new vector gauge boson, A', using the electron beam of the FEL scattering off a hydrogen gas target. APEX and HPS also propose searches for the A' using Jefferson Lab's Hall A and Hall B respectively. APEX has successfully completed and published a test run. Setup for both DarkLight and HPS are in progress. Thus Jefferson Lab is poised to venture into unexplored regions of dark matter parameter space.

Acknowlegements:

Notice: Authored by Jefferson Science Associates, LLC under U.S. DOE Contract No. DE-AC05-06OR23177. The U.S. Government retains a non-exclusive, paid-up, irrevocable, world-wide license to publish or reproduce this manuscript for U.S. Government purposes. This work supported by the Office of Naval Research Award N00014-06-1-1168, the Joint Technology Office, the Commonwealth of Virginia, the Air Force Research Laboratory, and by DOE under contract DE-AC05-060R23177.

References

1. www.jlab.org
2. http://wwwold.jlab.org/FEL/feldescrip.html
3. Benson S V, et al., 2010 *Nucl. Instr. and Meth. A* **12** 093.
4. **LIPSS Collaboration: Baker O K (Spokesperson)** and Slocum P *Yale Univ.*; **Afanasev A, (Co-Spokesperson)** *George Washington Univ. (Formerly at Hampton Univ.)*; Biallas G, Boyce J R, Robinson T, and Shinn, *Jefferson Lab*; Beard K B *Muons, Inc.*; Minarni M. *Universitas Riau (Formerly at Yale)*
5. Afanasev A, Baker O K, Beard K B. Biallas G, Boyce J R, Minarni M, Ramdon R, Shinn M, and Slocum P, 2008 *PhysRevLett* **101** 120401.
6. Sikivie P, 1983 *PhysRevLett* **51** 1415. Ansel'm A A, 1985, *Yad. Fiz.* **42** 1480. Van Bibber K, 1987, *PhysRevLett* **59** 759.
7. Afanasev A, Baker O K, Beard K B. Biallas G, Boyce J R, Minarni M, Ramdon R, Shinn M, and Slocum P, 2009 *Phys. Lett. B* **679** 317-320.
8. Ehret K, et al., 2010 *Phys. Lett B* **689** 149-155.
9. Baker O K, 2011, *7th Patras Workshop, Mykonos, Greece.*
10. DarkLight Collaboration: Alacorn R, *(Arizona State Univ.)*, Freytsis M, *(Berkeley)*, Benson S, Boyce J R, Douglas D, Ent R, Jordan K C, Neil G, Shinn M and Zang S, *(Jefferson Lab)*, Fiorito R, O'Shea P *(Univ. of Maryland)*, Balakrishnan P, Bertozzi W, Cowan R, Gilad S, **Fisher P (Spokesperson)**, Hays-Wehie J, Kahn Y, Kelleher A, **Milner R (Co-Spokesperson)**, Russell B, Thaler J, Thong S, Tschalär C *(MIT)* and Ovanesyan G. *(LANL)*.
11. Freytsis M, Ovanesyan G and Thaler J, 2010 *JHEP* **1001** 111. arXiv:0909.2862.
12. **HPS Collaboration:** Cooper M, Demarteau M, *(FNAL)*, Bueltmann S and Weinstein L, *(ODU)*, Grillo A, *(Univ. of California, Santa Cruz)*, **Holtrop M (Co-Spokesperson)**, Slifer K, Phillips S and Ebrahim E, *(Univ. of New Hampshire)*, Schuster P and Toro N *(Perimeter Institute)*, Essig R, Field C, Graham M, Haller R, **Jaros J (Co-Spokesperson)**, Kenney C, Maruyama T, Moffeit K, Nnelson T, Neal H, Odian A, Oriunno M, Partridge R and Walz D, *(SLAC)*, Boyarinov S, Burkert V, Deur A, Egiyan H, Freyberger A, Girod F-X, Kubarovsky V, **Stepanyan S, (Co-Spokesperson)** *(Jefferson*

Journal of Physics: Conference Series **384** (2012) 012008 doi:10.1088/1742-6596/384/1/012008

Lab), Fradi A, Guegan B, Guidal M, Niccolai S, Pisano S, Raulv E, Rosier P and Sokhan D, *(Institut de Physique Nuycleaire d'Orsay, France),* Khandaker M and Salgado C, *(Norfolk State Univ.),* Dashyan N, Gevorgvan N, Paremuzvan R and Voskanvan H, *(Yerevan Physics Institute, Armenia),* Battaglieri M and DeVitta R, *(INFN, Sezione di Genova, Italy).*

13. Holtrop M, http://conferences.jlab.org/boson2010 .
14. Ghaham M, http://conferences.jlab.org/boson2010 .
15. **APEX Collaboration: Essig R, Schuster P, Toro N and Wojtsekhowski B, (Co-Spokespersons,** Abrahamyan S, Afanasev A, Ahmed Z, Aliotta E, Allada K, Anez D, Armstrong D, Averett A, Barbieri A, Bartlett K, Beacham J, Beck S, Bjorken JD, Bono J, Bosted P, Boyce JR, Brindza P, Bubis N, Camsonne A, Chen O, Cranmer K, Curtis C, Chudakov E, Dalton M, de Jager C W, Deur A, Donagh J, Field C, Folts E, Gasparian A. Gavalya A, Gilad S, Gilman R, Glamazdin A, Goeckner-Wald N, Gomez J, Graham M, Hansen O, Higinbotham DW, Holmstrom T, Huang J, Iqbal S, Jaros J, Jensen E, Kelleher A, Khandaker M, Korover I, Kumbartzki G, LeRose JJ, Lindgren R, Liyanage N, Long E, Mammei J, Markowitz P, Maruyama T, Maxwell V, McDonald J, Meekins D, Michaels R, Mihovilovič M, Moffeit K, Nanda S, Nelyubin V, Norum BE, Odian A, Oriunno M, Partridge R, Paolone M, Piasetzky E, Pomerantz I, Puckett A, Punjabi V, Qiang Y, Ransome R, Riordan S, Roblin Y, Ron G, Saenboonruang K, Saha A, Sawatzky B, Segal J, Selvy L, Shahinyan A, Shneor R, Širca S, Subedi R, Sulkosky V, Stepanyan S, Waltz D, Weinstein L, Zhang J, Zhang Y, Zhao B, and the Hall A Collaboration.
16. Merkel, et al., 2010 *Phys. Rev. Lett* **106** 251802.
17. Abrahamyan S, et al., 2011 *Phys. Rev. Lett.* **107** 191804.

Journal of Physics: Conference Series **384** (2012) 012009 doi:10.1088/1742-6596/384/1/012009

Simultaneous Generation of WIMP Miracle-like Densities of Baryons and Dark Matter

John McDonald

Consortium for Fundamental Physics, Cosmology and Astroparticle Physics Group, Department of Physics, University of Lancaster, Lancaster LA1 4YB, UK

E-mail: j.mcdonald@lancaster.ac.uk

Abstract.

The observed density of dark matter is of the magnitude expected for a thermal relic weakly-interacting massive particle (WIMP). In addition, the observed baryon density is within an order of magnitude of the dark matter density. This suggests that the baryon density is physically related to a typical thermal relic WIMP dark matter density. We present a model which simultaneously generates thermal relic WIMP-like densities for both baryons and dark matter by modifying a large initial baryon asymmetry. Production of unstable scalars carrying baryon number at the LHC would be a clear signature of the model.

The ratio of the mass density of baryons to that of dark matter (the BDM ratio) is observed to be $\Omega_B/\Omega_{DM} \approx 1/5$. However, in most models the physics of baryogenesis and of dark matter production is physically unrelated. Why then is the density in baryons within an order of magnitude of that in dark matter? Either:

- A remarkable coincidence
- An anthropic selection mechanism
- The physics of the observed baryon asymmetry and dark matter densities is in some way related.

Therefore the BDM ratio may be a powerful clue to the correct particle physics theory.

Up to now there have been broadly two classes of model:

1). Direct mechanism: The dark matter particle and baryon number densities are directly related by a conserved charge,

$$\Rightarrow Q_{tot} = Q_B + Q_X \Rightarrow n_{cdm} \sim n_B \Rightarrow M_{DM} \sim m_n n_B/n_{cdm} \sim 1 - 10 \text{ GeV}$$

There are many models in this catagory, which typically predict asymmetric dark matter [1, 2, 3, 4, 5, 6, 7, 8]. Models also exist which break the simple mass relation $M_{DM} \sim 1 - 10$ GeV, see [9, 10, 11, 12, 13, 14]. (In [15] a new idea is discussed where the correct dark matter density is generated via its interaction with the baryon asymmetry.)

2). Indirect mechanism: The dark matter and baryon densities are related by *similar* but *seperate* physical mechanisms for their origin $\Rightarrow$ Less rigid relation between n_B and n_{cdm}; can have $n_B \gg n_{cdm} \Rightarrow$ larger M_{DM}. An example based on $d = 4$ Affleck-Dine leptogenesis is given in [16], where dark matter is due to a condensate of right-handed sneutrinos [1]

[1] The author is not aware of other examples in this class and would be interested to know if other models exist.

However, there are in fact *two* seperate coincidences. 1). Why are the baryon and dark matter densities similar to each other? 2). Why are they both similar to the "WIMP Miracle" density? (Most models do not address 2.)

The "WIMP Miracle" refers to the natural similarity of the observed dark matter density and the thermal relic density of particles with weak-scale masses and interactions. This is widely interpreted as a strong indication that dark matter is due to thermal relic WIMPs. For dark matter with constant $<\sigma v_{rel}>$, where σ is the non-relativistic annihilation cross-section, the observed dark matter density requires

$$<\sigma v_{rel}> \approx 10^{-9}\ \mathrm{GeV}^{-2}$$

The DM annihilation cross-section may be expresed in terms of the annihilation matrix element $\mathcal{M}$ as

$$<\sigma v_{rel}> = \frac{|\mathcal{M}|^2}{32\pi m_{DM}^2} \ . \tag{1}$$

The matrix element is dimensionless and typically has the form $|\mathcal{M}|^2 = g_{eff}^4$. where g_{eff} is an effective coupling constant. Therefore

$$<\sigma v_{rel}> = 1.6\times 10^{-9}\left(\frac{g_{eff}}{0.2}\right)^4\left(\frac{100\ \mathrm{GeV}}{m_{DM}}\right)^2\ \mathrm{GeV}^{-2}\ . \tag{2}$$

Therefore if M_{DM} is in the range 100 GeV – 1 TeV and if g_{eff} is not too much smaller than the weak interaction coupling $g\approx 0.65$ (which is natural as g_{eff} can have additional factors from mixing angles, mass ratios etc), then we get the right amount of thermal relic dark matter.

If the WIMP miracle does explain dark matter, and if we discount coincidence and antropic selection, then we need to explain why the baryon asymmetry is also similar to the WIMP miracle density. The similarity to the dark matter density then *follows* from this i.e. the observed baryon to dark matter ratio is actually due to a more fundamental relation between each of the densities and the WIMP miracle density.

In [17] and [18] we presented one approach to making this connection, based on a mechanism we call *baryomorphosis*. [Recently an alternative approach to achieving a WIMP-like baryon asymmetry, *WIMPy baryogenesis*, has been proposed in [19].]

This idea of baryomorphosis that the *observed* baryon density is determined by a process similar to thermal relic WIMP freeze-out. The model does not explain the baryon asymmetry, but modifies a large initial baryon asymmetry into a final thermal WIMP-like baryon asymmetry; hence "baryomorphosis" rather than "baryogenesis".

The ingredients of the baryomorphosis mechanism are:

- An initial baryon asymmetry, for example in a heavy particle Σ which is out of thermal equilibrium, which decays at a low temperature $T_d \lesssim O(100)$ GeV.
- Pairs of new scalar particles, ϕ_B and $\hat{\phi}_B$, called "annihilons", to which Σ decays. The annihilons have opposite Standard Model (SM) charges but, crucially, do not have opposite baryon numbers. ϕ_B and $\hat{\phi}_B$ are distinct particles which can have different masses, although for simplicity we will consider their masses to be equal in the following.
- A B-violating interaction which is broadly of weak interaction strength, via which ϕ_B and $\hat{\phi}_B$ annihilate, leaving a thermal WIMP-like baryon number density.
- A mechanism to transfer the baryon asymmetry from annihilons to conventional quarks.

This is illustrated schematically in Figure 1.

In [17] a simple model implementing the baryomorphosis mechanism was presented. In this model the annihilons annihilate to a complex boson $\hat{H}$ via a renormalizable B-violating interaction of the form

$$\mathcal{L}_{\phi_B\hat{\phi}_B\,ann} = \lambda_B\phi_B\hat{\phi}_B\hat{H}^{\dagger}\hat{H} \ + h.c.\ . \tag{3}$$

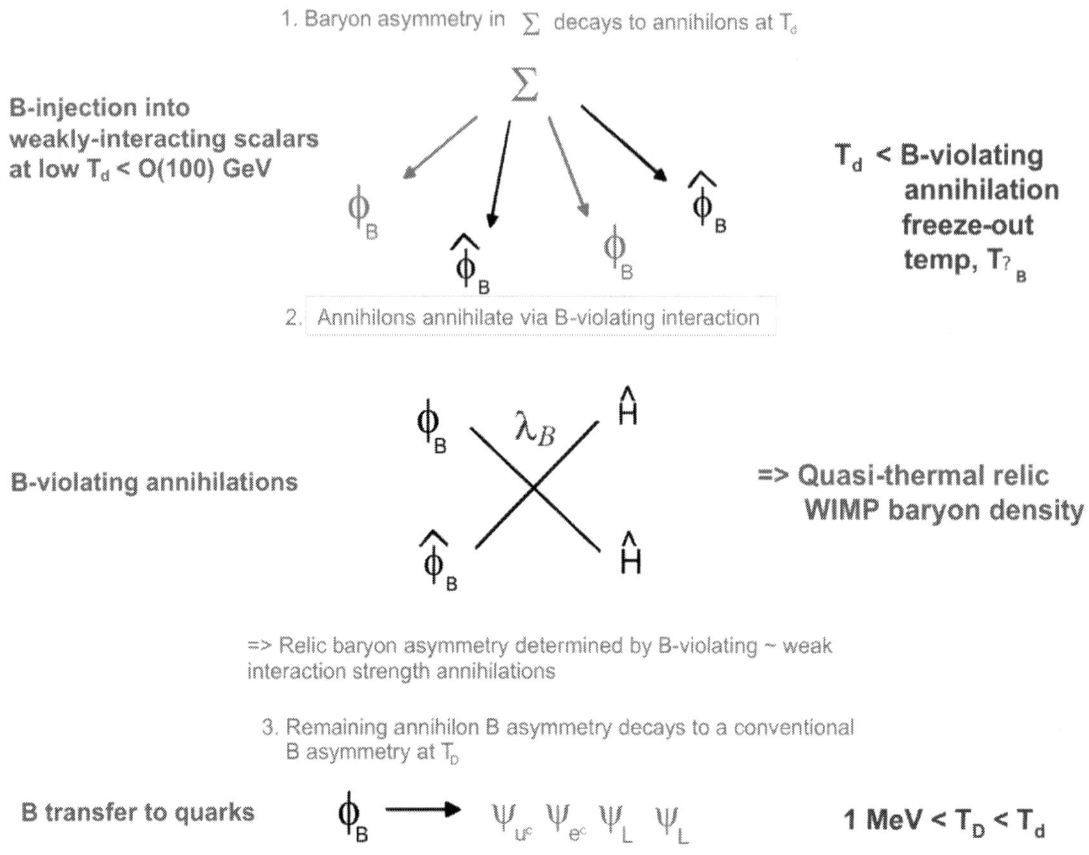

Figure 1.

Note that the $\hat{H}$ cannot be the Higgs boson, since the ϕ_B, $\hat{\phi}_B$ will mix once the Higgs VEV is introduced, resulting in a B-violating mass insertion which will wash out the annihilon asymmetry via scattering from thermal background quarks. Therefore the final state in the B-violating annihilation must have no VEV.

However, the original model in [17] has some features which do not have a symmetry explanation and must therefore simply be imposed:

- There is no symmetry to prevent B-violating mass mixing terms like $\phi_B\hat{\phi}_B$ ($\Rightarrow$ B washout).
- There is no suppression of renormalizable couplings of ϕ_B to SM fermions ($\Rightarrow$ ϕ_B decay too rapidly, before annihilating).

In addition, although it is not really problem, no WIMP dark matter candidate is specified.

In order to address these issues, in [18] a model was presented which uses a simple discrete symmetry to ensure no ϕ_B $\hat{\phi}_B$ mixing terms leading to baryon washout. Since additional discrete symmetries are necessary to evade B washout, and since additional symmetries are also necessary to stabilize dark matter, it is natural to identify as dark matter particles the additional fields necessary as final states in the ϕ_B $\hat{\phi}_B$ annihilation process. The model can be naturally combined with gauge singlet (or inert doublet [20]) dark matter to provide a unified model of baryomorphosis and scalar WIMP dark matter.

We introduce a pair of real singlet scalars s $\hat{s}$ plus a Z_2 discrete symmetry of the annihilons, Z_A,

$$Z_A : \phi_B \rightarrow \phi_B \, ; \;\; \hat{\phi}_B \rightarrow -\hat{\phi}_B \, ; \;\; s \rightarrow s \, ; \;\; \hat{s} \rightarrow -\hat{s} \, , \qquad (4)$$

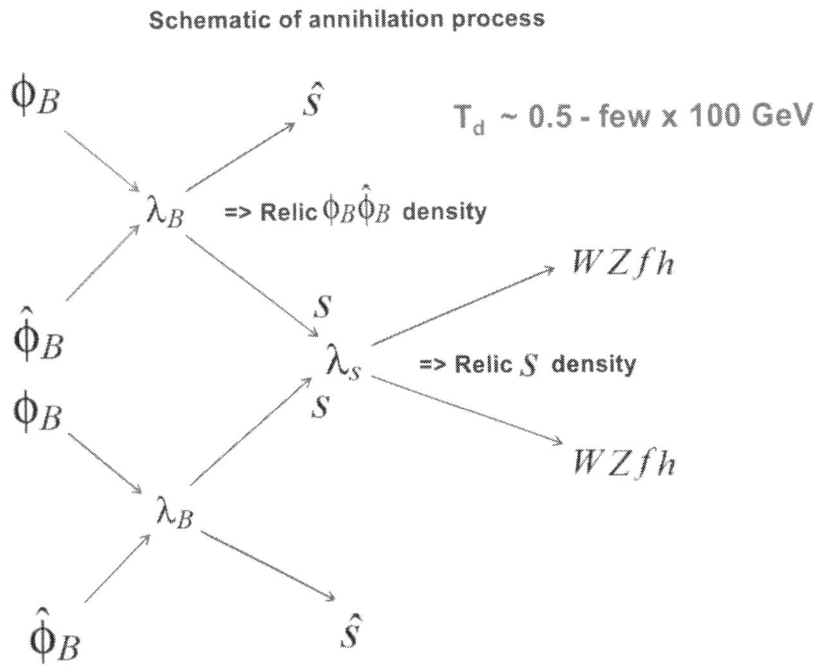

Figure 2.

with all SM fields invariant under Z_A. This eliminates the dangerous B-violating mass mixing terms $\phi_B\hat{\phi}_B$ and $\phi_B\hat{\phi}_B H^\dagger H$ but allows the B-violating annihilation process

$$\mathcal{L}_{\phi_B\hat{\phi}_B\, ann} = \lambda_B \phi_B \hat{\phi}_B s\hat{s} \ + h.\,c.\,. \quad (5)$$

Note that ϕ_B and $\hat{\phi}_B$ must carry gauge charges to prevent the dangerous terms $\phi_B\phi_B$ and $\hat{\phi}_B\hat{\phi}_B$, which are allowed under Z_A.

We need a sceond discrete symmetry to stabilize the dark matter, Z_S,

$$Z_S : s \longrightarrow -s\,;\ \ \hat{s} \longrightarrow -\hat{s}\,. \quad (6)$$

The $Z_A \times Z_S$ symmetry then allows the coupling to the SM:

$$\frac{\lambda_s}{2} s s H^\dagger H + \frac{\lambda_{\hat{s}}}{2} \hat{s}\hat{s} H^\dagger H\,. \quad (7)$$

This allows the s and $\hat{s}$ to annihilate down to a thermal WIMP-like dark matter density.

The formation of the final baryon asymmetry and dark matter density in this model is shown schematically in Figure 2. As before, the initial annihilon asymmetry is injected at a low temperature $T_d \sim 0.1 - 100$ GeV. The s and $\hat{s}$ subsequently annihilate down to thermal WIMP-like density if $T_d < T_s$, or to a standard thermal WIMP density if $T_d > T_s$, where $T_s \approx m_s/25$ is the freeze-out temperature of the s and $\hat{s}$ dark matter.

The annihilation cross-section for $\phi_B\,\hat{\phi}_B \longrightarrow s\,\hat{s}$ is

$$< \sigma v >_{\phi_B} = \frac{\lambda_B^2}{32\pi m_{\phi_B}^2}\left(1 - \frac{m_s^2}{m_{\phi_B}^2}\right)^{1/2}\,. \quad (8)$$

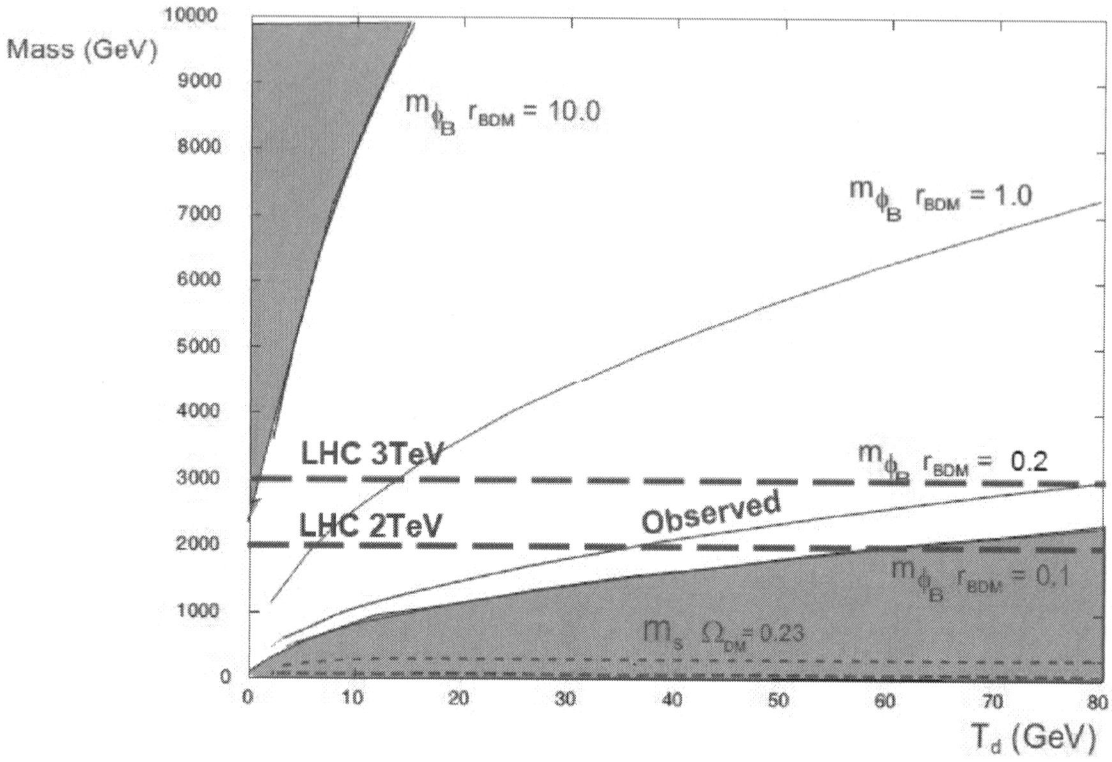

Figure 3.

The freeze-out number density at T_d is then

$$n_{\phi_B}(T_d) \approx \frac{H(T_d)}{<\sigma v>_{\phi_B}} \ . \tag{9}$$

(The $\hat{\phi}_B$ number density is the same when $m_{\phi_B} = m_{\hat{\phi}_B}$.) If ϕ_B later decays to baryon number $B(\phi_B)$ and $\hat{\phi}_B$ to $B(\hat{\phi}_B)$, the baryon asymmetry to dark matter ratio at present, $r_{BDM} \equiv \Omega_B/\Omega_{DM}$, is given by

$$r_{BDM} = 3(B(\phi_B) + B(\hat{\phi}_B))\frac{m_n}{\Omega_{DM}}\frac{g(T_\gamma)}{g(T_d)^{1/2}}\left(\frac{4\pi^3}{45M_{Pl}^2}\right)^{1/2}\frac{T_\gamma^3}{\rho_c}\frac{1}{T_d}\frac{1}{\langle\sigma v\rangle_{\phi_B}} \ . \tag{10}$$

Here $g(T)$ is the effective number of relativistic degrees of freedom, m_n is the nucleon mass, ρ_c the critical density, T_γ is the present photon temperature and $M_{Pl} = 1.22 \times 10^{19}$ GeV. The prefactor 3 accounts for the three colours of ϕ_B. Requiring that $\Omega_{DM} = 0.23$ then determines the annihilon mass

$$m_{\phi_B} = 2.81 \text{ TeV} \times g(T_d)^{1/4} r_{BDM}^{1/2}(B(\phi_B) + B(\hat{\phi}_B))^{-1/2}\left(\frac{T_d}{1 \text{ GeV}}\right)^{1/2}\lambda_B\left(1 - \frac{m_S^2}{m_{\phi_B}^2}\right)^{1/4} \ . \tag{11}$$

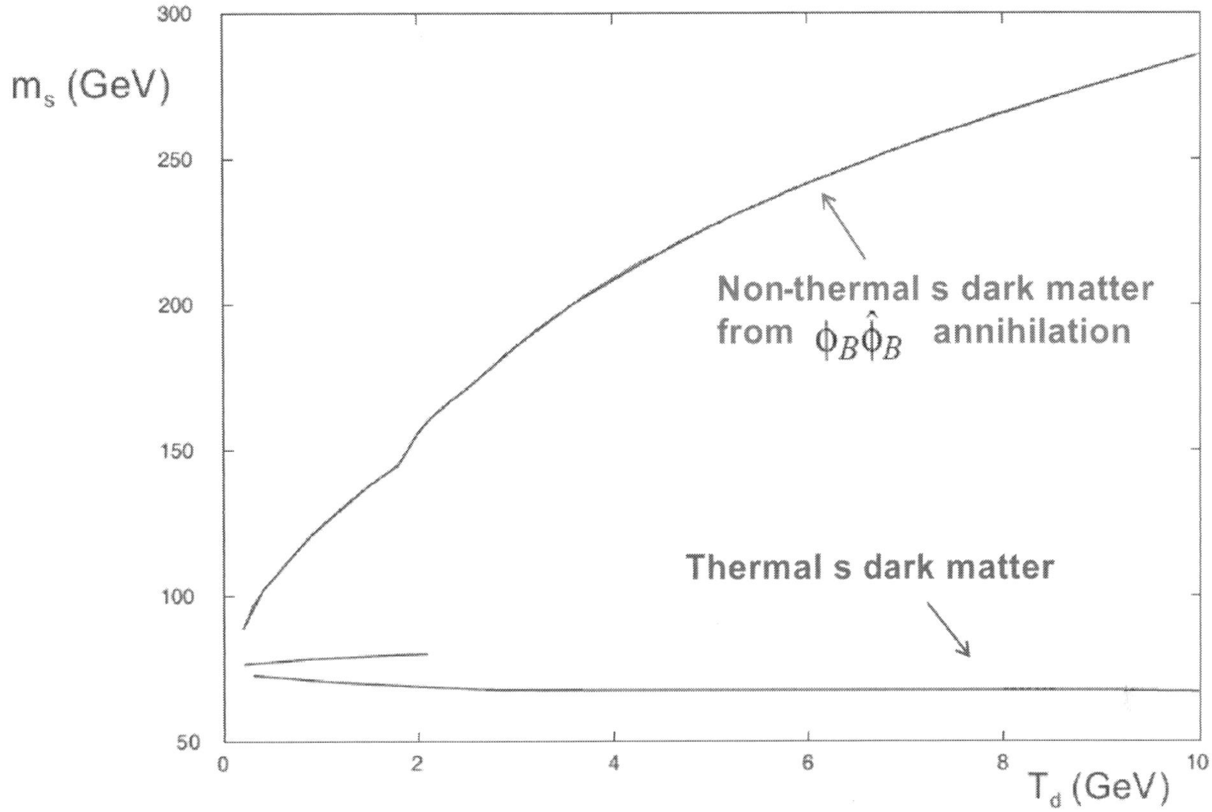

Figure 4.

In this we have assumed that ϕ_B and $\hat{\phi}_B$ are non-relativistic when the B-violating annihilation occurs. This is generally true for gauge-charged ϕ_B and $\hat{\phi}_B$, as discussed in [18]. It is also true in s and $\hat{s}$ annihilation provided $T_d \gtrsim 0.4$ GeV [18].

In Figure 3 we show the annihilon mass as a function of the B injection temperature T_d for different values of the baryon-to-dark matter ratio r_{BDM}. We have set $\lambda_B = \lambda_s = 0.1$, which are dimensionally natural values in particle physics models. Since the mystery of the baryon-to-dark matter ratio may be considered why they are within an order of magnitude of each other, we have shown r_{BDM} in the range 0.1 to 10. We see that a wide range of annihilon masses in the 100 GeV to 10 TeV range is compatible with r_{BDM} in the range 0.1 to 10 when T_d is in the range 0.1 GeV to 100 GeV. There is also an upper bound on T_d from the requirement that $T_d < T_{\phi_B}$, where T_{ϕ_B} is the freeze-out temperature of the B-violating annihilation process, $T_{\phi_B} \approx m_{\phi_B}/20$. T_{ϕ_B} equals a few hundred GeV for the range of m_{ϕ_B} considered in Figure 4.

Therefore we see that for a wide range of T_d, $O(0.1)$ GeV to $O(100)$ GeV, the baryon density is naturally within an order of magnitude of the dark matter density. (Note that a low T_d is essential for a baryon asymmetry to exist at all.)

In Figure 3 we have also shown lines for m_{ϕ_B} equal to 2 TeV and 3 TeV, corresponding to bounds that may be achieveable for coloured scalars at the LHC. The observed $r_{BDM} = 0.2$ favours annihilon masses less than ~ 3 TeV over the range of T_d from 0.1 to 100 GeV. Therefore, should the annihilons be coloured, there is a good prospect of producing them at the LHC.

In the case where s and $\hat{s}$ are degenerate in mass, both scalars are stable and contribute to dark matter.

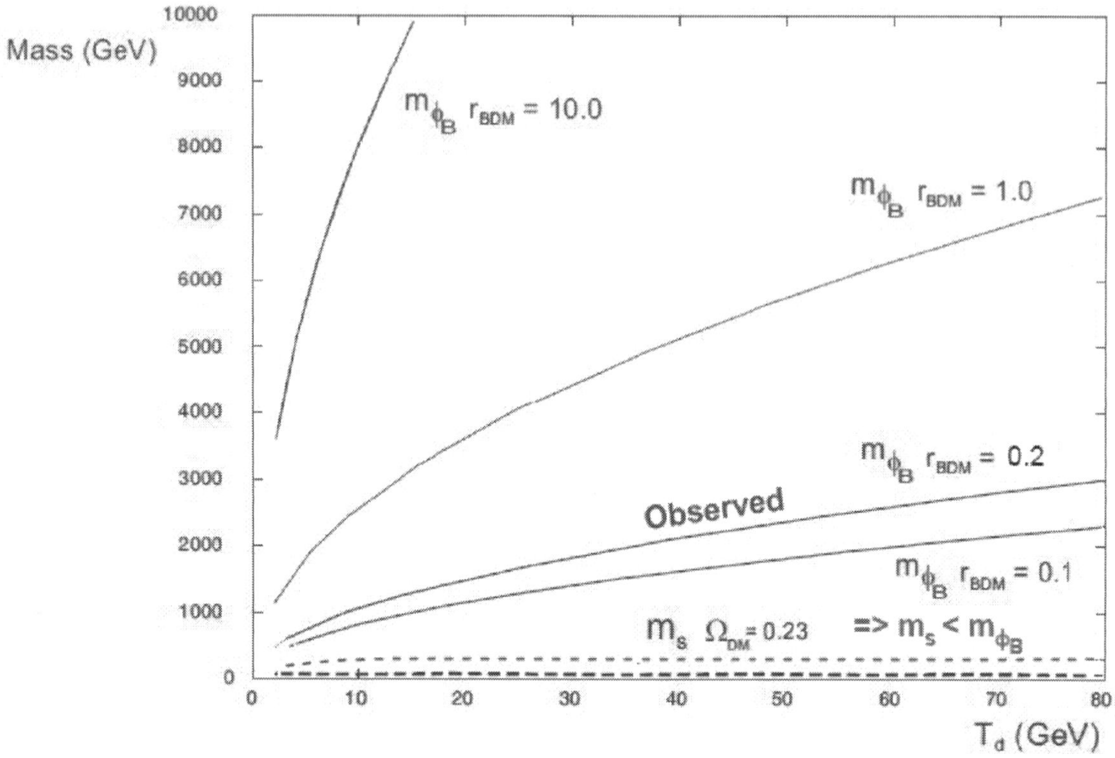

Figure 5.

A key requirement is that $m_s < m_{\phi_B}$, so that the $\phi_B\,\hat{\phi}_B$ annihilation process is kinematically allowed. If $T_d < T_s$, then the s and $\hat{s}$ dark matter density is non-thermal, coming from the s and $\hat{s}$ produced as final state particles in $\phi_B\,\hat{\phi}_B$ annihilation. This gives an enhancement of the density by T_s/T_d relative to the thermal relic s density,

$$\Omega_{DM} = \frac{2m_s}{\rho_c}\frac{g(T_\gamma)}{g(T_d)^{1/2}}\left(\frac{4\pi^3}{45M_{Pl}^2}\right)^{1/2}\frac{T_\gamma^3}{T_d}\frac{1}{\langle\sigma v\rangle_s} . \tag{12}$$

If $T_d > T_s$ then the dark matter is purely thermal relic in nature. In this case we can approximately replace T_d by T_s in Eq. (12). The resulting m_s is shown in Figure 4 for Higgs mass $m_h = 150$ GeV, $\lambda_S = 0.1$ and $\Omega_{DM} = 0.23$. In Figure 5 we show m_s together with m_{ϕ_B}. For all T_d the s mass is much less than the ϕ_B mass, as required for consistency of the model. In addition, for most T_d the s freeze-out temperature is less than T_d, so that s dark matter is in fact thermal relic in nature. Non-thermal s dark matter is found only at $T_d \lesssim 10$ GeV.

We conclude that a consistent baryomorphosis model can be constructed using only a simple $Z_2 \times Z_2$ symmetry.

So far we have produced a thermal WIMP-like baryon asymmetry in ϕ_B and $\hat{\phi}_B$ particles. This must still be transferred to conventional quarks. We will focus on the example of coloured annihilons, which have the best prospect of being produced at the LHC. We assume the annihilons transform under $SU(3)_c \times SU(2)_L$ as $\phi_B(\mathbf{3},\mathbf{1})$ and $\hat{\phi}_B(\overline{\mathbf{3}},\mathbf{1})$. In order to produce a thermal WIMP-like baryon asymmetry,

Journal of Physics: Conference Series **384** (2012) 012009 doi:10.1088/1742-6596/384/1/012009

the decay of ϕ_B, $\hat{\phi}_B$ must occur after the baryon asymmetry has frozen out, at $T_D < T_d$. This means that ϕ_B and $\hat{\phi}_B$ must be *very long-lived* particles:

$$1.5\,\mathrm{s} \gtrsim \tau \gtrsim 8 \times 10^{-11} \left(\frac{100\,\mathrm{GeV}}{T_d} \right)^2 s \,, \tag{13}$$

where the upper bound is from nucleosynthesis. This provides a key experimental signature of annihilons. The long annihilon life-time requires either an extremely small renormalizable Yukawa coupling $\lambda \phi_B \overline{\psi} \psi$ to SM fermions ψ with

$$\lambda \lesssim 1.2 \times 10^{-10} \left(\frac{T_d}{1\,\mathrm{GeV}} \right) \left(\frac{1\,\mathrm{TeV}}{m_{\phi_B}} \right)^{1/2} \,, \tag{14}$$

or a non-renormalizable coupling suppressed by a sufficiently large mass scale. The former possibility seems overtly unnatural, so we will consider the latter. However, in this case we need to explain why there are no renormalizable couplings leading to rapid ϕ_B decay.

The simplest way to achieve this is if ϕ_B has a large hypercharge e.g. $Y(\phi_B) = 5/3$. In this case ϕ_B and $\hat{\phi}_B$ can decay only via mass-suppressed non-renormalizable interactions. For example, for the case $\phi_B(\mathbf{3}, \mathbf{1}, 5/3)$ and $\hat{\phi}_B(\overline{\mathbf{3}}, \mathbf{1}, -5/3)$, we can form

$$\frac{1}{M^3} \phi_B \overline{d^c_R} d_R \overline{L^c_L} L_L \tag{15}$$

and

$$\frac{1}{M^3} \hat{\phi}_B \overline{d_R} e^c_R Q_L Q_L \,. \tag{16}$$

The mass M should then be in the range $10^6 - 10^8$ GeV to account for the low decay temperature T_D [17].

Note that for $\hat{\phi}_B$ to decay, we must assume that Z_A is slightly broken by the non-renormalizable operators. However, since these operators are suppressed by a large mass scale, this small breaking of Z_A will not introduce any dangerous mass mixing between ϕ_B and $\hat{\phi}_B$. If baryon number is conserved, then from Eq. (15) and Eq. (16) the baryon numbers[2] of ϕ_B and $\hat{\phi}_B$ would be $B(\phi_B) = -2/3$ and $B(\hat{\phi}_B) = -1/3$.

However, if we do not assume baryon number conservation, then there are other possible operators, for example

$$\frac{1}{M^3} \phi_B \left(\overline{e_R} Q_L \overline{e_R} L_L \right)^\dagger \,, \tag{17}$$

which allows ϕ_B to decay to a final state with $B = 1/3$. In this case the effective baryon number of ϕ_B and $\hat{\phi}_B$ will be determined by their dominant decay modes.

Observation of such unstable particles might be achieved via the decay of particles which are stopped in detectors (see e.g. [21]).

In summary, the key experimental feature of the baryomorphosis model are the annihilons ϕ_B, $\hat{\phi}_B$:

- Their mass must be 100 GeV - few TeV to produce weak-strength annihilations $\Rightarrow$ good prospect of producing at colliders.
- They have a long lifetime and decay to net B number.
- They may have B-violating decays.
- Their long lifetime suggests large hypercharge $|Y| \geq 5/3$.
- The best case is that of coloured annihilons: Can pair produce at LHC up to 2-3 TeV. (Similar to squark production.)

[2] ϕ_B and $\hat{\phi}_B$ also carry lepton number, $L(\phi_B) = -2$ and $L(\hat{\phi}_B) = 1$.

Conclusions

• It is possible to *modify* a large initial baryon asymmetry to be similar to a thermal relic WIMP mass density $\Rightarrow$ *Baryomorphosis*.

• Requires additional particles and discrete symmetries $\Rightarrow$ Can naturally combine with a WIMP dark matter candidate (gauge singlet scalars, inert doublets) $\Rightarrow$ Simultaneous generation of thermal WIMP-like baryon and dark matter densities.

• Therefore can understand both of the puzzles of the baryon density: why it is similar to the dark matter density and why it is similar to the "WIMP miracle" density.

• Need a low B-injection temperature, $T_d <$ few $\times$ 100 GeV.

• Generically requires pairs of new particles ("annihilons") with gauge interactions and with mass $\sim$ 100 GeV - few TeV $\Rightarrow$ Could be produced at the LHC.

• Two types of annihilon are necessary, with opposite gauge charge but different B.

• Annihilons have long lifetime and decay to baryon number, with possibly large hypercharge and B-violating decay modes.

To sum up, the WIMP miracle *can* account for DM without requiring either coincidence or anthropic selection to explain $\Omega_B/\Omega_{DM} = 1/5$.

Acknowledgements

This work is supported by the Lancaster-Manchester-Sheffield Consortium for Fundamental Physics under STFC grant ST/J000418/1.

References

[1] S.M.Barr, R.S.Chivukula and E.Fahri, *Phys. Lett.* **B241** (1990) 387; S.M.Barr, *Phys. Rev. D* **D44** (1991) 3062; D.B.Kaplan, *Phys. Rev. Lett.* **68** (1992) 741.

[2] N.Cosme, L.Lopez Honorez and M.H.G.Tytgat, *Phys. Rev. D* **D72** (2005) 043505; R. Kitano and I. Low, Phys. Rev. D **71**, 023510 (2005) [arXiv:hep-ph/0411133]. K. Agashe and G. Servant, JCAP **0502**, 002 (2005) [arXiv:hep-ph/0411254]; D. Hooper, J. March-Russell and S. M. West, Phys. Lett. B **605** (2005) 228 [arXiv:hep-ph/0410114]; M. Y. Khlopov, Pisma Zh. Eksp. Teor. Fiz. **83**, 3 (2006) [JETP Lett. **83**, 1 (2006)] [arXiv:astro-ph/0511796]. G. R. Farrar and G. Zaharijas, Phys. Rev. Lett. **96**, 041302 (2006) [arXiv:hep-ph/0510079]; S. Abel and V. Page, JHEP **0605**, 024 (2006) [arXiv:hep-ph/0601149]. H. An, S. L. Chen, R. N. Mohapatra and Y. Zhang, JHEP **1003**, 124 (2010) [arXiv:0911.4463 [hep-ph]].

[3] S.D.Thomas, *Phys. Lett.* **B356** (1995) 256.

[4] K.Enqvist and J.McDonald, *Nucl. Phys.* **B538** (1999) 321; K. Enqvist and J. McDonald, Phys. Lett. B **440**, 59 (1998) [arXiv:hep-ph/9807269]. L. Roszkowski and O. Seto, Phys. Rev. Lett. **98**, 161304 (2007) [arXiv:hep-ph/0608013]; O. Seto and M. Yamaguchi, Phys. Rev. D **75**, 123506 (2007) [arXiv:0704.0510 [hep-ph]]; F. Doddato and J. McDonald, JCAP **1106**, 008 (2011) [arXiv:1101.5328 [hep-ph]]; F. Doddato and J. McDonald, arXiv:1111.2305 [hep-ph].

[5] T. Cohen, D. J. Phalen, A. Pierce and K. M. Zurek, Phys. Rev. D **82**, 056001 (2010) [arXiv:1005.1655 [hep-ph]]; H. Davoudiasl, D. E. Morrissey, K. Sigurdson and S. Tulin, Phys. Rev. Lett. **105**, 211304 (2010) [arXiv:1008.2399 [hep-ph]]. J. Shelton and K. M. Zurek, Phys. Rev. D **82** (2010) 123512 [arXiv:1008.1997 [hep-ph]]. M. Blennow, B. Dasgupta, E. Fernandez-Martinez and N. Rius, JHEP **1103**, 014 (2011) [arXiv:1009.3159 [hep-ph]].

[6] T. Cohen and K. M. Zurek, Phys. Rev. Lett. **104**, 101301 (2010) [arXiv:0909.2035 [hep-ph]]; L. J. Hall, J. March-Russell and S. M. West, arXiv:1010.0245 [hep-ph]; B. Dutta and J. Kumar, Phys. Lett. B **699**, 364 (2011) [arXiv:1012.1341 [hep-ph]]; J. J. Heckman, S. -J. Rey, JHEP **1106** (2011) 120. [arXiv:1102.5346 [hep-th]]. M. L. Graesser, I. M. Shoemaker and L. Vecchi, arXiv:1103.2771 [hep-ph]; M. T. Frandsen, S. Sarkar and K. Schmidt-Hoberg, arXiv:1103.4350 [hep-ph]. S. D. McDermott, H. B. Yu and K. M. Zurek, arXiv:1103.5472 [hep-ph]; C. Kouvaris, P. Tinyakov, Phys. Rev. Lett. **107** (2011) 091301. [arXiv:1104.0382 [astro-ph.CO]]. C. Cheung, K. M. Zurek, [arXiv:1105.4612 [hep-ph]]; J. March-Russell, M. McCullough, [arXiv:1106.4319 [hep-ph]]; H. Davoudiasl, D. E. Morrissey, K. Sigurdson and S. Tulin, arXiv:1106.4320 [hep-ph]; C. Arina and N. Sahu, arXiv:1108.3967 [hep-ph]; K. Petraki, M. Trodden and R. R. Volkas, arXiv:1111.4786 [hep-ph].

[7] N. F. Bell, K. Petraki, I. M. Shoemaker and R. R. Volkas, arXiv:1105.3730 [hep-ph];

[8] B. von Harling, K. Petraki and R. R. Volkas, arXiv:1201.2200 [hep-ph].

[9] Y. Cai, M. A. Luty and D. E. Kaplan, arXiv:0909.5499 [hep-ph].

[10] M. R. Buckley and L. Randall, arXiv:1009.0270 [hep-ph].
[11] P. H. Gu, M. Lindner, U. Sarkar and X. Zhang, arXiv:1009.2690 [hep-ph].
[12] Y. Cui, L. Randall and B. Shuve, arXiv:1106.4834 [hep-ph].
[13] A. Falkowski, J. T. Ruderman and T. Volansky, JHEP **1105**, 106 (2011) [arXiv:1101.4936 [hep-ph]].
[14] R. Allahverdi, B. Dutta and K. Sinha, Phys. Rev. D **83**, 083502 (2011) [arXiv:1011.1286 [hep-ph]].
[15] F. D'Eramo, L. Fei and J. Thaler, arXiv:1111.5615 [hep-ph].
[16] J. McDonald, JCAP **0701**, 001 (2007) [arXiv:hep-ph/0609126].
[17] J. McDonald, Phys. Rev. D **83**, 083509 (2011) arXiv:1009.3227 [hep-ph].
[18] J. McDonald, arXiv:1108.4653 [hep-ph].
[19] Y. Cui, L. Randall and B. Shuve, arXiv:1112.2704 [hep-ph].
[20] L. Lopez Honorez, E. Nezri, J. F. Oliver and M. H. G. Tytgat, JCAP **0702**, 028 (2007) [arXiv:hep-ph/0612275].
[21] P. W. Graham, K. Howe, S. Rajendran and D. Stolarski, arXiv:1111.4176 [hep-ph].

The impact of XENON100 and the LHC on Supersymmetric Dark Matter

Keith A. Olive

William I. Fine Theoretical Physics Institute,
School of Physics and Astronomy, University of Minnesota,
Minneapolis, Minnesota 55455, USA

E-mail: olive@umn.edu

Abstract. The effect of 2010 and 2011 LHC data are discussed in connection to the potential for the direct detection of supersymmetric dark matter. The impact of the recent XENON100 results are contrasted to these predictions. Expectations for indirect detection are also discussed.

1. Introduction

The minimal supersymmetric Standard Model (MSSM) has over 100 undetermined parameters, which are mainly associated with the breaking of supersymmetry. However, it is often assumed that the soft supersymmetry-breaking parameters have some universality properties. These may include the universality of gaugino masses, $M_a = m_{1/2}$, trilinear supersymmetry-breaking mass parameters, $A_f = A_0$, and soft scalar masses. $m_{ij}^2 = \delta_{ij} m_0^2$. The simplified version of the MSSM in which universality in input at the grand unification scale is called the constrained MSSM (CMSSM) [1, 2, 3, 4, 5, 6, 7, 8]. The sparticle spectrum is run down to the electroweak scale and radiatively induces electroweak symmetry breaking (EWSB).

Minimization of the Higgs potential leads to two conditions at the weak scale which can be expressed as

$$\mu^2 = \frac{m_1^2 - m_2^2 \tan^2\beta + \frac{1}{2} m_Z^2 (1 - \tan^2\beta) + \Delta_\mu^{(1)}}{\tan^2\beta - 1 + \Delta_\mu^{(2)}}, \tag{1}$$

and

$$B\mu = (m_1^2 + m_2^2 + 2\mu^2) \sin 2\beta + \Delta_B \tag{2}$$

where μ is the Higgs mixing parameter, B is the associated supersymmetry-breaking bilinear mass, $\tan\beta$ is the ratio of the two Higgs vacuum expectation values, and Δ_B and $\Delta_\mu^{(1,2)}$ are loop corrections [9, 10, 11]. The combination $B\mu$ can be related to the Higgs pseudo-scalar mass, m_A. While one can choose to include B and μ (or m_A and μ) as free input parameters and calculate the two Higgs expectation values, or m_Z and $\tan\beta$, it is more common to use these equations to calculate μ and B upon assuming a value of $\tan\beta$ and the measured value of m_Z. Thus upon assuming radiative electroweak symmetry breaking, the CMSSM is a 4 parameter theory ($m_{1/2}, m_0, A_0$ and $\tan\beta$). In addition the sign of the μ term must also be specified.

An often considered generalization of the CMSSM allows for non-universal Higgs masses (NUHM) [12, 13, 14]. By allowing, both Higgs soft masses m_1 and m_2 to differ from m_0

(NUHM2), one can effectively choose both m_A and μ as free parameters as is seen from the electroweak conditions given above. One may also choose a subset of these models and take $m_1 = m_2 = m_0$ in which case either m_A or μ can be chosen in addition to the four CMSSM parameters (NUHM1).

It is also possible to consider a very constrained version of the MSSM (VCMSSM) [15] by applying the relation $B_0 = A_0 - m_0$ as expected from minimal supergravity [16]. In this case, μ and $\tan\beta$ are derived from the electroweak vacuum conditions and the theory has only three free parameters (and the sign of μ). True models based on minimal supergravity (mSUGRA) impose in addition the relation between the gravitino mass and m_0, namely $m_{3/2} = m_0$. In these models, it is often the gravitino which ends up as the lightest supersymmetric particle (LSP) and therefore the dark matter candidate.

For given values of $\tan\beta$, A_0, and $sgn(\mu)$, the regions of the CMSSM parameter space that yield an acceptable relic density and satisfy other phenomenological constraints may be displayed in the $(m_{1/2}, m_0)$ plane. In Fig. 1a, the dark (blue) shaded region corresponds to that portion of the CMSSM plane with $\tan\beta = 10$, $A_0 = 0$, and $\mu > 0$ such that the computed relic density yields the WMAP value [17] of

$$\Omega h^2 = 0.111 \pm 0.006. \tag{3}$$

The bulk region at relatively low values of $m_{1/2}$ and m_0, tapers off as $m_{1/2}$ is increased. At higher values of m_0, annihilation cross sections are too small to maintain an acceptable relic density and $\Omega_\chi h^2$ is too large. At large $m_{1/2}$, co-annihilation processes between the LSP and the next lightest sparticle (in this case the $\tilde{\tau}$) enhance the annihilation cross section and reduce the relic density. This occurs when the LSP and NLSP are nearly degenerate in mass. The dark (red) shaded region has $m_{\tilde{\tau}} < m_\chi$ and is excluded. The effect of coannihilations is to create an allowed band about 25-50 GeV wide in m_0 for $m_{1/2} \lesssim 950$ GeV, or $m_\chi \lesssim 400$ GeV, which tracks above the $m_{\tilde{\tau}_1} = m_\chi$ contour [18]. Also shown in the figure are some phenomenological constraints from the lack of detection of charginos [19], or Higgses [20] as well as constraints from $b \to s\gamma$ [21] and $g_\mu - 2$ [22]. The locations of these constraints are described in the caption.

At larger $m_{1/2}, m_0$ and $\tan\beta$, the relic neutralino density may be reduced by rapid annihilation through direct-channel H, A Higgs bosons, as seen in Fig. 1(b) [1, 3]. Finally, the relic density can again be brought down into the WMAP range at large m_0 in the 'focus-point' region close the boundary where EWSB ceases to be possible and the lightest neutralino χ acquires a significant higgsino component [23]. The start of the focus point region is seen in the upper left of Fig. 1b.

2. Mastercode: Markov-Chain Monte-Carlo

It is well established that Markov-Chain Monte-Carlo (MCMC) algorithms offer an efficient technique for sampling a large parameter space such as the CMSSM or its variants. MCMC has been utilized in the Mastercode [24] framework which incorporates a code for the electroweak observables based on [25] as well as the `SoftSUSY` [26], `FeynHiggs` [27], `SuFla` [28], `SuperIso` [29], `MicrOMEGAs` [30] and `SSARD` [31] codes, using the SUSY Les Houches Accord [32]. The MCMC technique is used to sample the SUSY parameter space, and thereby construct the χ^2 probability function, $P(\chi^2, N_{\rm dof})$. This accounts for the number of degrees of freedom, $N_{\rm dof}$, and thus provides a quantitative measure for the quality-of-fit such that $P(\chi^2, N_{\rm dof})$ can be used to estimate the absolute probability with which the CMSSM describes the experimental data.

The results of the mastercode analysis include the parameters of the best-fit points as well as the 68 and 95% C.L. regions found with default implementations of the phenomenological, experimental and cosmological constraints. These include precision electroweak data, the anomalous magnetic moment of the muon, $(g-2)_\mu$, B-physics observables, the bound on the lightest MSSM Higgs boson mass, M_h, and the cold dark matter (CDM) density inferred from astrophysical and cosmological data assuming that this is dominated by the relic density

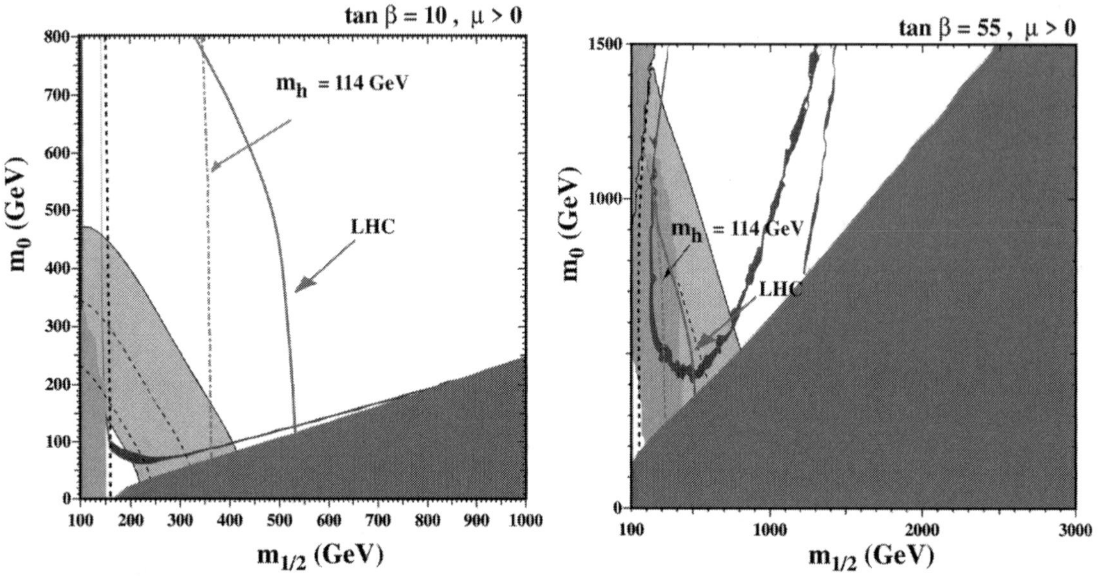

Figure 1. *The $(m_{1/2}, m_0)$ planes for (a) $\tan\beta = 10$ and $\mu > 0$, assuming $A_0 = 0, m_t = 173.1$ GeV and $m_b(m_b)^{\overline{MS}}_{SM} = 4.25$ GeV. The near-vertical (red) dot-dashed lines are the contours $m_h = 114$ GeV, and the near-vertical (black) dashed line is the contour $m_{\chi^\pm} = 104$ GeV. The medium (dark green) shaded region is excluded by $b \to s\gamma$, and the dark (blue) shaded area is the cosmologically preferred region. In the dark (brick red) shaded region, the LSP is the charged $\tilde{\tau}_1$. The region allowed by the E821 measurement of $g_\mu - 2$ at the 2-σ level, is shaded (pink) and bounded by solid black lines, with dashed lines indicating the 1-σ ranges. The curves marked LHC show the 95& CL exclusion region (to the left of the curves) for LHC sparticle searches at 1/fb. In (b), $\tan\beta = 55$. Here, in the upper left corner, the region with no EWSB is shaded dark pink.*

of the lightest neutralino, $\Omega_\chi h^2$. In addition one can include the constraint imposed by the experimental upper limit on the spin-independent DM scattering cross section $\sigma_p^{\rm SI}$. A purely frequentist analyses of the CMSSM was performed in [6, 7, 8, 33, 34, 35], in the NUHM1 in [8, 33, 34, 35], and in the VCMSSM/mSUGRA in [36, 33, 34].

In [7], a pre-LHC analysis of the CMSSM was performed. The 68% and 95% confidence-level (C.L.) regions in the $(m_{1/2}, m_0)$ plane of the CMSSM are shown in Fig. 2. Also shown for comparison are the physics reaches of ATLAS and CMS with 1/fb of integrated luminosity [37, 38]. The likelihood analysis assumed $\mu > 0$, as motivated by the sign of the apparent discrepancy in $g_\mu - 2$, but sampled all values of $\tan\beta$ and A_0: the experimental sensitivities were estimated assuming $\tan\beta = 10$ and $A_0 = 0$, but are probably not very sensitive to these assumptions. The global maximum of the likelihood function (indicated by the black dot) is at $m_{1/2} = 310$ GeV, $m_0 = 60$ GeV, $A_0 = 240$ GeV, $\tan\beta = 11$ and $\chi^2/N_{dof} = 20.4/19$ (37% probability). Note that the best-fit point lies well within the LHC discovery range, as does the 68% C.L. region. As we wil see, by the end of 2011, the LHC has met this reach (at 7 TeV center of mass energy) and as sparticles have yet to be discovered, this region is mostly excluded at 95% CL (see sections 4 and 5 below). A more detailed view of the $\Delta\chi^2$ function for the CMSSM is shown in Fig. 3. For other pre-LHC results see [39].

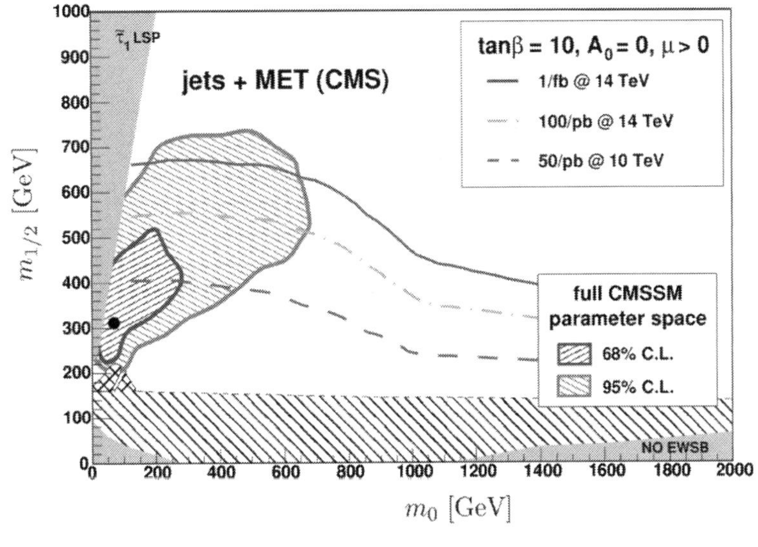

Figure 2. *The $(m_0, m_{1/2})$ plane in the CMSSM showing the regions favoured in a likelihood analysis at the 68% (blue) and 95% (red) confidence levels [7]. The best-fit point is shown by the black point. Also shown are the 5σ discovery contours for jet + missing E_T events at CMS with 1 fb^{-1} at 14 TeV, 100 pb^{-1} at 14 TeV and 50 pb^{-1} at 10 TeV centre-of-mass energy.*

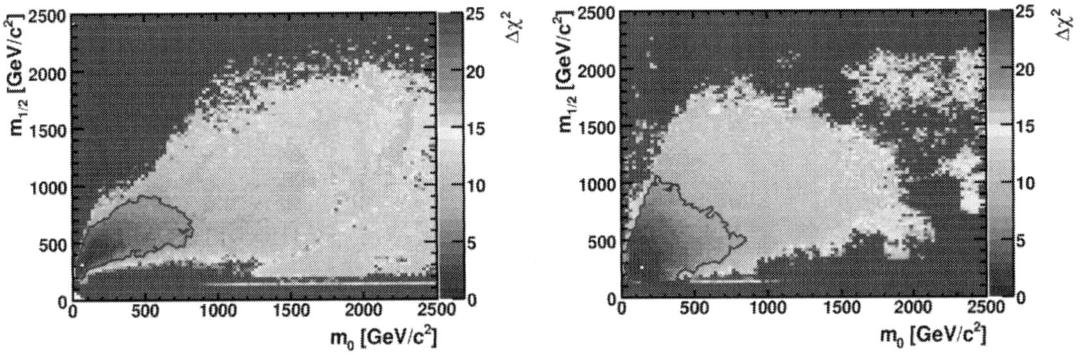

Figure 3. *The $\Delta\chi^2$ functions in the $(m_0, m_{1/2})$ planes for the CMSSM (left plot) and for the NUHM1 (right plot). The best fit points are indicated by a white dot and the 68% and 95% CL contours are the light and dark curves respectively [8].*

3. Results for the NUHM1

The cosmologically preferred regions move around in the $(m_{1/2}, m_0)$ plane if one abandons the universality assumptions of the CMSSM. As discussed above, if one allows the supersymmetry-breaking contributions to the Higgs masses to be non-universal (NUHM), the rapid-annihilation WMAP 'strip' can appear at different values of $\tan\beta$ and $m_{1/2}$, as seen in Fig. 4 [14]. In the left panel, we show an NUHM1 $(m_{1/2}, m_0)$ plane for $\tan\beta = 10, A_0 = 0$ and $\mu > 0$ with $m_A = 500$ GeV, and μ calculated using (1). In addition to the constraints discussed above, we also plot contours of $\mu = 300$, 500, 1000, and 1500 GeV (light pink). The thick green dot-dashed contour tracks the CMSSM parameters in the NUHM1 $(m_{1/2}, m_0)$ plane. The most prominent departure from the CMSSM is that the EWSB requirement constrains the plane at low m_0 rather than at large m_0. In this region (below the CMSSM contour), the fixed value m_A is larger than its corresponding value in the CMSSM, resulting in correspondingly larger m_1^2 and m_2^2 (smaller $|m_2^2|$, since $m_2^2 < 0$). We see from (1) that, the effect is to drive μ^2 smaller, and eventually negative. The stau LSP exclusion regions are qualitatively similar to those in the CMSSM, however there is a (black shaded) region of the plane where the lighter selectron is the LSP. In this case, the co-annihilation region connects the analogue of the focus point region with small μ discussed above and heavy Higgs funnel region which now exists at $\tan\beta = 10$ at $m_{1/2} \approx 550$ GeV and extends to large m_0. Some of the co-annihilation region also extends to larger $m_{1/2}$.

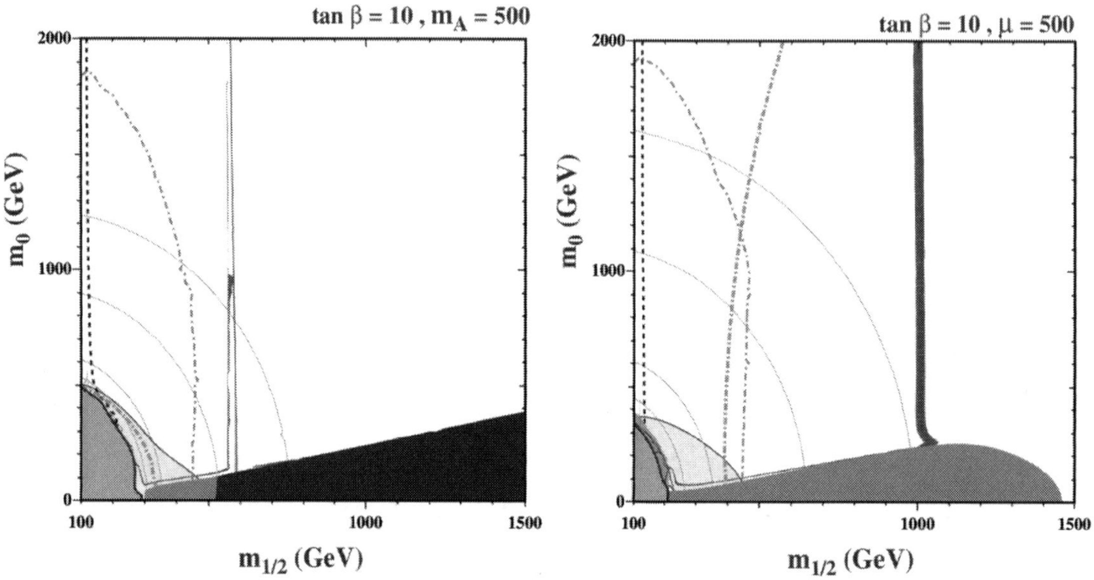

Figure 4. *The* $(m_{1/2}, m_0)$ *plane in the NUHM for* $\tan\beta = 10$*, a) with fixed* $m_A = 500$ *GeV and b) with fixed* $\mu = 500$ *GeV [14].*

In the right panel of Fig. 4, we show the NUHM1 $(m_{1/2}, m_0)$ plane with $\mu = 500$ GeV and m_A calculated using (2). At first glance, the $(m_{1/2}, m_0)$ plane with fixed μ has some similarities with those with fixed m_A. There are excluded regions at very low $(m_{1/2}, m_0)$ where the pseudoscalar Higgs mass squared is negative, corresponding to the absence of electroweak symmetry breaking, surrounded by four contours of fixed $m_A = 300$, 500, 1000, and 1500 GeV. At small values of m_0, extending out to large $m_{1/2}$, there are excluded $\tilde{\tau}$-LSP regions resembling those in the CMSSM. $b \to s\gamma$ excludes strips near the EWSB boundary. In addition to the coannihilation strip close to the $\tilde{\tau}$-LSP boundary, another strip close to the EWSB boundary, and the curved rapid-annihilation funnels that appear at low m_A, with strips of good relic density forming the funnel walls, there is a fourth, near-vertical strip, where the relic density is brought down into the WMAP range because of the large mixing between the bino and Higgsino components in the LSP. For smaller $m_{1/2} < 500$ GeV, the LSP is almost pure bino, and the relic density is too large except in the narrow strips mentioned previously. On the other hand, for larger $m_{1/2} > 1000$ GeV, the LSP is almost pure Higgsino, and the relic density falls below the WMAP range. It is also this change in the nature of the LSP that causes the boundary of the $\tilde{\tau}$LSP region to drop. Since the $\tilde{\tau}$ mass is affected only minimally by the value of μ, we find that $\tilde{\tau}$-LSP region terminates at some value of $m_{1/2}$ related primarily to μ. At large m_0 in panel (b) of Fig. 4, it is only in the 'crossover' strip that the relic density falls within the WMAP range. The CMSSM contour is a roughly vertical thick green dot-dashed line, the position of which is determined by the value of m_A that one would find from the electroweak vacuum conditions in the standard CMSSM.

The results of the Mastercode (pre-LHC) analysis for the NUHM is shown in the right panel of Fig. 3. The corresponding parameters of the best-fit NUHM1 point are $m_0 = 150$ GeV, $m_{1/2} = 270$ GeV, $A_0 = -1300$ GeV, $\tan\beta = 11$ and $m^2_{h_1} = m^2_{h_2} = -1.2 \times 10^6$ GeV^2 or, equivalently, $\mu = 1140$ GeV, yielding $\chi^2 = 18.4$ (corresponding to a similar fit probability to the CMSSM) and $M_h = 120.7$ GeV.

It is also possible to extract 1D likelihood functions for essentially any parameter of interest. Here, we show in Fig. 5 the $\Delta\chi^2$ function for the lightest neutralino mass in both the CMSSM and NUHM1. The left panel of Fig. 5 displays the likelihood function in the CMSSM. The solid line shows the result obtained when incorporating the LEP Higgs limit, while the dashed line corresponds to the case where the LEP Higgs constraint is removed. There is a sharp rise in the likelihood function at low values of $m_{\tilde{\chi}^0_1}$, which is caused by the limits from the direct searches for SUSY particles, but receives also contributions from BR($b \to s\gamma$) and other constraints. This sharp rise in the likelihood function persists when the LEP Higgs constraint is removed, but is shifted towards slightly lower values of $m_{\tilde{\chi}^0_1}$ in that case. The right panel of Fig. 5 shows the likelihood function for $m_{\tilde{\chi}^0_1}$ in the NUHM1, again with and without the LEP M_h constraint imposed.

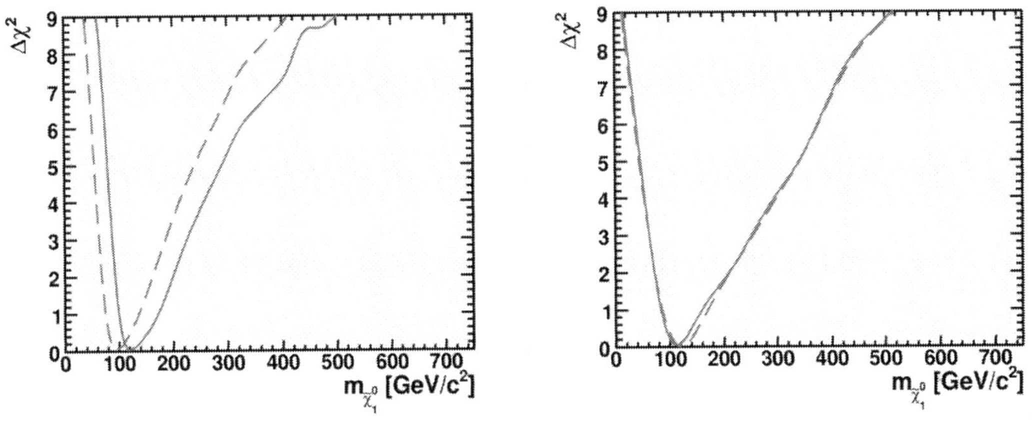

Figure 5. *The likelihood functions for $m_{\tilde{\chi}^0_1}$ in the CMSSM (left) and in the NUHM1 (right), both with (solid lines) and without (dashed lines) the LEP constraint on M_h [8].*

4. Early Results from the LHC

As 2010 came to a close, we obtained the first results from the LHC on sparticle searches. These include several new constraints on SUSY using an integrated luminosity of ~ 35/pb of data at 7 TeV. ATLAS has published the results of a search in multijet + $\not{E}_T$ channels (ATLAS 0L) [40] that has greater sensitivity in some regions to the types of gluino and squark pair-production events expected in the supersymmetric models discussed here than did the earlier ATLAS 1L search [41], and has also released results obtained by combining the one- and zero-lepton searches [42]. CMS has announced results from two other searches in multijet + $\not{E}_T$ channels that improve the CMS α_T sensitivity also to gluino and squark production in the models discussed here [43, 44].

In addition, the XENON100 Collaboration has recently released results from a search for direct spin-independent dark matter scattering with 100.9 live days of data using a fiducial target with a mass of 48 kg [45]. As we see later, this provides constraints on the parameter spaces of supersymmetric models that complement those provided by collider experiments.

The impact of the 35/pb LHC data on the CMSSM parameter space is rather dramatic [33, 34, 35] (see also [46]). In Fig. 6, we display the planes for the CMSSM (left) and NUHM1 (right) driven by the ATLAS 0L and CMS MHT constraints but also taking into account the other 2010 LHC constraints, as well as the XENON100 constraint [34]. The best-fit points are shown as green stars and 68 and 95% CL regions are shown as red and blue lines, respectively and correspond to $\Delta\chi^2 = 2.3$ (red) and 5.99 (blue) relative to the best fit points. The pre-LHC results, taken from [33], are displayed as 'snowflakes' and dotted lines, the post-2010-LHC/XENON100 results are displayed as full stars and solid lines [34].

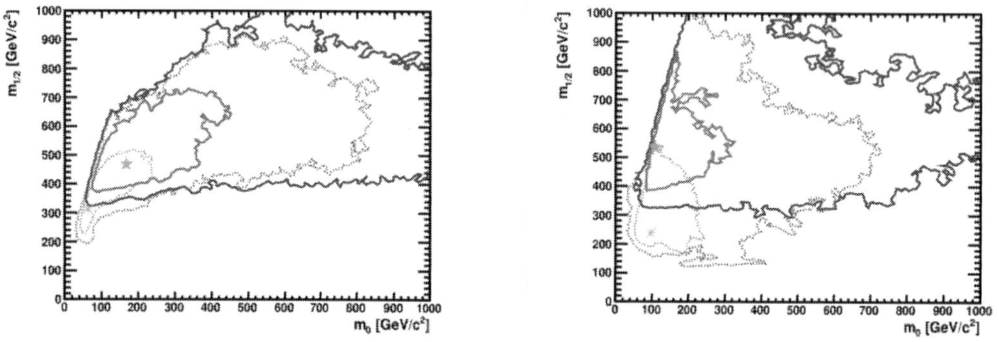

Figure 6. *The $(m_0, m_{1/2})$ planes in the CMSSM (left), and the NUHM1 (right). In each plane, the best-fit point after incorporation of the 2010 LHC and XENON100 constraints is indicated by a filled green star, and the pre-LHC fit by an open star. The 68 and 95% CL regions are indicated by red and blue contours, respectively, the solid lines including the 2010 LHC and XENON100 data, and the dotted lines showing the pre-LHC fits [33, 34].*

In the CMSSM and the NUHM, the direct 2010 LHC constraints push the best-fit values of $m_{1/2}$ to significantly higher values, as well as their 68 and 95% CL ranges. The effect of the LHC on the best-fit values of m_0 is smaller, though there is a significant increase in the CMSSM that is correlated with the increase in $m_{1/2}$. The positions of the best fit points pre/post LHC are collected in Table 1. The total value of χ^2 and the number of degrees of freedom (dof) along with the fit probably (p-value) are also given. The final column shows the predicted value of the Higgs mass at the best fit point. Note there is a slight shift in the best fit points for the pre-LHC data due to changes in the input data used in the Mastercode analysis. LHC2010 includes the 35/pb LHC data as well as the XENON100 data. While one would expect two additional dof's, one dof (the Higgs mass constraint) is lost since the best fit point is pushed past the previous LEP sensitivity.

5. The LHC @ 1/fb

By mid 2011, the LHC results for sparticle searches based on 1/fb of analyzed data were released by the ATLAS [47, 48], CMS [49, 50, 51] and LHCb Collaborations [52]. The absences of signals in the jets $+\not{E}_T$ searches disfavour the ranges of the model mass parameters $(m_0, m_{1/2})$ that had been favoured in our previous analyses of the CMSSM and NUHM1 [33, 34], and our current best fits have $m_0 \sim 150$ to 450 GeV and $m_{1/2} \sim 750$ GeV. Reconciling these larger values of $(m_0, m_{1/2})$ with $(g-2)_\mu$ favours values of $\tan\beta \sim 40$, though with a large uncertainty. Fig. 7 shows the positions of the 68 and 95% CL contours with solid curves corresponding to the 1/fb data as compared with the pre-LHC (and pre-XENON100) results (dashed) [35].

Model	Minimum χ^2/d.o.f.	Prob-ability	$m_{1/2}$ (GeV)	m_0 (GeV)	A_0 (GeV)	$\tan\beta$	M_h (GeV) (no LEP)
CMSSM pre-LHC	21.5/20	37%	360	90	−400	15	111.5
LHC 2010	25.2/21	24%	470	170	−780	22	115.7
$\mathrm{LHC}_{1/\mathrm{fb}}$	28.8/22	15%	780	450	−1100	41	119.1
$(g-2)_\mu$ neglected	21.3/20	38%	2000	1050	430	22	124.8
NUHM1 pre-LHC	20.8/18	29%	340	110	520	13	118.9
LHC 2010	24.5/20	22%	530	110	−370	27	117.9
$\mathrm{LHC}_{1/\mathrm{fb}}$	27.3/21	16%	730	150	−910	41	118.8
$(g-2)_\mu$ neglected	20.3/19	38%	2020	1410	2580	48	126.6

Table 1. *Comparison of the best-fit points found in the CMSSM and NUHM1 pre-LHC (including the upper limit on $BR(B_s \to \mu^+\mu^-)$ available then), the LHC 2010 result (including XENON100) and with the $\mathrm{LHC}_{1/\mathrm{fb}}$ data set (also including the XENON100 constraint) using the standard implementations of the $(g-2)_\mu$ constraint, followed by the case dropping $(g-2)_\mu$. The predictions for M_h do not include the constraint from the direct LEP Higgs search.*

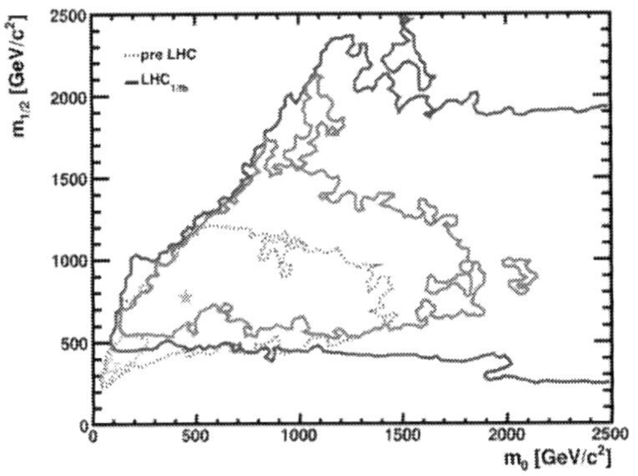

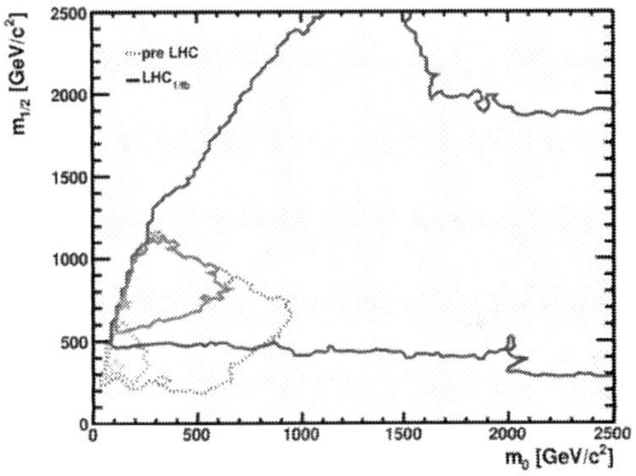

Figure 7. *The $(m_0, m_{1/2})$ planes in the CMSSM (left) and the NUHM1 (right). In each plane, the best-fit point after incorporation of the $\mathrm{LHC}_{1/\mathrm{fb}}$ constraints is indicated by a filled green star, and the pre-LHC fit [36] by an open star. The 68 and 95% CL regions are indicated in red and blue, respectively, the solid lines including the $\mathrm{LHC}_{1/\mathrm{fb}}$ data and the dotted lines showing the pre-LHC fits [35].*

The positions of the best fit points at 1/fb are tabulated in Table 1. We now find that the p-value for the CMSSM best-fit point is now $\sim 15\%$, and that for the NUHM1 is $\sim 16\%$. On the other hand, if the $(g-2)_\mu$ constraint is dropped much larger regions of the $(m_0, m_{1/2})$ and other parameter planes are allowed at the 68 and 95% CL, and these p-values increase to 38% in both models. For comparison, the p-value for the Standard Model (including $(g-2)_\mu$) is 9%.

In Fig. 8 we show the 68% and 95% CL contours in the $(m_{\tilde{\chi}^0_1}, \sigma_p^{\mathrm{SI}})$ planes for the CMSSM (left) and the NUHM1 (right). The solid lines are based on our global fits including the $\mathrm{LHC}_{1/\mathrm{fb}}$ constraints, whereas the dotted lines correspond to our previous fits using the pre-

LHC constraints. In both cases, we assume $\Sigma_{\pi N} = 50 \pm 14$ MeV [53] [1], and we include with the $\mathrm{LHC}_{1/\mathrm{fb}}$ data the XENON100 constraint on σ_p^{SI} [45]. We see that the $\mathrm{LHC}_{1/\mathrm{fb}}$ data tend to push $m_{\tilde{\chi}_1^0}$ to larger values, and that these are correlated with lower values of σ_p^{SI}, though with best-fit values still $\sim 10^{-45}$ cm^2 [35].

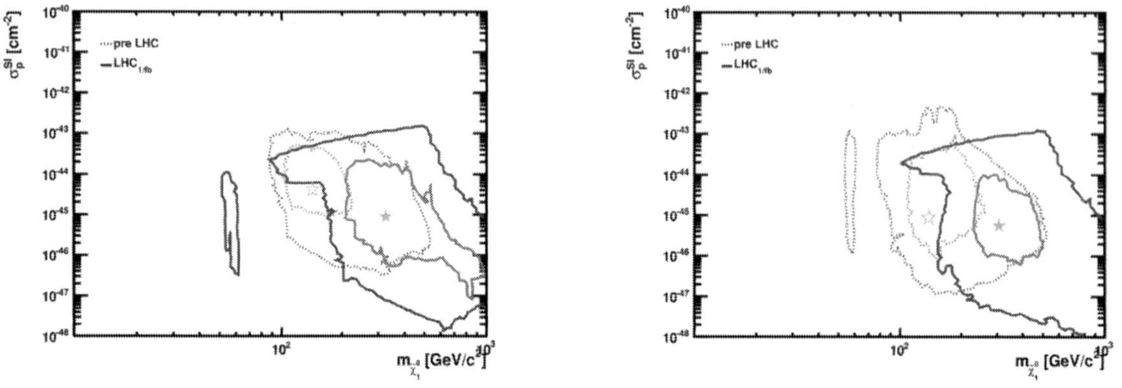

Figure 8. *The 68% and 95% CL contours (red and blue, respectively) in the CMSSM (left) and the NUHM1 (right). The solid lines are for fits including the XENON100 [45] and* $\mathrm{LHC}_{1/\mathrm{fb}}$ *data, whereas the dotted lines include only the pre-LHC data [35].*

6. The Higgs Search

At the close of 2011, the ATLAS and CMS Collaborations have released their official combination of the searches for a SM Higgs boson with the first $\sim 1 - 2.3$/fb of LHC luminosity at $E_{\mathrm{cm}} = 7$ TeV [54]. The combination excludes a SM Higgs boson with a mass between 141 and 476 GeV. Additionally, the ATLAS and CMS Collaborations have presented preliminary updates of their results with ~ 5/fb of data [55]. These results may be compatible with a SM-like Higgs boson around $M_h \simeq 125$ GeV, though CMS also report an excess at $M_h \simeq 119$ GeV in the ZZ^* channel.

It is interesting to note that based on the LHC 1/fb results, we have the predictions of the Higgs mass as seen in Table 1 of 119 and 125 GeV depending on whether $(g-2)_\mu$ is included in the analysis [35]. Specifically, the 1/fb fits included $M_h = 119.1^{+3.4}_{-2.9}$ GeV in the CMSSM and $M_h = 118.8^{+2.7}_{-1.1}$ GeV in the NUHM1 (which should be combined with an estimated theory error $\Delta M_h = \pm 1.5$ GeV). These two fits are based solely on the Higgs-*independent* searches including the $(g-2)_\mu$ constraint, i.e., they do not rely on the existing limits from LEP [20, 56], the Tevatron [57], or the LHC [48, 50]. These predictions increase to $M_h = 124.8^{+3.4}_{-10.5}$ GeV in the CMSSM and $126.6^{+0.7}_{-1.9}$ GeV in the NUHM1 if the $(g-2)_\mu$ constraint is dropped.

If indeed, the LHC has seen the Higgs at 125±1 (119±1) GeV, there are rather dramatic consequences for the supersymmetric parameter space [58] (see also [59]). Since in the CMSSM and NUHM1 the radiative corrections contributing to the value of M_h are sensitive primarily to $m_{1/2}$ and $\tan\beta$, and only to a lesser extent to m_0, we expect that the primary effect of imposing the M_h constraint should be to affect the preferred ranges of $m_{1/2}$ and $\tan\beta$, with a lesser effect on the preferred range of m_0. This effect is indeed seen in both panels of Fig. 9. We see that the 68% CL ranges of $m_{1/2}$ extend to somewhat larger values and with a wider spread than the pre-Higgs results, particularly in the NUHM1. However, the NUHM1 best-fit value of

[1] We recall the sensitivity of predictions for σ_p^{SI} to the uncertainty in $\Sigma_{\pi N}$ [34].

$m_{1/2}$ remains at a relatively low value of $\sim$ 800 GeV, whereas the best-fit value of $m_{1/2}$ in the CMSSM moves to $\sim$ 1900 GeV. This jump reflects the flatness of the likelihood function for $m_{1/2}$ between $\sim$ 700 GeV and $\sim$ 2 TeV.

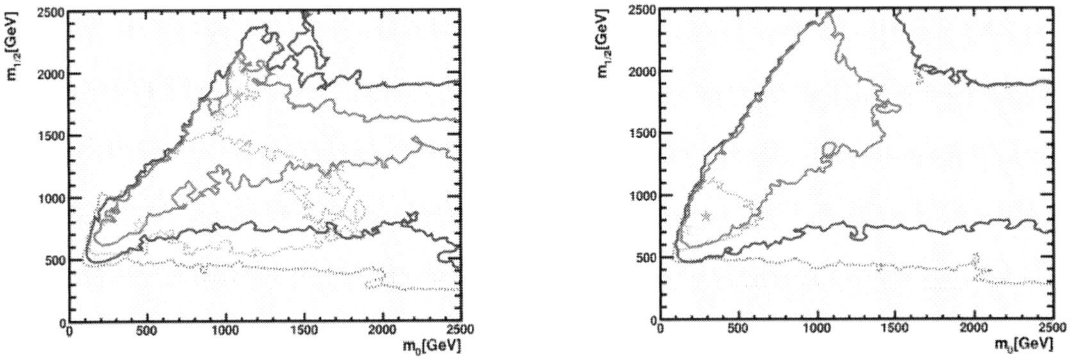

Figure 9. *The* $(m_0, m_{1/2})$ *planes in the CMSSM (left) and the NUHM1 (right). The 68 and 95% CL regions are indicated in red and blue, respectively, the solid lines including the hypothetical LHC measurement* $M_h = 125 \pm 1$ GeV *and allowing for a theoretical error* ± 1.5 GeV *[58], and the dotted lines showing the contours found previously in [35] without this* M_h *constraint. Here the open green stars denote the pre-Higgs best-fit points, whereas the solid green stars indicate the new best-fit points.*

In Fig. 10 we show results for the preferred regions in the $(m_{\tilde{\chi}^0_1}, \sigma_p^{\rm SI})$ plane. As seen in Fig. 10, the fact that larger values of $m_{1/2}$ and hence $m_{\tilde{\chi}^0_1}$ are favoured by the larger values of M_h implies that at the 68% CL the preferred range of $\sigma_p^{\rm SI}$ is significantly lower when $M_h \simeq 125$ GeV, when compared to our previous best fit with $M_h = 119$ GeV, rendering direct detection of dark matter significantly more difficult. Again, this effect on $m_{\tilde{\chi}^0_1}$ is more pronounced in the CMSSM, whereas in the NUHM1 the value of $m_{\tilde{\chi}^0_1}$ for the best-fit point changes only slightly.

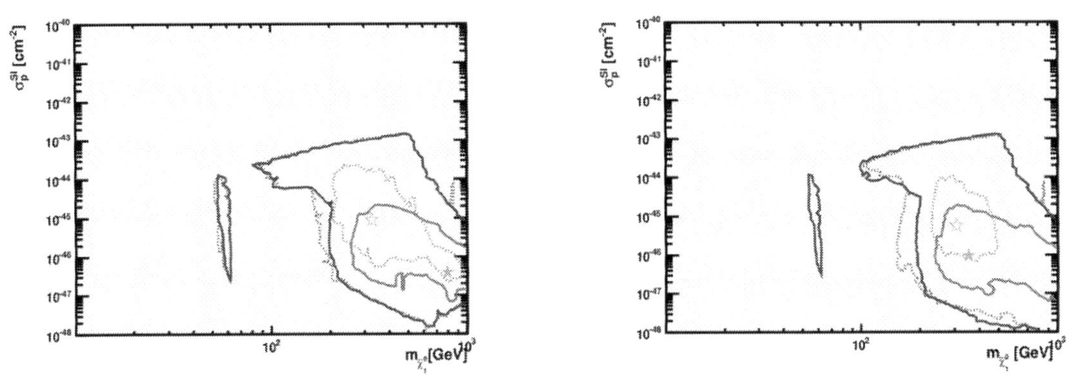

Figure 10. *The* $(m_{\tilde{\chi}^0_1}, \sigma_p^{\rm SI})$ *planes in the CMSSM (left) and the NUHM1 (right), for* $M_h \simeq$ 125 GeV. *The notations and significations of the contours are the same as in Fig. 8 [58].*

If instead, we assume an alternative potential LHC measurement $M_h = 119 \pm 1$ GeV, which corresponds to the CMS ZZ^* signal and our earlier predictions including the $(g-2)_\mu$ constraint, we obtain the $(m_0, m_{1/2})$ planes shown in Fig. 11. Since this assumed LHC value of M_h coincides with the previous best-fit values in both the CMSSM and NUHM1, the best-fit points in these models (indicated by the green stars in Fig. 11) are unaffected by the imposition of the putative LHC constraint. The effect of the hypothetical measurement restricting the range in $m_{1/2}$ is indeed seen in both panels of Fig. 11, though for the 68% CL contour (shown in red) it is much more pronounced for the CMSSM than for the NUHM1.

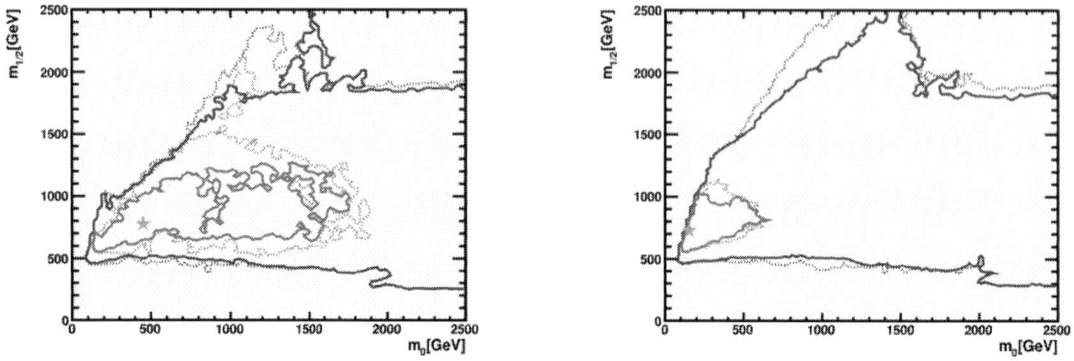

Figure 11. *The $(m_0, m_{1/2})$ planes in the CMSSM (left) and the NUHM1 (right) assuming a hypothetical measurement of $M_h = 119{\pm}1$ GeV. The notations and significations of the contours are the same as in Fig. 7 [58].*

In this case, in both the CMSSM and the NUHM1 there is little impact on the 95% CL regions nor on the 68% CL region in the NUHM1 in the (m_χ, σ_p^{SI}) plane. The only substantial change, as can be seen in Fig. 12, appears in the 68% CL region of the CMSSM, where now values of $m_{\tilde{\chi}_1^0} \gtrsim 500$ GeV and $\sigma_p^{SI} 10^{-46}\mathrm{cm}^{-2}$ are disfavoured after the inclusion of a Higgs-boson mass measurement at 119 GeV.

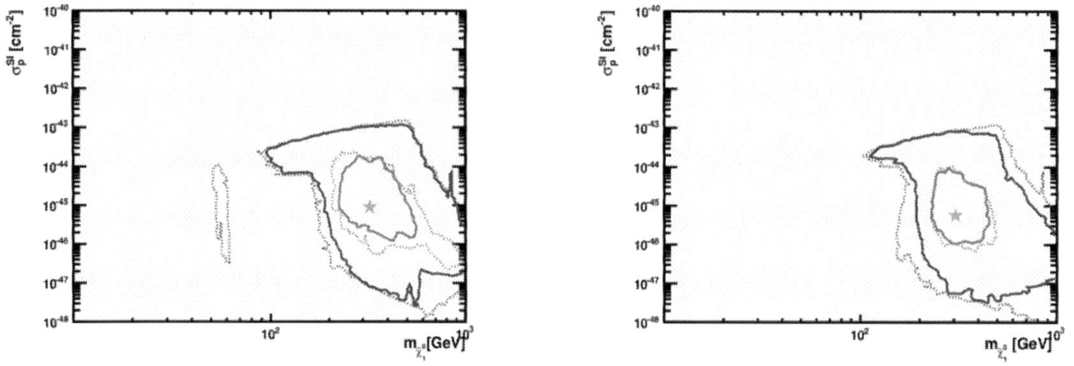

Figure 12. *The $(m_{\tilde{\chi}_1^0}, \sigma_p^{SI})$ planes in the CMSSM (left) and the NUHM1 (right), for $M_h \simeq$ 119 GeV. The notations and significations of the contours are the same as in Fig. 11 [58].*

The past year has shown an immense amount of activity. We have seen direct detection experiments (XENON100 [45]) for the first time have a direct impact on supersymmetric parameter space making these data a necessary input to a global likelihood analysis. The

LHC constraints have moved at a frightening pace. As reviewed here, starting with the 35/pb data, our notion of the best fit point in the CMSSM and indeed our prospects for low energy supersymmetry are greatly different as we start 2012 compared with the pre-LHC era.

Acknowledgments

I would like to thank all the members of Mastercode collaboration whose work went into the summary presented here. This work was supported in part by DOE grant DE–FG02–94ER–40823 at the University of Minnesota.

References

[1] M. Drees and M. M. Nojiri, Phys. Rev. D **47** (1993) 376 [arXiv:hep-ph/9207234]; H. Baer and M. Brhlik, Phys. Rev. D **53** (1996) 597 [arXiv:hep-ph/9508321]; Phys. Rev. D **57** (1998) 567 [arXiv:hep-ph/9706509]; H. Baer, M. Brhlik, M. A. Diaz, J. Ferrandis, P. Mercadante, P. Quintana and X. Tata, Phys. Rev. D **63** (2001) 015007 [arXiv:hep-ph/0005027].

[2] G. L. Kane, C. F. Kolda, L. Roszkowski and J. D. Wells, Phys. Rev. D **49** (1994) 6173 [arXiv:hep-ph/9312272]; J. R. Ellis, T. Falk, K. A. Olive and M. Schmitt, Phys. Lett. B **388** (1996) 97 [arXiv:hep-ph/9607292]; Phys. Lett. B **413** (1997) 355 [arXiv:hep-ph/9705444]; J. R. Ellis, T. Falk, G. Ganis, K. A. Olive and M. Schmitt, Phys. Rev. D **58** (1998) 095002 [arXiv:hep-ph/9801445]; V. D. Barger and C. Kao, Phys. Rev. D **57** (1998) 3131 [arXiv:hep-ph/9704403]; J. R. Ellis, T. Falk, G. Ganis and K. A. Olive, Phys. Rev. D **62** (2000) 075010 [arXiv:hep-ph/0004169].

[3] J. R. Ellis, T. Falk, G. Ganis, K. A. Olive and M. Srednicki, Phys. Lett. B **510** (2001) 236 [arXiv:hep-ph/0102098].

[4] V. D. Barger and C. Kao, Phys. Lett. B **518** (2001) 117 [arXiv:hep-ph/0106189]; L. Roszkowski, R. Ruiz de Austri and T. Nihei, JHEP **0108** (2001) 024 [arXiv:hep-ph/0106334]; A. Djouadi, M. Drees and J. L. Kneur, JHEP **0108** (2001) 055 [arXiv:hep-ph/0107316]; U. Chattopadhyay, A. Corsetti and P. Nath, Phys. Rev. D **66** (2002) 035003 [arXiv:hep-ph/0201001]; J. R. Ellis, K. A. Olive and Y. Santoso, New Jour. Phys. **4** (2002) 32 [arXiv:hep-ph/0202110]; H. Baer, C. Balazs, A. Belyaev, J. K. Mizukoshi, X. Tata and Y. Wang, JHEP **0207** (2002) 050 [arXiv:hep-ph/0205325]; R. Arnowitt and B. Dutta, arXiv:hep-ph/0211417.

[5] J. R. Ellis, K. A. Olive, Y. Santoso and V. C. Spanos, Phys. Lett. B **565** (2003) 176 [arXiv:hep-ph/0303043]; H. Baer and C. Balazs, JCAP **0305**, 006 (2003) [arXiv:hep-ph/0303114]; A. B. Lahanas and D. V. Nanopoulos, Phys. Lett. B **568**, 55 (2003) [arXiv:hep-ph/0303130]; U. Chattopadhyay, A. Corsetti and P. Nath, Phys. Rev. D **68**, 035005 (2003) [arXiv:hep-ph/0303201]; C. Munoz, Int. J. Mod. Phys. A **19**, 3093 (2004) [arXiv:hep-ph/0309346]; R. Arnowitt, B. Dutta and B. Hu, arXiv:hep-ph/0310103. J. Ellis and K. A. Olive, arXiv:1001.3651 [astro-ph.CO].

[6] O. Buchmueller *et al.*, Phys. Lett. B **657** (2007) 87 [arXiv:0707.3447 [hep-ph]].

[7] O. Buchmueller *et al.*, JHEP **0809** (2008) 117 [arXiv:0808.4128 [hep-ph]].

[8] O. Buchmueller *et al.*, Eur. Phys. J. C **64** (2009) 391 [arXiv:0907.5568 [hep-ph]].

[9] V. D. Barger, M. S. Berger and P. Ohmann, Phys. Rev. D **49** (1994) 4908 [arXiv:hep-ph/9311269].

[10] W. de Boer, R. Ehret and D. I. Kazakov, Z. Phys. C **67** (1995) 647 [arXiv:hep-ph/9405342].

[11] M. Carena, J. R. Ellis, A. Pilaftsis and C. E. Wagner, Nucl. Phys. B **625** (2002) 345 [arXiv:hep-ph/0111245].

[12] J. Ellis, K. Olive and Y. Santoso, Phys. Lett. B **539**, 107 (2002) [arXiv:hep-ph/0204192]; J. R. Ellis, T. Falk, K. A. Olive and Y. Santoso, Nucl. Phys. B **652**, 259 (2003) [arXiv:hep-ph/0210205].

[13] H. Baer, A. Mustafayev, S. Profumo, A. Belyaev and X. Tata, Phys. Rev. D **71**, 095008 (2005) [arXiv:hep-ph/0412059]; H. Baer, A. Mustafayev, S. Profumo, A. Belyaev and X. Tata, *JHEP* **0507** (2005) 065, hep-ph/0504001.

[14] J. R. Ellis, K. A. Olive and P. Sandick, Phys. Rev. D **78**, 075012 (2008) [arXiv:0805.2343 [hep-ph]].

[15] J. R. Ellis, K. A. Olive, Y. Santoso and V. C. Spanos, Phys. Lett. B **573** (2003) 162 [arXiv:hep-ph/0305212], and Phys. Rev. D **70** (2004) 055005 [arXiv:hep-ph/0405110].

[16] J. Polonyi, *Generalization Of The Massive Scalar Multiplet Coupling To The Supergravity*, Hungary Central Inst Res - KFKI-77-93; E. Cremmer, B. Julia, J. Scherk, P. van Nieuwenhuizen, S. Ferrara and L. Girardello, Phys. Lett. B **79**, 231 (1978); E. Cremmer, B. Julia, J. Scherk, S. Ferrara, L. Girardello and P. van Nieuwenhuizen, Nucl. Phys. B **147**, 105 (1979); R. Barbieri, S. Ferrara and C. A. Savoy, Phys. Lett. B **119**, 343 (1982); A. H. Chamseddine, R. L. Arnowitt and P. Nath, Phys. Rev. Lett. **49**, 970 (1982).

[17] E. Komatsu *et al.* [WMAP Collaboration], Astrophys. J. Suppl. **192** (2011) 18 [arXiv:1001.4538 [astro-ph.CO]].

[18] J. Ellis, T. Falk, and K.A. Olive, Phys. Lett. **B444** (1998) 367 [arXiv:hep-ph/9810360]; J. Ellis, T. Falk,

K.A. Olive, and M. Srednicki, *Astr. Part. Phys.* **13** (2000) 181 [Erratum-ibid. **15** (2001) 413] [arXiv:hep-ph/9905481].

[19] Joint LEP 2 Supersymmetry Working Group, *Combined LEP Chargino Results, up to 208 GeV*, http://lepsusy.web.cern.ch/lepsusy/www/inos_ moriond01/charginos_pub.html.

[20] R. Barate *et al.* [ALEPH, DELPHI, L3, OPAL Collaborations: the LEP Working Group for Higgs boson searches], Phys. Lett. B **565**, 61 (2003) [arXiv:hep-ex/0306033]; D. Zer-Zion, *Prepared for 32nd International Conference on High-Energy Physics (ICHEP 04), Beijing, China, 16-22 Aug 2004*; LHWG-NOTE-2004-01, ALEPH-2004-008, DELPHI-2004-042, L3-NOTE-2820, OPAL-TN-744, http://lephiggs.web.cern.ch/LEPHIGGS/papers/August2004_MSSM/index.html.

[21] S. Chen *et al.* [CLEO Collaboration], Phys. Rev. Lett. **87** (2001) 251807 [arXiv:hep-ex/0108032]; P. Koppenburg *et al.* [Belle Collaboration], Phys. Rev. Lett. **93** (2004) 061803 [arXiv:hep-ex/0403004]. B. Aubert *et al.* [BaBar Collaboration], arXiv:hep-ex/0207076; E. Barberio *et al.* [Heavy Flavor Averaging Group (HFAG)], arXiv:hep-ex/0603003.

[22] [The Muon g-2 Collaboration], *Phys. Rev. Lett.* **92** (2004) 161802, hep-ex/0401008; G. Bennett et al. [The Muon g-2 Collaboration], *Phys. Rev.* **D 73** (2006) 072003 [arXiv:hep-ex/0602035].

[23] J. L. Feng, K. T. Matchev and T. Moroi, *Phys. Rev. D* **61** (2000) 075005 [arXiv:hep-ph/9909334].

[24] For more information and updates, please see http://cern.ch/mastercode/.

[25] S. Heinemeyer *et al.*, JHEP **0608** (2006) 052 [arXiv:hep-ph/0604147]; S. Heinemeyer, W. Hollik, A. M. Weber and G. Weiglein, JHEP **0804** (2008) 039 [arXiv:0710.2972 [hep-ph]].

[26] B. C. Allanach, Comput. Phys. Commun. **143** (2002) 305 [arXiv:hep-ph/0104145].

[27] G. Degrassi, S. Heinemeyer, W. Hollik, P. Slavich and G. Weiglein, Eur. Phys. J. C **28** (2003) 133 [arXiv:hep-ph/0212020]; S. Heinemeyer, W. Hollik and G. Weiglein, Eur. Phys. J. C **9** (1999) 343 [arXiv:hep-ph/9812472]; S. Heinemeyer, W. Hollik and G. Weiglein, Comput. Phys. Commun. **124** (2000) 76 [arXiv:hep-ph/9812320]; M. Frank *et al.*, JHEP **0702** (2007) 047 [arXiv:hep-ph/0611326]; See http://www.feynhiggs.de .

[28] G. Isidori and P. Paradisi, Phys. Lett. B **639** (2006) 499 [arXiv:hep-ph/0605012]; G. Isidori, F. Mescia, P. Paradisi and D. Temes, Phys. Rev. D **75** (2007) 115019 [arXiv:hep-ph/0703035], and references therein.

[29] F. Mahmoudi, Comput. Phys. Commun. **178** (2008) 745 [arXiv:0710.2067 [hep-ph]]; Comput. Phys. Commun. **180** (2009) 1579 [arXiv:0808.3144 [hep-ph]]; D. Eriksson, F. Mahmoudi and O. Stal, JHEP **0811** (2008) 035 [arXiv:0808.3551 [hep-ph]].

[30] G. Belanger, F. Boudjema, A. Pukhov and A. Semenov, Comput. Phys. Commun. **176** (2007) 367 [arXiv:hep-ph/0607059]; Comput. Phys. Commun. **149** (2002) 103 [arXiv:hep-ph/0112278]; Comput. Phys. Commun. **174** (2006) 577 [arXiv:hep-ph/0405253].

[31] Information about this code is available from K. A. Olive: it contains important contributions from T. Falk, A. Ferstl, G. Ganis, A. Mustafayev, J. McDonald, K. A. Olive, P. Sandick, Y. Santoso and M. Srednicki.

[32] P. Skands *et al.*, JHEP **0407** (2004) 036 [arXiv:hep-ph/0311123]; B. Allanach *et al.*, Comput. Phys. Commun. **180** (2009) 8 [arXiv:0801.0045 [hep-ph]].

[33] O. Buchmueller *et al.*, Eur. Phys. J. C **71** (2011) 1634 [arXiv:1102.4585 [hep-ph]].

[34] O. Buchmueller *et al.*, Eur. Phys. J. C **71** (2011) 1722 [arXiv:1106.2529 [hep-ph]].

[35] O. Buchmueller, R. Cavanaugh, A. De Roeck, M. J. Dolan, J. R. Ellis, H. Flacher, S. Heinemeyer and G. Isidori *et al.*, arXiv:1110.3568 [hep-ph].

[36] O. Buchmueller *et al.*, Eur. Phys. J. C **71** (2011) 1583 [arXiv:1011.6118 [hep-ph]].

[37] "ATLAS detector and physics performance. Technical design report. Vol. 2,", CER-LHCC-99-15.

[38] G. L. Bayatian *et al.* [CMS Collaboration], J. Phys. G **34**, 995 (2007).

[39] For a sampling of other pre-LHC analyses, see: E. A. Baltz and P. Gondolo, JHEP **0410** (2004) 052 [arXiv:hep-ph/0407039]; B. C. Allanach and C. G. Lester, Phys. Rev. D **73** (2006) 015013 [arXiv:hep-ph/0507283]; R. R. de Austri, R. Trotta and L. Roszkowski, JHEP **0605** (2006) 002 [arXiv:hep-ph/0602028]; R. Lafaye, T. Plehn, M. Rauch and D. Zerwas, Eur. Phys. J. C **54** (2008) 617 [arXiv:0709.3985 [hep-ph]]; S. Heinemeyer, X. Miao, S. Su and G. Weiglein, JHEP **0808** (2008) 087 [arXiv:0805.2359 [hep-ph]]; R. Trotta, F. Feroz, M. P. Hobson, L. Roszkowski and R. Ruiz de Austri, JHEP **0812** (2008) 024 [arXiv:0809.3792 [hep-ph]]; P. Bechtle, K. Desch, M. Uhlenbrock and P. Wienemann, Eur. Phys. J. C **66** (2010) 215 [arXiv:0907.2589 [hep-ph]].

[40] G. Aad *et al.* [Atlas Collaboration], Phys. Lett. B **701**, 186 (2011) [arXiv:1102.5290 [hep-ex]].

[41] G. Aad *et al.* [Atlas Collaboration], Phys. Rev. Lett. **106**, 131802 (2011) [arXiv:1102.2357 [hep-ex]].

[42] G. Aad *et al.* [ATLAS Collaboration], http://cdsweb.cern.ch/record/1345745/files/ATLAS-CONF-2011-064.pdf.

[43] V. Khachatryan *et al.* [CMS Collaboration], Phys. Lett. B **698**, 196 (2011) [arXiv:1101.1628 [hep-ex]].

[44] V. Khachatryan *et al.* [CMS Collaboration], http://cdsweb.cern.ch/record/1343076/files/SUS-10-005-pas.pdf; see also

https://twiki.cern.ch/twiki/bin/view/CMSPublic/PhysicsResultsSUS.
[45] E. Aprile *et al.* [XENON100 Collaboration], Phys. Rev. Lett. **107**, 131302 (2011) [arXiv:1104.2549 [astro-ph.CO]].
[46] D. Feldman, K. Freese, P. Nath, B. D. Nelson and G. Peim, Phys. Rev. D **84**, 015007 (2011) [arXiv:1102.2548 [hep-ph]]; B. C. Allanach, Phys. Rev. D **83**, 095019 (2011) [arXiv:1102.3149 [hep-ph]]; S. Scopel, S. Choi, N. Fornengo and A. Bottino, Phys. Rev. D **83**, 095016 (2011) [arXiv:1102.4033 [hep-ph]]; P. Bechtle *et al.*, arXiv:1102.4693 [hep-ph]; B. C. Allanach, T. J. Khoo, C. G. Lester and S. L. Williams, JHEP **1106**, 035 (2011) [arXiv:1103.0969 [hep-ph]]; S. Akula, N. Chen, D. Feldman, M. Liu, Z. Liu, P. Nath and G. Peim, Phys. Lett. B **699**, 377 (2011) [arXiv:1103.1197 [hep-ph]]; M. J. Dolan, D. Grellscheid, J. Jaeckel, V. V. Khoze and P. Richardson, JHEP **1106**, 095 (2011) [arXiv:1104.0585 [hep-ph]]; S. Akula, D. Feldman, Z. Liu, P. Nath and G. Peim, Mod. Phys. Lett. A **26**, 1521 (2011) [arXiv:1103.5061 [hep-ph]]; M. Farina, M. Kadastik, D. Pappadopulo, J. Pata, M. Raidal and A. Strumia, Nucl. Phys. B **853**, 607 (2011) [arXiv:1104.3572 [hep-ph]]; S. Profumo, Phys. Rev. D **84**, 015008 (2011) [arXiv:1105.5162 [hep-ph]]; T. Li, J. A. Maxin, D. V. Nanopoulos and J. W. Walker, arXiv:1106.1165 [hep-ph]; N. Bhattacharyya, A. Choudhury and A. Datta, Phys. Rev. D **84**, 095006 (2011) [arXiv:1107.1997 [hep-ph]]; A. Fowlie, A. Kalinowski, M. Kazana, L. Roszkowski and Y. L. S. Tsai, arXiv:1111.6098 [hep-ph].
[47] G. Aad *et al.* [ATLAS Collaboration], arXiv:1109.6572 [hep-ex].
[48] ATLAS Collaboration,
http://cdsweb.cern.ch/record/1383835/files/ATLAS-CONF-2011-132.pdf.
[49] S. Chatrchyan *et al.* [CMS Collaboration], arXiv:1109.2352 [hep-ex].
[50] CMS Collaboration,
http://cdsweb.cern.ch/record/1378096/files/HIG-11-020-pas.pdf.
[51] S. Chatrchyan *et al.* [CMS Collaboration], arXiv:1107.5834 [hep-ex].
[52] R. Aaij *et al.* [LHCb Collaboration], Phys. Lett. B **699** (2011) 330 [arXiv:1103.2465 [hep-ex]].
[53] A. Bottino, F. Donato, N. Fornengo and S. Scopel, Astropart. Phys. **13** (2000) 215 [arXiv:hep-ph/9909228]; E. Accomando, R. L. Arnowitt, B. Dutta and Y. Santoso, Nucl. Phys. B **585** (2000) 124 [arXiv:hep-ph/0001019]; J. R. Ellis, K. A. Olive, Y. Santoso and V. C. Spanos, Phys. Rev. D **71** (2005) 095007 [arXiv:hep-ph/0502001]; J. R. Ellis, K. A. Olive and C. Savage, Phys. Rev. D **77** (2008) 065026 [arXiv:0801.3656 [hep-ph]]; J. Gasser, H. Leutwyler and M. E. Sainio, Phys. Lett. B **253** (1991) 252; M. Knecht, PiN Newslett. **15** (1999) 108 [arXiv:hep-ph/9912443]; M. E. Sainio, PiN Newslett. **16** (2002) 138 [arXiv:hep-ph/0110413]; B. Borasoy and U. G. Meissner, Annals Phys. **254** (1997) 192 [arXiv:hep-ph/9607432]; M. M. Pavan, I. I. Strakovsky, R. L. Workman and R. A. Arndt, PiN Newslett. **16** (2002) 110 [arXiv:hep-ph/0111066]; see also T. Strauch *et al.*, arXiv:1011.2415 [nucl-ex], as interpreted in V. Baru, C. Hanhart, M. Hoferichter, B. Kubis, A. Nogga and D. R. Phillips, arXiv:1003.4444 [nucl-th].
[54] The ATLAS and CMS Collaborations,
https://cdsweb.cern.ch/record/1399607/files/HIG-11-023-pas.pdf.
[55] CMS Collaboration,
http://cdsweb.cern.ch/record/1406347/files/HIG-11-032-pas.pdf; ATLAS Collaboration,
http://cdsweb.cern.ch/record/1406358/files/ATLAS-CONF-2011-163.pdf.
[56] S. Schael *et al.* [ALEPH, DELPHI, L3, OPAL Collaborations and LEP Working Group for Higgs boson searches], Eur. Phys. J. C **47** (2006) 547 [arXiv:hep-ex/0602042].
[57] http://tevnphwg.fnal.gov/ and references therein.
[58] O. Buchmueller, R. Cavanaugh, A. De Roeck, M. J. Dolan, J. R. Ellis, H. Flacher, S. Heinemeyer and G. Isidori *et al.*, arXiv:1112.3564 [hep-ph].
[59] H. Baer, V. Barger and A. Mustafayev, arXiv:1112.3017 [hep-ph]; J. L. Feng, K. T. Matchev and D. Sanford, arXiv:1112.3021 [hep-ph]; S. Heinemeyer, O. Stal and G. Weiglein, arXiv:1112.3026 [hep-ph]; A. Arbey, M. Battaglia, A. Djouadi, F. Mahmoudi and J. Quevillon, arXiv:1112.3028 [hep-ph]; P. Draper, P. Meade, M. Reece and D. Shih, arXiv:1112.3068 [hep-ph]; M. Kadastik, K. Kannike, A. Racioppi and M. Raidal, arXiv:1112.3647 [hep-ph]; N. Karagiannakis, G. Lazarides and C. Pallis, arXiv:1201.2111 [hep-ph].

7th International Workshop on the Dark Side of The Universe (DSU 2011) IOP Publishing
Journal of Physics: Conference Series **384** (2012) 012011 doi:10.1088/1742-6596/384/1/012011

Probing reheating via direct detection of the inflationary gravitational wave background and its implications for dark matter candidate

Sachiko Kuroyanagi

Institute for Cosmic Ray Research, University of Tokyo, Chiba 277-8582, Japan

E-mail: skuro@icrr.u-tokyo.ac.jp

Abstract. Future direct detection gravitational wave experiments would provide vital information on the early Universe. It may enable us to determine the reheating temperature by detecting a signature of reheating on the spectrum of the inflationary gravitational wave background. We investigate the detectability of the reheating signature and discuss its implications for particle physics, especially on the scenario where gravitinos are dark matter candidates.

1. Introduction

Inflation generically predicts the cosmological gravitational wave background originating from quantum fluctuations in the spacetime metric. If it is directly detected by future experiments such as DECIGO/BBO [1, 2, 3], it will be a powerful tool to extract information about the early Universe. In particular, information on the Hubble expansion rate is imprinted on the shape of the gravitational wave background spectrum and it may enable us to observe the epoch of reheating [4, 5, 6, 7, 8].

The effect of reheating is seen as a change of the frequency dependence of the spectrum, which is induced by the transition of the expansion rate at the end of reheating. We estimate the detectability of the reheating signature, by examining whether the experiment can distinguish the change of the characteristic frequency dependence. Also, we discuss the implications of a possible lower bound of the reheating temperature for particle physics, which is expected from the future direct detection by DECIGO/BBO.

2. Gravitational wave background from inflation

Gravitational waves in the expanding Universe can be described as a transverse-traceless part of the metric perturbation in a Friedmann Robertson-Walker background, $ds^2 = -dt^2 + a^2(t)(\delta_{ij}+h_{ij})dx^i dx^j$. Applying the metric of this space-time to the perturbed Einstein equation, $\delta G_{ij} = 8\pi G\delta T_{ij}$, and Fourier transforming the equation, we obtain the equation of motion for the gravitational wave of the mode $\mathbf{k}$

$$\ddot{h}_{\mathbf{k}} + 3H\dot{h}_{\mathbf{k}} + \frac{k^2}{a^2}h_{\mathbf{k}} = \Pi_{\mathbf{k}}, \tag{1}$$

where the over-dot describes the time derivative, H is the Hubble parameter, and $\Pi_{\mathbf{k}}$ is the Fourier transform of the transverse-traceless part of the anisotropic stress. The effect of reheating appears through the second term via the Hubble expansion rate.

If we assume reheating is proceeded by perturbative decay of the inflaton field into light fermions, the background evolution of the Universe during inflation and reheating can be described by the following equations,

$$\ddot{\phi} + (3H + \Gamma)\dot{\phi} + \frac{\partial V(\phi)}{\partial \phi} = 0, \tag{2}$$

$$\dot{\rho}_r + 4H\rho_r = \Gamma\rho_\phi, \tag{3}$$

$$H^2 = \frac{8\pi}{3m_{\rm Pl}^2}(\rho_\phi + \rho_r), \tag{4}$$

where Γ is the decay rate of the scalar field into radiation, $V(\phi)$ is the potential of the inflaton scalar field, ρ_r is the energy density of the radiation, and the energy density of the scalar field is given as $\rho_\phi = \dot{\phi}^2/2 + V(\phi)$. The reheating temperature $T_{\rm R}$, defined at the end of reheating, can be related to the decay rate Γ as

$$T_{\rm R} \simeq g_{*,{\rm R}}^{-(1/4)} \left(\frac{45}{8\pi^3}\right)^{1/4} (m_{\rm Pl}\Gamma)^{1/2}, \tag{5}$$

where $g_{*,{\rm R}} \sim 100$ is the effective number of relativistic degrees of freedom at the end of reheating and $m_{\rm Pl} = G^{-1/2}$ denotes the Planck mass.

From Eqs. (2), (3) and (4), one can find the Hubble expansion rate behaves like matter-dominated Universe during reheating and becomes radiation-dominated after reheating. This transition gives rise to change in frequency dependencies of the spectrum. In Fig 1, the spectrum is shown in terms of the density parameter per logarithmic wave number, defined as $\Omega_{\rm GW} \equiv (d\rho_{\rm GW}/d\ln k)/\rho_c$, where the critical density of the Universe is defined as $\rho_c \equiv 3H^2/8\pi G$ and $\rho_{\rm GW}$ denotes the energy density of the gravitational waves, $\rho_{\rm GW} = \langle(\partial_\tau h_{ij})^2 + (\vec{\nabla} h_{ij})^2\rangle/(64\pi G a^2)$. The matter-dominated Universe gives the dependence of $\Omega_{\rm GW} \propto f^{-2}$, and the radiation-dominated Universe gives f^0. As seen from the figure, if the reheating temperature is $\sim 10^7$GeV, future gravitational wave experiments would be able to detect the knee-like feature in the spectrum. The change of the frequency dependence arises around the frequency of

$$f_{\rm R} \simeq 0.26 \left(\frac{T_{\rm R}}{10^7{\rm GeV}}\right) \left(\frac{g_{*,{\rm R}}}{110}\right)^{1/2} \left(\frac{g_{*s,{\rm R}}}{110}\right)^{-1/3} {\rm Hz}, \tag{6}$$

where g_{*s} is the effective number of relativistic degrees of freedom for the entropy density. Since the signature arises at different frequency depending on the reheating temperature, it may even be possible to constraint the reheating temperature by measuring the position where the change of the frequency dependence arises.

3. Prospects for detectability of the reheating signature

Using the Fisher matrix formalism, we present an example of the expected future constraints on reheating temperature in Fig. 2, where the fiducial parameters are chosen as $r = 0.1$ and $T_{\rm R} = 10^7$ GeV. Each ellipse represents the 2σ errors on the reheating temperature $T_{\rm R}$ and the tensor-to-scalar ratio r, expected from 1, 3 and 10 years of observation with the sensitivity of BBO. Here, the amplitude of the gravitational wave background is parametrized in terms of the tensor-to-scalar ratio, defined as the ratio of the amplitudes of the scalar and tensor power spectra $r \equiv \Delta^2_{h,{\rm prim}}(k_0)/\Delta^2_{\mathcal{R},{\rm prim}}(k_0)$, where $k_0 = 0.002\,{\rm Mpc}^{-1}$ is usually taken to be the scale of the CMB.

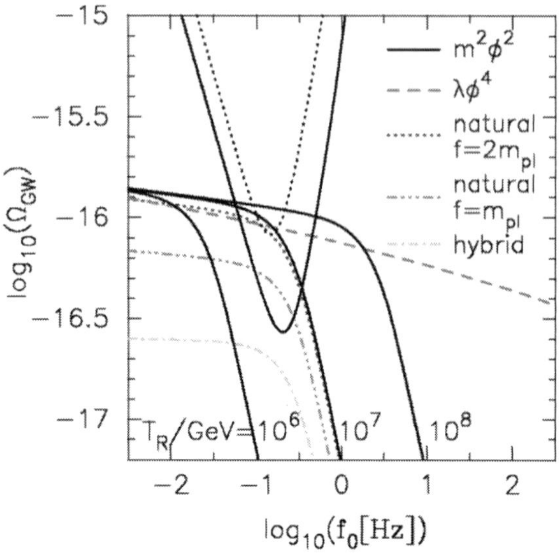

Figure 1. Spectra of the inflationary gravitational wave background for different inflation models [8]; chaotic ($m^2\phi^2$, $\lambda\phi^4$), natural and hybrid inflation for typical values of parameters. The sensitivity curves of FP-DECIGO (dotted) and BBO (solid) experiments are also shown. The spectra are calculated assuming $T_{\mathrm{R}} = 10^6$GeV, 10^7GeV and 10^8GeV for the $m^2\phi^2$ model, respectively.

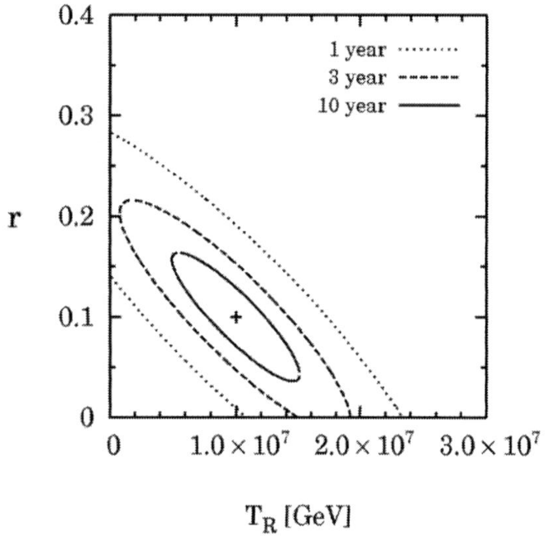

Figure 2. The 2σ confidence level contours in the $T_{\mathrm{R}} - r$ plane for 1-year (dotted line), 3-year (dashed line) and 10-year (solid line) observation by BBO [9]. The fiducial parameters are set as $r = 0.1$ and $T_{\mathrm{R}} = 10^7$GeV, which is shown by a cross mark.

In Fig. 3, we show the accessible parameter space where the reheating signature is detected at greater than 2σ level with 3-year observation by BBO. As seen from the figure, if $T_{\mathrm{R}} \sim 10^7$GeV, the reheating temperature can be determined with more than 2σ accuracy. We also show the region where the inflationary gravitational wave background can be detected with a signal-to-noise ratio higher than 5. However, in this case, the reheating signature is outside the frequency band of sensitivity so we can only obtain a lower bound of the reheating temperature, such as $T_{\mathrm{R}} > 10^7$GeV.

4. Implication for dark matter candidate

As shown in the previous section, direct detection of the inflationary gravitational wave background may be able to provide constraints on reheating temperature. In the following, we consider the implication of the lower bound of the reheating temperature, on gravitino cosmology in the context of chaotic/natural inflation [8].

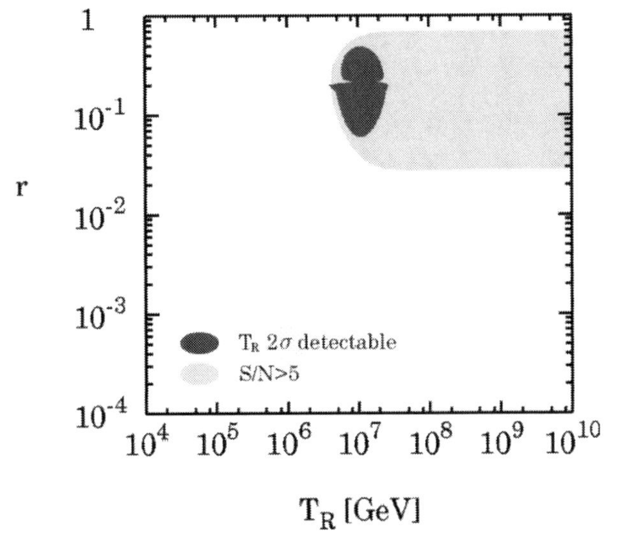

Figure 3. Accessible parameter space in the $T_R - r$ space [9]. In the blue shaded region, reheating signature would be detectable with a 2σ confidence level by 3-year observation of BBO. In the light-blue shaded region, the inflationary gravitational wave background would be detected with a signal-to-noise ratio higher than 5.

In the case where the gravitinos are stable and produced by the thermal scatterings at the reheating, the present-day abundance of the gravitinos is given by [10]

$$\Omega_{3/2}^{TH}h^2 \simeq 0.27 \; \frac{T_{\rm R}}{10^8{\rm GeV}} \; \frac{M_{3/2}}{1{\rm GeV}}^{-1}, \tag{7}$$

where we have set the gluino mass $M_{\tilde{g}} = 1{\rm TeV}$. Since the total abundance should satisfy the bound $\Omega_{3/2}h^2 = \Omega_{3/2}^{\rm TH}h^2 + \Omega_{3/2}^{X}h^2 \leq \Omega_{dm}h^2 \simeq 0.134$, if $T_{\rm R} > T_{\rm R}^{\rm gw}$ is indicated from gravitational wave experiments, we get the lower bound on the gravitino mass,

$$M_{3/2} > 0.21{\rm GeV}\left(\frac{T_{\rm R}^{\rm gw}}{10^7{\rm GeV}}\right)\left(\frac{\Omega_{dm}h^2}{0.134}\right)^{-1}. \tag{8}$$

If the gravitinos are produced nonthermally via the decay of heavy scalar fields X, the abundance of the gravitinos produced by the X decay is given by

$$\Omega_{3/2}^{X}h^2 \simeq 6.8 \times 10^1 d_{3/2}^2 \; \frac{M_X}{10^{10}{\rm GeV}}^{1/2} \; \frac{M_{3/2}}{1{\rm GeV}}, \tag{9}$$

where s_0 is the present entropy density. In this case, the lower bound on the reheating temperature from gravitational wave experiments, $T_{\rm R} > T_{\rm R}^{\rm gw}$, places the upper bound on the mass of the X particle,

$$M_X < 6 \times 10^4{\rm GeV}\left(\frac{T_{\rm R}^{\rm gw}}{10^7{\rm GeV}}\right)^{-2}\left(\frac{\Omega_{dm}h^2}{0.134}\right)^4 d_{3/2}^{-4}. \tag{10}$$

5. Conclusion

Gravitational waves generated during inflation have potential to be a powerful observational tool to probe the early Universe. If detected, they surely provide generous information not only on inflation but also reheating. We considered the case of $T_{\rm R} \sim 10^7{\rm GeV}$, where the characteristic feature produced at the end of reheating is observed in the spectrum as a dip at $\sim 10^{-1}$Hz. If inflation generates large amplitude of the spectrum, the reheating temperature can be determined by future space-based gravitational wave detectors with more than 2σ accuracy. We also considered the case of $T_{\rm R} > 10^7$GeV, which would be able to confirmed by detection of the inflationary gravitational background. Taking into account of both thermal and nonthermal production of gravitinos, we showed that lower bound on the reheating temperature provides a lower bound on the gravitino mass and an upper bound on the moduli mass. These constraints could be a unique source of information for dark matter physics and complement with astrophysical and particle collider experiments.

References

[1] S. Phinney *et al.*, *The big bang observer: direct detection of gravitational waves from the birth of the Universe to the present*, NASA Mission Concept Study.
[2] N. Seto, S. Kawamura and T. Nakamura, Phys. Rev. Lett. **87**, 221103 (2001) [arXiv:astro-ph/0108011].
[3] S. Kawamura *et al.*, Class. Quant. Grav. **23**, S125 (2006).
[4] N. Seto, J. Yokoyama, J. Phys. Soc. Jap. **72**, 3082-3086 (2003). [gr-qc/0305096].
[5] K. Nakayama, S. Saito, Y. Suwa, J. Yokoyama, Phys. Rev. **D77**, 124001 (2008). [arXiv:0802.2452 [hep-ph]].
[6] K. Nakayama, S. Saito, Y. Suwa, J. Yokoyama, JCAP **0806**, 020 (2008). [arXiv:0804.1827 [astro-ph]].
[7] S. Kuroyanagi, T. Chiba, N. Sugiyama, Phys. Rev. **D79**, 103501 (2009). [arXiv:0804.3249 [astro-ph]].
[8] S. Kuroyanagi, T. Chiba, N. Sugiyama, Phys. Rev. **D83**, 043514 (2011). [arXiv:1010.5246 [astro-ph.CO]].
[9] S. Kuroyanagi, K. Nakayama and S. Saito, Phys. Rev. D **84**, 123513 (2011) [arXiv:1110.4169 [astro-ph.CO]].
[10] M. Bolz, A. Brandenburg and W. Buchmuller, Nucl. Phys. B **606**, 518 (2001) [Erratum-ibid. B **790**, 336 (2008)] [arXiv:hep-ph/0012052].

Dark Matter and Higgs Mass in the CMSSM with Yukawa Quasi-Unification

N Karagiannakis[1], **G Lazarides**[1], **C Pallis**[2]

[1] Physics Division, School of Technology, Aristotle University of Thessaloniki, Thessaloniki 54124, Greece

[2] Department of Physics, University of Cyprus, P.O. Box 20537, CY-1678 Nicosia, CYPRUS

E-mail: nikar@auth.gr, lazaride@eng.auth.gr, kpallis@auth.gr

Abstract. We present an updated analysis of the constrained minimal supersymmetric standard model with $\mu > 0$ supplemented by an 'asymptotic' Yukawa coupling quasi-unification condition, which allows an acceptable b-quark mass. Imposing constraints from the cold dark matter abundance in the universe, B physics, the muon anomalous magnetic moment, and the mass m_h of the lightest neutral CP-even Higgs boson, we find that the lightest neutralino cannot act as a cold dark matter candidate. This is mainly because the upper bound on the lightest neutralino relic abundance from cold dark matter considerations, despite the fact that this abundance is drastically reduced by neutralino-stau coannihilations, is incompatible with the recent data on the branching ratio of $B_s \to \mu^+\mu^-$. Allowing for a different particle, such as the axino or the gravitino, to be the lightest supersymmetric particle and, thus, constitute the cold dark matter in the universe, we find that the predicted m_h's in our model favor the range $(119 - 126)$ GeV.

1. Introduction

The well-known *constrained minimal supersymmetric standard model* (CMSSM) [1, 2, 3, 4], which is a highly predictive version of the *minimal supersymmetric standard model* (MSSM) based on universal boundary conditions for the soft *supersymmetry* (SUSY) breaking parameters, can be further restricted by being embedded in a SUSY *grand unified theory* (GUT) with a gauge group containing $SU(4)_c$ and $SU(2)_R$. This can lead [5] to 'asymptotic' *Yukawa unification* (YU) [6], i.e. the exact unification of the third generation Yukawa coupling constants h_t, h_b, and h_τ of the top quark, the bottom quark, and the tau lepton, respectively, at the SUSY GUT scale $M_{\rm GUT}$. The simplest GUT gauge group which contains both $SU(4)_c$ and $SU(2)_R$ is the *Pati-Salam* (PS) group $G_{\rm PS} = SU(4)_c \times SU(2)_L \times SU(2)_R$ [7, 8] – for YU within $SO(10)$, see Refs. [9, 10].

However, given the experimental values of the top-quark and tau-lepton masses (which, combined with YU, naturally restrict $\tan\beta \sim 50$), the CMSSM supplemented by the assumption of YU yields unacceptable values of the b-quark mass m_b for both signs of the parameter μ. This is due to the presence of sizable SUSY corrections [11] to m_b (about 20%), which arise [11, 12] from sbottom-gluino (mainly) and stop-chargino loops and have the sign of μ – with the standard sign convention of Ref. [13]. The predicted tree-level $m_b(M_Z)$, which turns out to be close to the upper edge of its 95% *confidence level* (c.l.) experimental range, receives, for $\mu > 0$ [$\mu < 0$], large positive [negative] corrections which drive it well above [a little below] the allowed range.

Consequently, for both signs of μ, YU leads to an unacceptable $m_b(M_Z)$ with the $\mu < 0$ case being much less disfavored.

The usual strategy to resolve this discrepancy is the introduction of several kinds of nonuniversalities in the scalar [9, 10] and/or gaugino [14, 15] sector of MSSM with an approximate preservation of YU. On the contrary, in Ref. [16], concrete SUSY GUT models based on the PS gauge group are constructed which naturally yield a moderate deviation from exact YU and, thus, can allow acceptable values of the b-quark mass for both signs of μ within the CMSSM. In particular, the Higgs sector of the simplest PS model [7, 8] is extended so that the electroweak Higgs fields are not exclusively contained in a $SU(2)_L \times SU(2)_R$ bidoublet superfield, but receive subdominant contributions from other representations too. As a consequence, a moderate violation of YU is naturally obtained, which can allow an acceptable b-quark mass even with universal boundary conditions. It is also remarkable that the resulting extended SUSY PS models support new successful versions [17] of hybrid inflation based solely on renormalizable superpotential terms.

These models provide us with a set of 'asymptotic' Yukawa quasi-unification conditions, which replace exact YU and can be applicable [16, 18, 19] for both signs of the MSSM parameter μ. We focus here on the $\mu > 0$ case since $\mu < 0$ is strongly disfavored by the constraint arising from the deviation δa_μ of the measured value of the muon anomalous magnetic moment a_μ from its predicted value $a_\mu^{\rm SM}$ in the *standard model* (SM). Indeed, $\mu < 0$ is defended only at 3σ by the calculation of $a_\mu^{\rm SM}$ based on the τ-decay data which is presented in Ref. [20], whereas there is a stronger and stronger tendency at present to prefer the e^+e^--annihilation data for the calculation of $a_\mu^{\rm SM}$, which favor the $\mu > 0$ regime. Moreover, in Ref. [21], it was claimed that, after some improvements, the τ-based result shifts considerably towards the e^+e^--based one.

Let us recall that, in this case, the suitable 'asymptotic' Yukawa quasi-unification condition applied [16, 18, 19] is

$$h_t : h_b : h_\tau = |1 + c| : |1 - c| : |1 + 3c|. \tag{1}$$

This relation depends on a single parameter c, which is taken, for simplicity, to be real and lying in the range $0 < c < 1$. With fixed masses for the fermions of the third generation, we can determine the parameters c and $\tan\beta$ so that Eq. (1) is satisfied. In contrast to the original version of the CMSSM [2, 3, 4], therefore, $\tan\beta$ is not a free parameter, but can be restricted, within our set-up, via Eq. (1) to relatively large values. The remaining free parameters of our model are the universal soft SUSY breaking parameters defined at $M_{\rm GUT}$, i.e.

$$M_{1/2}, \quad m_0, \quad \text{and} \quad A_0, \tag{2}$$

where the symbols above denote the common gaugino mass, scalar mass, and trilinear scalar coupling constant, respectively. These parameters can be restricted by employing a number of experimental and cosmological requirements as in Refs. [16, 18, 19] and most recently in Refs. [22, 23]. In this talk, we review the results of Ref. [22] implementing the following improvements:

- We do not take into account the upper bound on $m_{\tilde\chi}$ implied by the lower bound on δa_μ from the τ-based calculation of Ref. [20] raising, thereby, the upper bound on $m_{\tilde\chi}$ from the muon anomalous magnetic moment – see Sec. 2.
- We employ the recently released data on the branching ratio of $B_s \to \mu^+\mu^-$ [24] and the mass m_h of the lightest CP-even Higgs boson [25]. As a consequence, our predictions in Ref. [22] for m_h and the role of the lightest neutralino as *cold dark matter* (CDM) particle have been significantly altered.

All the cosmological and phenomenological requirements which we considered in our investigation are exhibited in detail in Sec. 2. Restrictions on the parameter space of our model are derived in Sec. 3 and our conclusions are summarized in Sec. 4.

2. Cosmological and Phenomenological Constraints

In our investigation, we integrate the two-loop renormalization group equations for the gauge and Yukawa coupling constants and the one-loop ones for the soft SUSY breaking parameters between $M_{\rm GUT}$ and a common SUSY threshold $M_{\rm SUSY} \simeq (m_{\tilde{t}_1} m_{\tilde{t}_2})^{1/2}$ ($\tilde{t}_{1,2}$ are the stop mass eigenstates) determined in consistency with the SUSY spectrum. At $M_{\rm SUSY}$, we impose radiative electroweak symmetry breaking, evaluate the SUSY spectrum employing the publicly available calculator `SOFTSUSY` [26], and incorporate the SUSY corrections to the b and τ mass [12]. The corrections to the τ-lepton mass m_τ (almost 4%) lead [16, 18] to a small decrease of $\tan\beta$. From $M_{\rm SUSY}$ to M_Z, the running of gauge and Yukawa coupling constants is continued using the SM renormalization group equations.

The parameter space of our model can be restricted by using a number of phenomenological and cosmological constraints. We calculate them using the latest version of the publicly available code `micrOMEGAs` [27]. We now briefly discuss these requirements – for similar recent analyses, see Ref. [28] for the CMSSM or Refs. [14, 29] for the MSSM with YU.

SM Fermion Masses. The masses of the fermions of the third generation play a crucial role in the determination of the evolution of the Yukawa coupling constants. For the b-quark mass, we adopt as an input parameter in our analysis the $\overline{\rm MS}$ b-quark mass, which at 1σ is [30]

$$m_b\,(m_b)^{\overline{\rm MS}} = 4.19^{+0.18}_{-0.06}\ {\rm GeV}. \tag{3}$$

This range is evolved up to M_Z using the central value $\alpha_s(M_Z) = 0.1184$ [30] of the strong fine structure constant at M_Z and then converted to the $\overline{\rm DR}$ scheme in accordance with the analysis of Ref. [31]. We obtain, at 95% c.l.,

$$2.745 \lesssim m_b(M_Z)/{\rm GeV} \lesssim 3.13 \tag{4}$$

with the central value being $m_b(M_Z) = 2.84$ GeV. For the top-quark mass, we use the central pole mass (M_t) as an input parameter [32]:

$$M_t = 173\ {\rm GeV} \quad\Rightarrow\quad m_t(m_t) = 164.6\ {\rm GeV} \tag{5}$$

with $m_t(m_t)$ being the running mass of the t quark. We also take the central value $m_\tau(M_Z) = 1.748$ GeV [31] of the $\overline{\rm DR}$ tau-lepton mass at M_Z.

Cold Dark Matter Considerations. According to the WMAP results [33], the 95% c.l. range for the CDM abundance is

$$\Omega_{\rm CDM} h^2 = 0.1126 \pm 0.0072. \tag{6}$$

In the context of the CMSSM, the lightest neutralino $\tilde{\chi}$ can be the *lightest supersymmetric particle* (LSP) and, thus, naturally arises as a CDM candidate. In this case, the requirement that the $\tilde{\chi}$ relic abundance $\Omega_{\tilde{\chi}} h^2$ does not exceed the 95% c.l. upper bound from Eq. (6), i.e.

$$\Omega_{\tilde{\chi}} h^2 \lesssim 0.12, \tag{7}$$

strongly restricts the parameter space of the CMSSM. This is because $\Omega_{\tilde{\chi}} h^2$ increases, in general, with $m_{\tilde{\chi}}$ and so an upper bound on $m_{\tilde{\chi}}$ can be derived from Eq. (7). The calculation of $\Omega_{\tilde{\chi}} h^2$ in `micrOMEGAs` includes accurately thermally averaged exact tree-level cross sections of all the possible (co)annihilation processes [3, 34], treats poles [4, 16, 35] properly, and uses one-loop QCD and SUSY QCD corrections [11, 16, 36] to the Higgs decay widths and couplings to fermions. It should, though, be noted that the restrictions induced by Eq. (7) can be evaded if we adopt one (or a combination) of the following scenarios:

- The cosmological evolution before *Big Bang nucleosynthesis* (BBN) deviates from the standard one [37, 38]. Since $\tilde{\chi}$ within the CMSSM is essentially a pure bino, the scenario which fits better this case is the low reheat temperature scenario with the decoupling of $\tilde{\chi}$ occurring before reheating. This scenario, however, is disfavored since it requires a very low reheat temperature $\sim (1-5)$ GeV. We will, thus, assume that the decoupling of the neutralino from the cosmic fluid occurs during the conventional radiation dominated era.
- The lightest neutralino is not the LSP and, thus, the relic density of another SUSY particle [39], which is the LSP, is to account for $\Omega_{\rm CDM}h^2$. This particle could be the gravitino [40] or the axino [41, 42]. The case of gravitino is tightly restricted in the CMSSM due to the BBN constraints imposed during the decay of the *lightest ordinary supersymmetric particle* (LOSP) to it. On the other hand, axino CDM [42] is, in general, possible once its mass and the reheat temperature are chosen appropriately. In such a case, $\tilde{\chi}$ may play the role of the LOSP and can contribute to the non-thermal production of the LSP. In particular, its contribution to the relic density of the LSP is equal to $\Omega_{\tilde{\chi}}h^2$ times the ratio of the LSP mass to $m_{\tilde{\chi}}$ and, thus, $\Omega_{\tilde{\chi}}h^2$'s exceeding the bound in Eq. (7) can be perfectly acceptable.

The Branching Ratio $\mathrm{BR}\,(b\to s\gamma)$ *of* $b\to s\gamma$. The most recent experimental world average for $\mathrm{BR}(b\to s\gamma)$ is known [43] to be $(3.52\pm0.23\pm0.09)\times10^{-4}$ and its updated SM prediction is $(3.15\pm0.23)\times10^{-4}$ [44]. Combining in quadrature the experimental and theoretical errors involved, we obtain the following constraints on this branching ratio at 95% c.l.:

$$2.84\times10^{-4}\lesssim\mathrm{BR}\,(b\to s\gamma)\lesssim4.2\times10^{-4}. \tag{8}$$

The computation of $\mathrm{BR}\,(b\to s\gamma)$ in the `micrOMEGAs` package presented in Ref. [36] includes [45] *next-to-leading order* (NLO) QCD corrections to the charged Higgs boson ($H^\pm$) contribution, the $\tan\beta$ enhanced contributions, as well as resummed NLO SUSY QCD corrections. The $H^\pm$ contribution interferes constructively with the SM contribution, whereas the SUSY contribution interferes destructively with the other two contributions for $\mu>0$. The SM plus the $H^\pm$ and SUSY contributions initially increases with $m_{\tilde{\chi}}$ and yields a lower bound on $m_{\tilde{\chi}}$ from the lower bound in Eq. (8). (For higher values of $m_{\tilde{\chi}}$, it starts mildly decreasing.)

The Branching Ratio $\mathrm{BR}\,(B_s\to\mu^+\mu^-)$ *of* $B_s\to\mu^+\mu^-$. The rare decay $B_s\to\mu^+\mu^-$ occurs via Z penguin and box diagrams in the SM and, thus, its branching ratio is highly suppressed. The SUSY contribution, though, originating [46, 47] from neutral Higgs bosons in chargino-, $H^\pm$-, and $W^\pm$-mediated penguins behaves as $\tan^6\beta/m_A^4$ (m_A is the mass of the CP-odd Higgs boson A) and hence is particularly important for large $\tan\beta$'s, especially after the new more stringent 95% c.l. upper bound

$$\mathrm{BR}\,(B_s\to\mu^+\mu^-)\lesssim1.08\times10^{-8} \tag{9}$$

recently reported by CMS and LHCb [24]. This new bound significantly reduces the previous bound [48], which we had adopted in Ref. [22]. The bound in Eq. (9) implies a lower bound on $m_{\tilde{\chi}}$ since $\mathrm{BR}\,(B_s\to\mu^+\mu^-)$ decreases as $m_{\rm LSP}$ increases.

The Branching Ratio $\mathrm{BR}\,(B_u\to\tau\nu)$ *of* $B_u\to\tau\nu$. The purely leptonic decay $B_u\to\tau\nu$ proceeds via $W^\pm$- and $H^\pm$-mediated annihilation processes. The SUSY contribution, contrary to the SM one, is not helicity suppressed and depends on the mass $m_{H^\pm}$ of the charged Higgs boson since it behaves [47, 49] as $\tan^4\beta/m_{H^\pm}^4$. The ratio $\mathrm{R}\,(B_u\to\tau\nu)$ of the CMSSM to the SM branching ratio of $B_u\to\tau\nu$ increases with $m_{\tilde{\chi}}$ and approaches unity. It is to be consistent with the following 95% c.l. range [43]:

$$0.52\lesssim\mathrm{R}\,(B_u\to\tau\nu)\lesssim2.04\,. \tag{10}$$

A lower bound on $m_{\tilde{\chi}}$ can be derived from the lower bound in this inequality.

Journal of Physics: Conference Series **384** (2012) 012012 doi:10.1088/1742-6596/384/1/012012

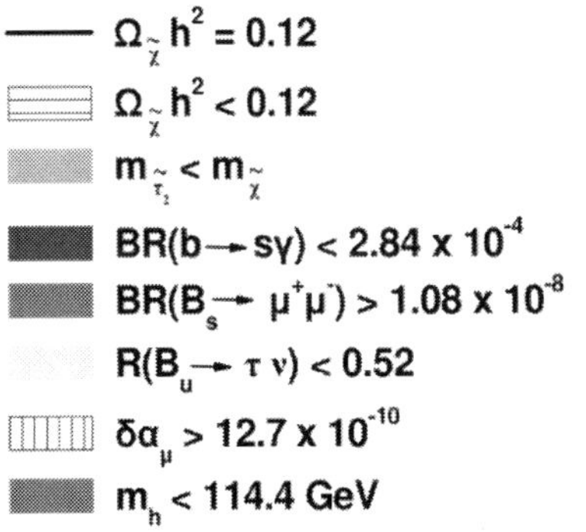

Figure 1. Summary of the conventions adopted in Figs. 2 and 3 for the various restrictions on the model parameters.

Muon Anomalous Magnetic Moment. The quantity δa_μ, which is defined in Sec. 1, can be attributed to SUSY contributions arising from chargino-sneutrino and neutralino-smuon loops. The relevant calculation is based on the formulas of Ref. [50]. The absolute value of the result decreases as $m_{\tilde{\chi}}$ increases and its sign is positive for $\mu > 0$. On the other hand, the calculation of $a_\mu^{\rm SM}$ is not yet completely stabilized mainly because of the ambiguities in the calculation of the hadronic vacuum-polarization contribution. According to the evaluation of this contribution in Ref. [20], there is still a discrepancy between the findings based on the e^+e^--annihilation data and the ones based on the τ-decay data – however, in Ref. [21], it was claimed that this discrepancy can be considerably ameliorated. Taking into account the more reliable calculation based on the e^+e^- data and the experimental measurements [51] of a_μ, we obtain the following 95% c.l. range – cf. Ref. [52]:

$$12.7 \times 10^{-10} \lesssim \delta a_\mu \lesssim 44.7 \times 10^{-10}. \tag{11}$$

A lower [upper] bound on $m_{\tilde{\chi}}$ can be derived from the upper [lower] bound in Eq. (11). As it turns out, only the upper bound on $m_{\tilde{\chi}}$ is relevant in our case. Taking into account the aforementioned computational instabilities and the common practice [28], we consider this bound only as an optional constraint.

Collider Bounds. For our analysis, the only relevant collider bound is the 95% c.l. LEP bound [53] on the lightest CP-even neutral Higgs boson mass

$$m_h \gtrsim 114.4~{\rm GeV}, \tag{12}$$

which gives a lower bound on $m_{\tilde{\chi}}$. However, we should keep in mind that recent data from ATLAS and CMS [25] provide a 99% c.l. upper bound $m_h \lesssim 128$ GeV and a hint in favor of the range (125 ± 1) GeV. Allowing for a theoretical error of ± 1.5 GeV and adding in quadrature the experimental and theoretical uncertainties, we construct the 1σ range of interest [54]:

$$123.2 \lesssim m_h/{\rm GeV} \lesssim 126.8. \tag{13}$$

The calculation of m_h in the package `SOFTSUSY` [26] includes the full one-loop SUSY corrections and some zero-momentum two-loop corrections [55]. The results are well tested [56] against other spectrum calculators.

3. Restrictions on the SUSY Parameters

Imposing the requirements described above, we can delineate the allowed parameter space of our model. Throughout our investigation, we consider the central values for the SM parameters

M_t, $m_b(M_Z)$, $m_\tau(M_Z)$, and $\alpha_s(M_Z)$. We adopt the following conventions for the various lines and regions in the relevant figures (Figs. 2 and 3) – see Fig. 1:

- On the solid black line, Eq. (7) is saturated.
- The horizontally hatched region is allowed by Eq. (7).
- In the light gray region, the lightest stau $\tilde{\tau}_2$ is lighter than $\tilde{\chi}$.
- The dark gray region is excluded by the lower bound in Eq. (8).
- The gray region is excluded by Eq. (9).
- The yellow region is excluded by the lower bound in Eq. (10).
- The vertically hatched region is favored by the lower bound in Eq. (11).
- The red region is excluded by Eq. (12).

Note that the upper bounds in Eqs. (8), (10), and (11) do not restrict the parameters of our model. The region with $\tilde{\tau}_2$ lighter than $\tilde{\chi}$ can not be excluded if the LSP is a neutral sparticle other than $\tilde{\chi}$. One should, though, make sure that the decay of $\tilde{\tau}_2$ to the LSP does not destroy the predictions of the standard BBN [40, 41].

We present the restrictions from all the requirements imposed in the $M_{1/2} - m_0$ plane for $A_0/M_{1/2} = 0$, 1, -1, and -2 in Fig. 2. From the relevant data, we observe that the lower bound in Eq. (10) is fulfilled for the mass of the CP-odd Higgs boson $m_A \gtrsim 520$ GeV and almost independently of the other parameters. Note also that, for $A_0/M_{1/2} = -1$ and -2, the bound in Eq. (12) is violated for $M_{1/2} < 400$ GeV and, consequently, does not appear in the relevant diagrams. It is obvious that, for all the $A_0/M_{1/2}$'s considered in Fig. 2, we are left with no region allowed by all the restrictions of Sec. 2. This is due to the fact that the constraint in Eq. (7), which necessarily holds if $\tilde{\chi}$ is the LSP, is nowhere fulfilled simultaneously with the bound in Eq. (9). Note, finally, that the upper bound on $M_{1/2}$ from the lower bound in Eq. (11) is also nowhere satisfied simultaneously with the bound in Eq. (9) for the values taken for $A_0/M_{1/2}$ in Fig. 2. However, this conflict is less serious since, as we already explained in Sec. 2, the lower bound in Eq. (11) is considered here only as an optional constraint.

The constraint in Eq. (7) is, in general, satisfied in two well-defined distinct regions in the diagrams of Fig. 2, which are:

- The region to the left of the almost vertical part of the line corresponding to the upper bound on $M_{1/2}$ from Eq. (7), where the neutralino annihilation via the s-channel exchange of a CP-odd Higgs boson A is by far the dominant (co)annihilation process. However, this region is excluded by the constraints in Eqs. (9) and (10). On the other hand, it is well known – see e.g. Refs. [4, 16] – that this region is extremely sensitive to variations of $m_b(M_Z)$. Indeed, we find that, as $m_b(M_Z)$ decreases, the A-boson mass m_A increases and approaches $2m_{\tilde{\chi}}$. The A-pole neutralino annihilation is then enhanced and $\Omega_{\tilde{\chi}}h^2$ is drastically reduced causing an increase of the upper bound on $M_{1/2}$. However, even if we reduce $m_b(M_Z)$, we do not find any A-pole neutralino annihilation region which is allowed by the requirements in Eqs. (9) and (10).
- The narrow region which lies just above the light gray area with $\tilde{\tau}_2$ lighter than the neutralino, where bino-stau coannihilations [3, 34] take over leading to a very pronounced reduction of $\Omega_{\tilde{\chi}}h^2$. A large portion of this region survives after the application of all the other requirements of Sec. 2 except for that in Eq. (9). To get a better understanding of this region, we can replace the parameter m_0 by the relative mass splitting $\Delta_{\tilde{\tau}_2} = (m_{\tilde{\tau}_2} - m_{\tilde{\chi}})/m_{\tilde{\chi}}$ between $\tilde{\chi}$ and the lightest stau, which controls the strength of bino-stau coannihilations. The coannihilation region then approximately corresponds to $\Delta_{\tilde{\tau}_2} = 0 - 0.25$. It is evident from Fig. 2 that the slope of the boundary line with $\Delta_{\tilde{\tau}_2} = 0$ increases as $A_0/M_{1/2}$ moves away from zero in both directions. Note that this slope in our model turns out to be larger

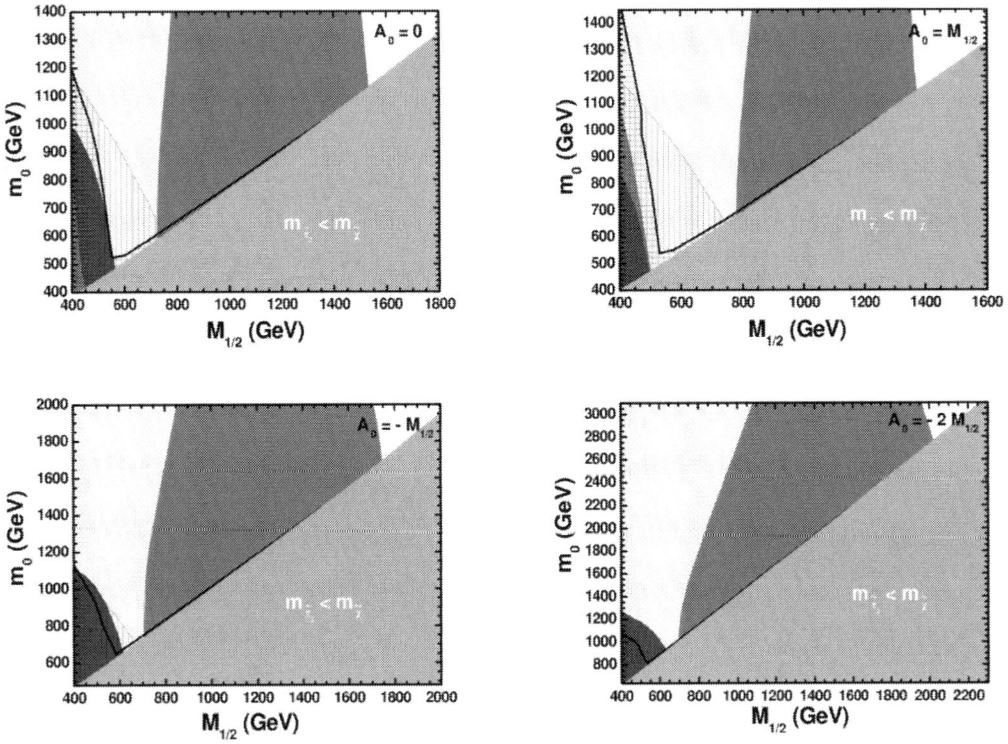

Figure 2. The restrictions in the $M_{1/2} - m_0$ plane for various values of $A_0/M_{1/2}$ indicated in the graphs. The conventions adopted are described in Fig. 1.

than the one obtained in other versions of the CMSSM – cf. Ref. [3] – with lower values of $\tan\beta$. As a consequence, small variations of m_0 or $M_{1/2}$ lead, in our model, to more drastic variations in $\Delta_{\tilde{\tau}_2}$.

To investigate further whether the incompatibility between the constraints in Eqs. (7) and (9) extends to all possible $A_0/M_{1/2}$'s, we focus on the coannihilation regime and construct the regions allowed by all the restrictions of Sec. 2 in the $M_{1/2} - A_0/M_{1/2}$ plane for $\Delta_{\tilde{\tau}_2} = 0$. We depict our results in Fig. 3. The choice $\Delta_{\tilde{\tau}_2} = 0$ ensures the maximal possible reduction of $\Omega_{\tilde{\chi}}h^2$ due to the $\tilde{\chi} - \tilde{\tau}_2$ coannihilation. So, for a given value of $A_0/M_{1/2}$, the maximal $M_{1/2}$ or $m_{\tilde{\chi}}$ allowed by Eq. (7), which holds under the assumption that the neutralino is the LSP, corresponds to $\Delta_{\tilde{\tau}_2} = 0$. We find that, for $A_0/M_{1/2} < 0$, processes with $\tilde{\tau}_2\tilde{\tau}_2^*$ in the initial state and $W^\pm W^\mp$, $W^\pm H^\mp$ in the final one become more efficient (with a total contribution to the effective cross section of about 14 to 22% as $A_0/M_{1/2}$ decreases from 0 to -2) and so coannihilation is strengthened and $m_{\tilde{\chi}}$'s larger than in the $A_0/M_{1/2} > 0$ case are allowed by Eq. (7). The overall maximal $M_{1/2} \simeq 1575$ GeV or $m_{\tilde{\chi}} \simeq 722$ GeV allowed by Eq. (7) is encountered at $A_0/M_{1/2} \simeq -2$ yielding $\mathrm{BR}(B_s \to \mu^+\mu^-) = 1.82 \times 10^{-8}$. Comparing the above upper bound on $M_{1/2}$ with the corresponding one in Fig. 3 of Ref. [22] (represented by a solid and two dotted black lines), we observe that here the bound is considerably enhanced in the region of low as well as the region of large values of $A_0/M_{1/2}$ since we do not consider the constraint from the lower bound on δa_μ from the τ-based calculation. However, it always remains smaller than the lower bound on $M_{1/2}$ derived from Eq. (9) – note that for $3.7 \lesssim A_0/M_{1/2} \lesssim 3.9$ the overall lower bound on $M_{1/2}$ is derived from Eq. (10). Indeed, the smallest lower bound on $M_{1/2} = 1306$ GeV or $m_{\tilde{\chi}} \simeq 590$ GeV is found at $A_0/M_{1/2} \simeq 2$ yielding $\Omega_{\tilde{\chi}}h^2 = 0.15$. Note that

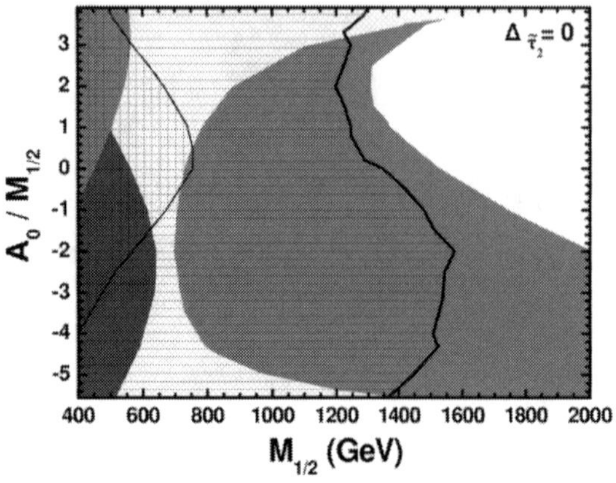

Figure 3. The restrictions in the $M_{1/2} - A_0/M_{1/2}$ plane for $\Delta_{\tilde{\tau}_2} = 0$ following the conventions of Fig. 1.

increasing $\Delta_{\tilde{\tau}_2}$ within the range $0 - 0.25$ does not alter the boundaries of the various constraints in any essential way, except the solid line which is displaced to the left shrinking, thereby, the area allowed by Eq. (7) considerably. Needless to say that the more stringent optional upper bound on $M_{1/2}$ from the lower bound in Eq. (11) is also not compatible with the constraint in Eq. (9). Consequently, for every A_0, there is no range of parameters simultaneously allowed by all the constraints and, therefore, $\tilde{\chi}$ can be now excluded as a CDM particle in our model.

The exclusion, in our model, of $\tilde{\chi}$ as a CDM candidate, resulting from the incompatibility between Eqs. (7) and (9), is further strengthened if one tries to reconcile Eqs. (7) and (13). Indeed, the tension between the neutralino CDM and the new data on m_h is quite generic within the CMSSM since the fulfillment of Eq. (13) requires a very heavy SUSY spectrum, which leads to conflict with Eq. (7) – cf. Refs. [54, 57]. It would be interesting to investigate this issue in our model, which yields large values of $\tan\beta$ and so Eq. (13) can be possibly satisfied with a lighter SUSY spectrum relative to other versions of the CMSSM with lower $\tan\beta$'s. Our results are presented in Fig. 4, where we draw m_h (bold lines) versus $m_{\tilde{\chi}}$ for $\Delta_{\tilde{\tau}_2} \simeq 0$ and $A_0/M_{1/2} = -2$, -1, 0, 1, and 3.9. In the table included in Fig. 4, we also list the minimal $m_{\tilde{\chi}}$'s, $m_{\tilde{\chi}}|_{\rm min}$, for which the inequality in Eq. (9) and the lower bound in Eq. (10) are satisfied for given values of $A_0/M_{1/2}$ as well as the corresponding ranges of $\Omega_{\tilde{\chi}}h^2$ as $m_{\tilde{\chi}}$ varies from $m_{\tilde{\chi}}|_{\rm min}$ to about 1 TeV. Since we take $\Delta_{\tilde{\tau}_2} \simeq 0$, the derived $\Omega_{\tilde{\chi}}h^2$ takes its minimal possible value. In this plot, we also depict by a dotted line the value of $m_{\tilde{\chi}}|_{\rm min}$ for all possible $A_0/M_{1/2}$'s and $\Delta_{\tilde{\tau}_2} \simeq 0$. This line terminates at $A_0/M_{1/2} \simeq 3.9$ since beyond this value the stability of the electroweak vacuum fails. The overall minimal $m_h \simeq 119$ GeV is encountered at $A_0/M_{1/2} \simeq 2$ and $m_{\tilde{\chi}} \simeq 590$ GeV. It is interesting to note that m_h increases with $m_{\tilde{\chi}}$ (or $M_{1/2}$) and as A_0 decreases and, eventually, this mass enters inside the gray region in Fig. 4, which is preferred, at 1σ, by the recent LHC searches. However, it is obvious from Fig. 4 and the values of $\Omega_{\tilde{\chi}}h^2$ in the table included in this figure that Eqs. (7) and (13) are incompatible even in our model.

Departure from $\Delta_{\tilde{\tau}_2} = 0$ is not expected to alter drastically our predictions as regards the value of m_h since m_h depends crucially on $M_{1/2}$, but only mildly on m_0. Moreover, one can deduce from the slop of the left boundary of the gray regions in Fig. 2 that, increasing $\Delta_{\tilde{\tau}_2}$, smaller $M_{1/2}$'s and, therefore, slightly lighter Higgs masses are permitted by Eq. (9). Note that, since $\tilde{\chi}$ cannot be the LSP, both signs of $\Delta_{\tilde{\tau}_2}$ are possible. In particular $\Delta_{\tilde{\tau}_2} > 0$ [$\Delta_{\tilde{\tau}_2} < 0$] corresponds to $\tilde{\chi}$ [$\tilde{\tau}_2$] being the LOSP. The critical case $\Delta_{\tilde{\tau}_2} \simeq 0$ gives the minimal possible relic abundance of the LOSP in both cases due to the coannihilation effect and, therefore, the maximal possible mass of the LSP if this is produced mainly non-thermally – see Refs. [40, 41, 42].

Having in mind mostly the latter possibility, we proceed in the presentation of our predictions for the sparticle and the Higgs boson spectrum of our model, which may be observable at the

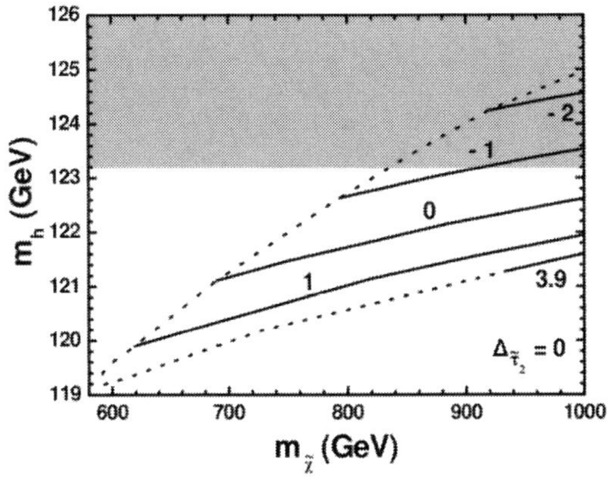

$A_0/M_{1/2}$	$m_{\tilde\chi}\|_{\rm min}$	$\Omega_{\tilde\chi}h^2$
-2	919 GeV	$0.22-0.32$
-1	793 GeV	$0.17-0.31$
0	688 GeV	$0.15-0.38$
1	619 GeV	$0.16-0.44$
3.9	935 GeV	$0.31-0.34$

Figure 4. The variation of m_h as a function of $m_{\tilde\chi}$ for $\Delta_{\tilde\tau_2}\simeq 0$ and various $A_0/M_{1/2}$'s indicated on the curves and in the table included in this figure. In this table, listed are also the minimal $m_{\tilde\chi}$'s for which the inequality in Eq. (9) and the lower bound in Eq. (10) are satisfied and the corresponding ranges of $\Omega_{\tilde\chi}h^2$ as $m_{\tilde\chi}$ increases from $m_{\tilde\chi}|_{\rm min}$ to about 1 TeV. The minimal m_h for all the values of $A_0/M_{1/2}$ is also depicted by a dotted line. The part of the region of Eq. (13) preferred by the recent LHC data which lies in the panel is painted light gray.

LHC. In Table 1, we list the model input and output parameters, the masses in GeV of the sparticles – neutralinos $\tilde\chi$, $\tilde\chi_2^0$, $\tilde\chi_3^0$, $\tilde\chi_4^0$, charginos $\tilde\chi_1^\pm$, $\tilde\chi_2^\pm$, gluinos $\tilde g$, squarks $\tilde t_1$, $\tilde t_2$, $\tilde b_1$, $\tilde b_2$, $\tilde u_L$, $\tilde u_R$, $\tilde d_L$, $\tilde d_R$, and sleptons $\tilde\tau_1$, $\tilde\tau_2$, $\tilde\nu_\tau$, $\tilde e_L$, $\tilde e_R$, $\tilde\nu_e$ – and the Higgs bosons (h, H, $H^\pm$, A), and the values of the various low energy observables for $\Delta_{\tilde\tau_2}\simeq 0$, $A_0/M_{1/2}=-3$, -2, -1, 0, and 2 and for the minimal $M_{1/2}$ allowed by Eq. (9) in each case. We consider the squarks and sleptons of the two first generations as degenerate. From the values of the various observable quantities, we see that the bound in Eq. (7) and the optional lower bound in Eq. (11) are violated. So, the lightest neutralino cannot be the LSP. It is also very interesting to observe that the predicted values of m_h lie close or even inside the range in Eq. (13) favored by Ref. [25] – cf. Refs. [54, 57].

The deviation from YU can be estimated by defining [19] the relative splittings δh_b and δh_τ at $M_{\rm GUT}$ through the relations:

$$\delta h_b\equiv\frac{h_b-h_t}{h_t}=-\frac{2c}{1+c}=-\delta h_\tau\equiv\frac{h_t-h_\tau}{h_t}. \qquad (14)$$

Along the dotted line of Fig. 4, the ranges of the parameters c, δh_τ, δh_b, and $\tan\beta$ are

$$0.148\lesssim c\lesssim 0.16,\quad 0.26\lesssim\delta h_\tau=-\delta h_b\lesssim 0.28,\quad 56.2\lesssim\tan\beta\lesssim 56.9.$$

Let us underline that, although the required deviation from YU is not so small, the restrictions from YU are not completely lost since $\tan\beta$ remains large – close to 60 – and that the deviation from exact YU is generated within well-motivated SUSY GUTs described in Ref. [16].

It is worth emphasizing that our results do not invalidate the $\tilde\chi$ candidacy for a CDM particle in all versions of the CMSSM with Yukawa quasi-unification. This is because the (monoparametric) condition of Eq. (1), which we considered here, is only a simplified case of the Yukawa quasi-unification conditions shown in Eq. (15) of Ref. [16], which depend on one real and two complex parameters. Actually, the investigation of the viability of $\tilde\chi$ as a CDM candidate within the CMSSM with Yukawa quasi-unification conditions more complicated than the one in Eq. (1) derived from the GUT models of Ref. [16] is under consideration. Alternatively, our present model may be perfectly consistent with data if we avoid the restriction from Eq. (7) by assuming that the LSP is the axino [41, 42] with mass a little lower than $m_{\tilde\chi}$ and that the reheat temperature is adequately low.

Journal of Physics: Conference Series **384** (2012) 012012 doi:10.1088/1742-6596/384/1/012012

Table 1. Input and output parameters, masses of the sparticles and Higgs bosons, and values of the low energy observables for $\Delta_{\tilde{\tau}_2} \simeq 0$, five values of $A_0/M_{1/2}$, and the minimal $M_{1/2}$.

Input parameters					
$A_0/M_{1/2}$	-3	-2	-1	0	2
c	0.1589	0.1592	0.1585	0.153	0.1475
$M_{1/2}$/GeV	2355.02	2019.6	1744.65	1530.3	1317.65
m_0/GeV	4292.06	2760.01	1691.01	1132.04	1542.88
Output parameters					
$\tan\beta$	57.1	56.9	56.3	56.2	56.2
$100\delta h_\tau(M_{\rm GUT})$	27.4	27.4	27.3	26.5	25.7
μ/GeV	3704.64	2755	2059	1588	1250
Masses in GeV of sparticles and Higgs bosons					
$\tilde{\chi}$	1089.2	926.5	794.2	692.5	595.2
$\tilde{\chi}_2^0$	2087.2	1777.7	1524.2	1325.4	1130.1
$\tilde{\chi}_3^0$	3688.5	2747.6	2057.3	1588.9	1254.8
$\tilde{\chi}_4^0$	3789.7	2750.3	2062.7	1600.5	1279.9
$\tilde{\chi}_1^\pm$	3690.1	2750.6	2062.8	1600.4	1279.6
$\tilde{\chi}_2^\pm$	2087.3	1777.8	1524.3	1325.5	1130.2
$\tilde{g}$	5190.9	4454	5190.9	3388.6	2981.2
$\tilde{t}_1$	4336.3	3608.2	3094.3	2752.2	2567.8
$\tilde{t}_2$	3593.2	3084.5	2709.5	2449.2	2303.1
$\tilde{b}_1$	4514.1	3653.4	3097.4	2747.9	2575.7
$\tilde{b}_2$	4310.6	3561.2	3005.7	2644.1	2508.8
$\tilde{u}_L$	6215.9	4786	3815.1	3231.7	3051.1
$\tilde{u}_R$	6047.6	4624.8	3661.4	3090.5	2941.3
$\tilde{d}_L$	6216.3	4786.5	3815.8	3232.5	3052
$\tilde{d}_R$	6026.1	4604.4	3641.9	3072.7	2927.7
$\tilde{\tau}_1$	3447.3	2413.2	1721.0	1354.4	1436.4
$\tilde{\tau}_2$	1089.9	927.1	794.4	692.5	595.7
$\tilde{\nu}_\tau$	3443.9	2407.7	1712.4	1343.3	1430.2
$\tilde{e}_L$	4582.6	3085.1	2070.4	1544.9	1789.6
$\tilde{e}_R$	4389.7	2869.5	1821.4	1278.5	1626.4
$\tilde{\nu}_e$	4581.6	3083.7	2068.6	1542.6	1787.5
h	126.17	124.3	122.68	121.15	119.30
H	1463.72	1334.6	1181.53	1012.4	730.78
$H^\pm$	1466.38	1337.6	1185.04	1016.55	736.91
A	1463.99	1334.9	1182.00	1013	732
Low energy observables					
$10^4{\rm BR}\,(b\to s\gamma)$	3.25	3.23	3.22	3.23	3.35
$10^8{\rm BR}\,(B_s\to\mu^+\mu^-)$	1.08	1.08	1.08	1.08	1.08
${\rm R}\,(B_u\to\tau\nu)$	0.929	0.915	0.893	0.856	0.736
$10^{10}\delta a_\mu$	0.565	1.09	2.04	3.27	3.4
$\Omega_{\tilde{\chi}}h^2$	0.301	0.219	0.167	0.152	0.151

4. Conclusions

We performed a revised scan of the parameter space of the CMSSM with $\mu > 0$ applying a suitable Yukawa quasi-unification condition predicted by the SUSY GUT model of Ref. [16], which has been constructed in order to remedy the b-quark mass problem arising from exact YU and universal boundary conditions. We took into account updated constraints from collider and cosmological data. These constraints originate from the CDM abundance in the universe, B physics ($b \to s\gamma$, $B_s \to \mu^+\mu^-$, and $B_u \to \tau\nu$), δa_μ, and m_h. Although the neutralino-stau coannihilations drastically reduce the neutralino relic abundance and, thus, enhance the upper bound on $m_{\tilde{\chi}}$ implied by the assumption that the neutralino is a CDM particle, they do not quite succeed to bring it to an acceptable level compatible with the lower bound on $m_{\tilde{\chi}}$ induced by BR$(B_s \to \mu^+\mu^-)$. Therefore, – contrary to our findings in Ref. [22] – $\tilde{\chi}$ is excluded as CDM particle by the combination of the constraints from BR$(B_s \to \mu^+\mu^-)$ and CDM. As a consequence, the model can become consistent with observations only if the LSP is a SUSY particle other than the neutralino. This could be the axino or the gravitino and can account for the present CDM abundance in the universe. It is interesting to note that, in this case, the lowest predicted m_h is enhanced and gets closer to the range favored by the recent preliminarily results announced by LHC.

Acknowledgments

We would like to thank A. Djouadi, N. Mahmoudi, K.A. Olive, P. Paradisi, Q. Shafi, and J. Wells for useful discussions. This work was supported by the European Union under the Marie Curie Initial Training Network 'UNILHC' PITN-GA-2009-237920 and the Greek Ministry of Education, Lifelong Learning and Religious Affairs and the Operational Program: Education and Lifelong Learning 'HERACLITOS II'.

References

[1] A.H. Chamseddine, R.L. Arnowitt, and P. Nath, *Phys. Rev. Lett.* **49**, 970 (1982); P. Nath, R.L. Arnowitt, and A.H. Chamseddine, *Nucl. Phys.* **B227**, 121 (1983); L.J. Hall, J.D. Lykken, and S. Weinberg, *Phys. Rev. D* **27**, 2359 (1983).

[2] R. Arnowitt and P. Nath, *Phys. Rev. Lett.* **69**, 725 (1992); G.G. Ross and R.G. Roberts, *Nucl. Phys.* **B377**, 571 (1992); V.D. Barger, M.S. Berger, and P. Ohmann, *Phys. Rev. D* **49**, 4908 (1994); G.L. Kane, C.F. Kolda, L. Roszkowski, and J.D. Wells, *ibid.* **49**, 6173 (1994).

[3] J.R. Ellis, T. Falk, and K.A. Olive, *Phys. Lett. B* **444**, 367 (1998); J.R. Ellis, T. Falk, K.A. Olive, and M. Srednicki, *Astropart. Phys.* **13**, 181 (2000); **15**, 413(E) (2001).

[4] A.B. Lahanas, D.V. Nanopoulos, and V.C. Spanos, *Phys. Rev. D* **62**, 023515 (2000); J.R. Ellis, T. Falk, G. Ganis, K.A. Olive, and M. Srednicki, *Phys. Lett. B* **510**, 236 (2001).

[5] G. Lazarides and C. Panagiotakopoulos, *Phys. Lett. B* **337**, 90 (1994); S. Khalil, G. Lazarides, and C. Pallis, *ibid.* **508**, 327 (2001).

[6] B. Ananthanarayan, G. Lazarides, and Q. Shafi, *Phys. Rev. D* **44**, 1613 (1991); *Phys. Lett. B* **300**, 245 (1993).

[7] I. Antoniadis and G.K. Leontaris, *Phys. Lett. B* **216**, 333 (1989).

[8] R. Jeannerot, S. Khalil, G. Lazarides, and Q. Shafi, *J. High Energy Phys.* **10**, 012 (2000).

[9] H. Baer, M.A. Diaz, J. Ferrandis, and X. Tata, *Phys. Rev. D* **61**, 111701 (2000); H. Baer, M. Brhlik, M.A. Diaz, J. Ferrandis, P. Mercadante, P. Quintana, and X. Tata, *ibid.* **63**, 015007 (2000); D. Auto, H. Baer, C. Balázs, A. Belyaev, J. Ferrandis, and X. Tata, *J. High Energy Phys.* **06**, 023 (2003); D. Auto, H. Baer, A. Belyaev, and T. Krupovnickas, *ibid.* **10**, 066 (2004).

[10] T. Blažek, R. Dermíšek, and S. Raby, *Phys. Rev. Lett.* **88**, 111804 (2002); *Phys. Rev. D* **65**, 115004 (2002); R. Dermíšek, S. Raby, L. Roszkowski, and R. Ruiz de Austri, *J. High Energy Phys.* **04**, 037 (2003).

[11] M.S. Carena, M. Olechowski, S. Pokorski, and C.E.M. Wagner, *Nucl. Phys.* **B426**, 269 (1994); R. Hempfling, *Phys. Rev. D* **49**, 6168 (1994); L.J. Hall, R. Rattazzi, and U. Sarid, *ibid.* **50**, 7048 (1994).

[12] D.M. Pierce, J.A. Bagger, K.T. Matchev, and R. Zhang, *Nucl. Phys.* **B491**, 3 (1997); M.S. Carena, D. Garcia, U. Nierste, and C.E.M. Wagner, *ibid.* **B577**, 88 (2000).

[13] P.Z. Skands *et al.*, *J. High Energy Phys.* **07**, 036 (2004).

[14] S.F. King and M. Oliveira, *Phys. Rev. D* **63**, 015010 (2001); I. Gogoladze, R. Khalid, and Q. Shafi, *ibid.* **79**,

115004 (2009); **80**, 095016 (2009); I. Gogoladze, R. Khalid, S. Raza, and Q. Shafi, *J. High Energy Phys.* **12**, 055 (2010).

[15] U. Chattopadhyay and P. Nath, *Phys. Rev. D* **65**, 075009 (2002); U. Chattopadhyay, A. Corsetti, and P. Nath, *ibid.* **66**, 035003 (2002); C. Pallis, *Nucl. Phys.* **B678**, 398 (2004).

[16] M.E. Gómez, G. Lazarides, and C. Pallis, *Nucl. Phys.* **B638**, 165 (2002).

[17] R. Jeannerot, S. Khalil, and G. Lazarides, *J. High Energy Phys.* **07**, 069 (2002); G. Lazarides and A. Vamvasakis, *Phys. Rev. D* **76**, 083507 (2007); **76**, 123514 (2007); G. Lazarides, I.N.R. Peddie, and A. Vamvasakis, *ibid.* **78**, 043518 (2008); G. Lazarides, arXiv:1006.3636.

[18] M.E. Gómez, G. Lazarides, and C. Pallis, *Phys. Rev. D* **67**, 097701 (2003).

[19] G. Lazarides and C. Pallis, hep-ph/0404266; hep-ph/0406081.

[20] M. Davier, A. Hoecker, B. Malaescu, and Z. Zhang, *Eur. Phys. J. C* **71**, 1515 (2011).

[21] F. Jegerlehner and R. Szafron, *Eur. Phys. J. C* **71**, 1632 (2011).

[22] N. Karagiannakis, G. Lazarides, and C. Pallis, *Phys. Lett. B* **704**, 43 (2011).

[23] S. Dar, I. Gogoladze, Q. Shafi, and C.S. Un, *Phys. Rev. D* **84**, 085015 (2011).

[24] CMS and LHCb Collaborations, CMS-PAS-BPH-11-019, LHCb-CONF-2011-047, `http://cdsweb.cern.ch/record/1374913/files/BPH-11-019-pas.pdf`.

[25] Update on the Standard Model Higgs searches in ATLAS and CMS, talks by F. Gianotti and G. Tonelli, 13/12/2011, CERN, `http://indico.cern.ch/conferenceDisplay.py?confId=164890`.

[26] B.C. Allanach, *Computer Physics Commun.* **143**, 305 (2002).

[27] G. Belanger, F. Boudjema, A. Pukhov, and A. Semenov, `http://lapth.in2p3.fr/micromegas`; G. Belanger, F. Boudjema, P. Brun, A. Pukhov, S. Rosier-Lees, P. Salati, and A. Semenov, *Comput. Phys. Commun.* **182**, 842 (2011).

[28] R. Trotta, F. Feroz, M.P. Hobson, L. Roszkowski, and R. Ruiz de Austri, *J. High Energy Phys.* **12**, 024 (2008); A. Belyaev, S. Dar, I. Gogoladze, A. Mustafayev, and Q. Shafi, arXiv:0712.1049 [hep-ph]; O. Buchmueller *et al.*, *Eur. Phys. J. C* **64**, 391 (2009); Y. Akrami, P. Scott, J. Edsjo, J. Conrad, and L. Bergstrom, *J. High Energy Phys.* **04**, 057 (2010); L. Roszkowski, R. Ruiz de Austri, and R. Trotta, *Phys. Rev. D* **82**, 055003 (2010); O. Buchmueller *et al.*, arXiv:1110.3568.

[29] H. Baer, S. Kraml, S. Sekmen, and H. Summy, *J. High Energy Phys.* **03**, 056 (2008); **10**, 079 (2008); H. Baer, S. Kraml, and S. Sekmen, *ibid.* **09**, 005 (2009); H. Baer, S. Kraml, A. Lessa, and S. Sekmen, *ibid.* **02**, 055 (2010); H. Baer, S. Kraml, A. Lessa, S. Sekmen, and H. Summy, *Phys. Lett. B* **685**, 72 (2010).

[30] K. Nakamura *et al.* [Particle Data Group], *J. Phys. G* **37**, 075021 (2010).

[31] H. Baer, J. Ferrandis, K. Melnikov, and X. Tata, *Phys. Rev. D* **66**, 074007 (2002); K. Tobe and J.D. Wells, *Nucl. Phys.* **B663**, 123 (2003).

[32] Tevatron Electroweak Working Group [CDF and D0 collaborations], arXiv:0903.2503; T. Aaltonen *et al.* [CDF Collaboration], *Phys. Rev. Lett.* **105**, 252001 (2010).

[33] E. Komatsu *et al.* [WMAP Collaboration], Astrophys. J. Suppl. **192**, 18 (2011); `http://lambda.gsfc.nasa.gov/product/map`.

[34] M.E. Gómez, G. Lazarides, and C. Pallis, *Phys. Rev. D* **61**, 123512 (2000); *Phys. Lett. B* **487**, 313 (2000).

[35] T. Nihei, L. Roszkowski, and R. Ruiz de Austri, *J. High Energy Phys.* **05**, 063 (2001); *ibid.* **07**, 024 (2002).

[36] G. Bélanger, F. Boudjema, A. Pukhov, and A. Semenov, *Comput. Phys. Commun.* **174**, 577 (2006).

[37] G.F. Giudice, E.W. Kolb, and A. Riotto, *Phys. Rev. D* **64**, 023508 (2001); N. Fornengo, A. Riotto, and S. Scopel, *ibid.* **67**, 023514 (2003); G.B. Gelmini and P. Gondolo, *ibid.* **74**, 023510 (2006); G. Barenboim and J.D. Lykken, *J. High Energy Phys.* **12**, 005 (2006); A.B. Lahanas, N.E. Mavromatos and D.V. Nanopoulos, *Phys. Lett. B* **649**, 83 (2007); M. Drees, H. Iminniyaz, and M. Kakizaki, *Phys. Rev. D* **76**, 103524 (2007).

[38] C. Pallis, *Astropart. Phys.* **21**, 689 (2004); *J. Cosmol. Astropart. Phys.* **10**, 015 (2005); *Nucl. Phys.* **B751**, 129 (2006); hep-ph/0610433.

[39] K. Matchev, hep-ph/0402088; E.A. Baltz, astro-ph/0412170; G. Lazarides, hep-ph/0601016; M. Taoso, G. Bertone, and A. Masiero, *J. Cosmol. Astropart. Phys.* **03**, 022 (2008).

[40] J.R. Ellis, K.A. Olive, Y. Santoso, and V.C. Spanos, *Phys. Lett. B* **588**, 7 (2004); K. Jedamzik, K.-Y. Choi, L. Roszkowski, and R. Ruiz de Austri, *J. Cosmol. Astropart. Phys.* **07**, 007 (2006); J. Pradler and F.D. Steffen, *Phys. Lett. B* **666**, 181 (2008).

[41] L. Covi, L. Roszkowski, R. Ruiz de Austri, and M. Small, *J. High Energy Phys.* **06**, 003 (2004); K.-Y. Choi, L. Covi, J.E. Kim, and L. Roszkowski, arXiv:1108.2282.

[42] H. Baer and H. Summy, *Phys. Lett. B* **666**, 5 (2008); H. Baer, M. Haider, S. Kraml, S. Sekmen, and H. Summy, *J. Cosmol. Astropart. Phys.* **02**, 002 (2009); H. Baer, A.D. Box, and H. Summy, *J. High Energy Phys.* **10**, 023 (2010).

[43] E. Barberio *et al.* [Heavy Flavor Averaging Group], arXiv:0808.1297.

[44] F. Borzumati and C. Greub, *Phys. Rev. D* **58**, 074004 (1998); M. Misiak *et al.*, *Phys. Rev. Lett.* **98**, 022002 (2007).

[45] M. Ciuchini, G. Degrassi, P. Gambino, and G.F. Giudice, *Nucl. Phys.* **B527**, 21 (1998); G. Degrassi, P. Gambino, and G.F. Giudice, *J. High Energy Phys.* **12**, 009 (2000); M.E. Gómez, T. Ibrahim, P. Nath, and S. Skadhauge, *Phys. Rev. D* **74**, 015015 (2006).
[46] P.H. Chankowski and L. Slawianowska, *Phys. Rev. D* **63**, 054012 (2001); C.S. Huang, W. Liao, Q.S. Yan, and S.H. Zhu, *ibid.* **63**, 114021 (2001); **64**, 059902(E) (2001); C. Bobeth, T. Ewerth, F. Kruger, and J. Urban, *ibid.* **64**, 074014 (2001); A. Dedes, H.K. Dreiner, and U. Nierste, *Phys. Rev. Lett.* **87**, 251804 (2001); J.R. Ellis, K.A. Olive, and V.C. Spanos, *Phys. Lett. B* **624**, 47 (2005).
[47] F. Mahmoudi, *Comput. Phys. Commun.* **180**, 1579 (2009).
[48] T. Aaltonen *et al.* [CDF Collaboration], *Phys. Rev. Lett.* **100**, 101802 (2008).
[49] G. Isidori and P. Paradisi, *Phys. Lett. B* **639**, 499 (2006); G. Isidori, F. Mescia, P. Paradisi, and D. Temes, *Phys. Rev. D* **75**, 115019 (2007).
[50] S.P. Martin and J.D. Wells, *Phys. Rev. D* **64**, 035003 (2001).
[51] G.W. Bennett *et al.* [Muon $g-2$ Collaboration], *Phys. Rev. D* **73**, 072003 (2006).
[52] K. Hagiwara, R. Liao, A.D. Martin, D. Nomura, and T. Teubner, *J. Phys. G* **38**, 085003 (2011).
[53] The LEP Collaborations ALEPH, DELPHI, L3, OPAL, The LEP Working Group for Higgs Boson Searches, *Eur. Phys. J. C* **47**, 547 (2006).
[54] O. Buchmueller *et al.*, arXiv:1112.3564.
[55] G. Degrassi, P. Slavich, and F. Zwirner, *Nucl. Phys.* **B611**, 403 (2001); A. Brignole, G. Degrassi, P. Slavich, and F. Zwirner, *ibid.* **B631**, 195 (2002); *ibid.* **B643**, 79 (2002); A. Dedes, G. Degrassi, and P. Slavich, *ibid.* **B672**, 144 (2003).
[56] B.C. Allanach, S. Kraml, and W. Porod, *J. High Energy Phys.* **03**, 045 (2003); B.C. Allanach, A. Djouadi, J.L. Kneur, W. Porod, and P. Slavich, *ibid.* **09**, 044 (2004).
[57] D. Albornoz Vasquez, G. Belanger, R.M. Godbole, and A. Pukhov, arXiv:1112.2200; I. Gogoladze, Q. Shafi, and C.S. Un, arXiv:1112.2206.

Journal of Physics: Conference Series **384** (2012) 012013 doi:10.1088/1742-6596/384/1/012013

The CRESST II Dark Matter Search

Leo Stodolsky[1], G Angloher[1], M Bauer[3], I Bavykina[1], A Bento[1,5], C Bucci[4], C Ciemniak[2], G Deuter[3], F v Feilitzsch[2], D Hauff[1], P Huff[1], C Isaila[2], J Jochum[3], M Kiefer[1], M Kimmerle[3], J C Lanfranchi[2], S Pfister[2], F Petricca[1], W Potzel[2], F Pröbst[1], F Reindl[1], S Roth[2], K Rottler[3], C Sailer[3], K Schäffner[1], J Schmaler[1], S Scholl[3], W Seidel[1], M v Sivers[2], C Strandhagen[3], R Strauss[2], A Tanzke[1], I Usherov[3], S Wawoczny[2], M Willers[2] and A Zöller[2]

[1] Max-Planck-Institut für Physik, Föhringer Ring 6, D-80805 München, Germany
[2] Physik-Department E15, Technische Universität München, D-85747 Garching, Germany
[3] Eberhard-Karls-Universität Tübingen, D-72076 Tübingen, Germany
[4] INFN, Laboratori Nazionali del Gran Sasso, I-67010 Assergi, Italy
[5] CI, Physics Department, University of Coimbra, P-3004 516 Coimbra, Portugal

E-mail: `les@mpp.mpg.de`

Abstract. Direct Dark Matter detection with cryodetectors is briefly discussed, with particular mention of the possibility of the identification of the recoil nucleus. Preliminary results from the CREEST II Dark Matter search, with 730 kg-days of data, are presented. Major backgrounds and methods of identifying and dealing with them are indicated.

1. Introduction

CRESST II is a cryogenic Dark Matter search operating in the Gran Sasso laboratory and is a collaboration between the Max-Planck-Institute, the Technical University of Munich, the University of Tübingen, and the Laboratori Nazionale del Gran Sasso.

CRESST is distinguished by the presence of two cryogenic readout channels. One is for heat or phonons. This provides a very good measurment of the total energy of an event. The other channel is for the scintillation light produced in the target material, which is the scintillating crystal $CaWO_4$. This light signal is used to greatly reduce the electron-photon backgrounds. For the nuclear recoils it also provides some information on which nucleus is recoiling, which can play an important role in the analysis.

2. Why Cryodetectors?

The proposal to look for direct detection of dark matter was stimulated by the suggestion of using cryodetectors for weak processes [1] and it might be helpful to recall the motivation for using cryodetectors. We are looking for 'WIMP'-induced nuclear recoils. The big problem in detecting this process is the small energy expected for the recoil.

As one sees from the example illustrated in Fig 1 with a 50 GeV WIMP, the recoil energy spectrum drops radically with energy. And the situation is even more dramatic with light WIMPs, as we see in Fig 2 with a 10 GeV WIMP.

Journal of Physics: Conference Series **384** (2012) 012013 doi:10.1088/1742-6596/384/1/012013

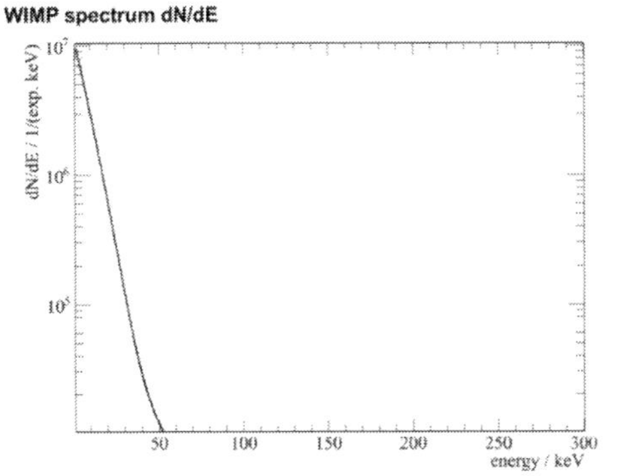

Figure 1. Recoil energy spectrum expected in CRESST ($CaWO_4$) for a M=50 GeV WIMP.

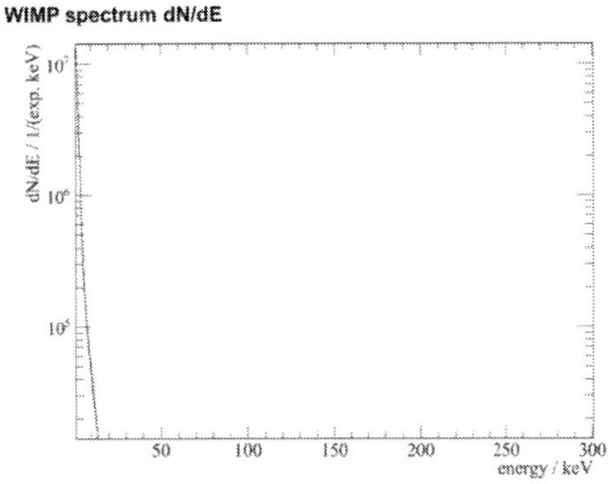

Figure 2. Recoil energy spectrum expected in CRESST ($CaWO_4$) for a M=10 GeV WIMP.

It is evident that an optimal dark matter detector should be able to see and resolve small ($\leq keV$) energies. However, most classical particle detection methods are barely able to get down to this range, and when employed for dark matter detection are operating on the ragged edge of their capabilities. For this reason many detector projects, not employing low temperatures, are effectively operating way out on the tail of the distributions, or in a regime where a special understanding of the detector is necessary. This difficulty, as we see from the figures, is especially present for light WIMPs.

2.1. Energy Threshold and Resolution

On the other hand cryodetectors are well adapted to this problem. The reason may be understood in terms of energy scales [2]. In conventional detectors, whether using liquids, gases, or solid state devices, via ionization or scintillation, the detection process starts by activating or ejecting an electron in some way. As is familiar from atomic or solid state physics, the energy unit for such processes is the electron volt. On the other hand with cryo-devices we are dealing with much smaller energies. For superconductivity for example, the typical energy unit is the energy to break a Cooper pair, which is $\sim 1^o K \sim 10^{-4} eV$ for classical superconductors and can be even less, as for the tungsten thermometers in CRESST. Thus one has considerably more and finer excitations for our few keV, implying a much higher accuracy in the final measurement.

Fig 3 shows the principle of a CRESST detector. A superconducting film thermometer, held at the superconducting-normal transition, has a resistance which is sensitive to very small temperature changes ΔT. These changes are read out by Squid electronics as in Fig 4.

The beautiful energy resolution of these detectors is illustrated [3] in Fig 5. One notes, around 46 KeV, the onset of a $\gamma - \beta$ feature due to the presence of a small ^{210}Pb impurity. This onset is very sharp, showing the fine energy resolution. The other features of the spectrum are also understood.

There are two points to be noted here. One is the low energy *threshold*, which allows one to get down to the main part of the expected recoil energy spectrum. Secondly, there is the very good energy *resolution*. It will be appreciated from Figs 1 and 2 that, due to the very rapid variation with energy, even a small error in an energy determination can lead to a very big error in the rate to be expected. Thus even in the absence of a positive signal, where we are just setting upper limits, a good understanding of the energy resolution is necessary to quantitatively determine what WIMP parameters are being excluded. This is especially true for light WIMPs.

Journal of Physics: Conference Series **384** (2012) 012013 doi:10.1088/1742-6596/384/1/012013

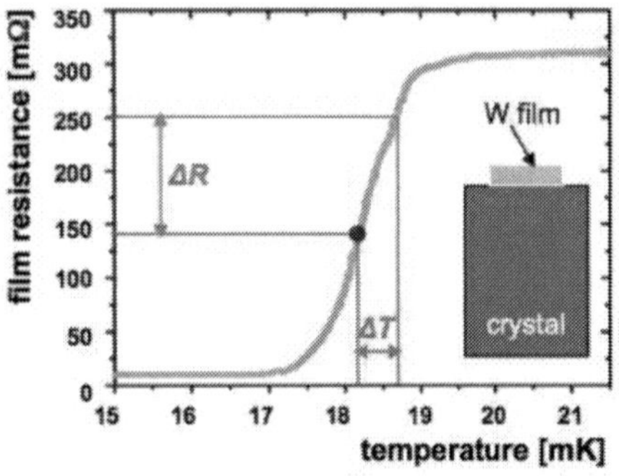

Figure 3. Principle of a CRESST detector with a superconducting thermometer deposited on a crystal ($CaWO_4$).

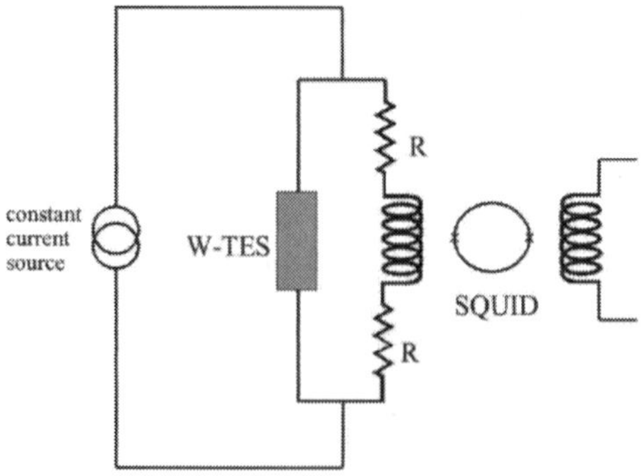

Figure 4. Readout circuit for the superconducting thermometer (TES).

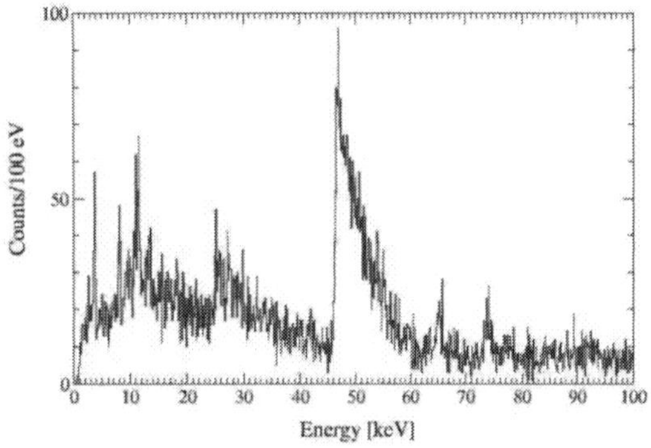

Figure 5. Spectrum from a CRESST detector illustrating the very good energy resolution, as exhibited by the features due to small impurities.

2.2. Multi-element targets

There is another advantage to cryodetectors. The cryo-technique is well adapted to studying different nuclear targets at the same time; that is, in parallel in the same setup. For example, in CRESST we have oxygen, calcium, and tungsten nuclei present simultaneously and their recoils are read out at the same time by the same system. In addition, the superconducting thermometer, or perhaps another cryo-sensor, can be applied onto various materials. These features are in strong contrast to detectors using only one element, such as a noble gas or liquid, which are designed around the properties of that one element. While such detectors have certain advantages, the multi-target aspect is definitely missing.

This ability to compare different nuclei gives valuable extra information and could play an important role in verifying a positive signal. Given the suggestion of a positive signal, there will always be the suspicion that it is due to some unsuspected background, and this fear may not be unfounded, given the very low rates involved and the rather unspecific nature of the WIMP signal. It would therefore be very helpful to have some features of the data which are characteristic of the sought-for WIMP signal.

A good, and one of the few, possibilities for something characteristic would be the comparison of the properties of the signal on different nuclei. This should vary in a definite way as we look at different nuclei, something not in general true for most backgrounds.

2.2.1. Recoil Spectrum One of the simplest features concerns the shape of the recoil energy spectrum. Since the incoming WIMP flux is evidently the same for all our target nuclei, the

Table 1. The differential scattering rate on various nuclei in the diffractive 'black disc' limit, at E_r =20 keV and E_r =30 keV. The same rate for a coherently scattering WIMP at $E_r = 20\,keV$ for masses 10 and 50 GeV is also shown. One notes different patterns of A behavior for neutrons and WIMPs. All values are per nucleus and normalized to that for oxygen. From ref [4].

Element	*A*	neutron E_r =20 keV	neutron E_r =30 keV	WIMP M=10 GeV	WIMP M=50 GeV
O	16	1	1	1	1
F	19	1.5	1.5	1.3	1.8
Na	23	2.2	2.2	1.6	3.3
Si	28	3.4	3.3	1.8	6.7
Ar	40	7.0	6.4	1.1	19
Ca	40	7.0	6.4	1.1	19
Ge	74	19	13	~0	93
I	127	20	5.1	~0	200
Xe	132	18	3.9	~0	240
W	184	2.6	1.6	~0	230

shape of the recoil energy spectrum should vary in a well defined way from nucleus to nucleus. This shape is just given by the mass and velocity spectrum of the incoming WIMPs, which is the same for all nuclei, and the mass and form factor of the nucleus. In addition to the mass, the form factor is the only quantity varying with the nucleus. It is however, a well known quantity, and for the small recoils where most of the rate is concentrated, depends on only one parameter, the radius of the nucleus.

Thus the shape of the recoil spectra on different nuclei should be in agreement with each other, once a mass for the WIMP is assumed, and observation of the correct behavior would go a long way towards making a WIMP signal convincing.

We stress the shape of the recoil spectrum and not its absolute level or total rate. The absolute rate can of course vary from nucleus to nucleus, depending on the quantum numbers of the WIMP and the number of neutrons and protons in the nucleus, and whether we have coherent or spin-dependent interactions. In fact such absolute variations could be used to disentangle the composition of the WIMP[5]. But the shape is governed by the simple factors just mentioned.

2.3. *Annual variation*

Similarly the annual variation effect[6], coming from the different velocities the earth has with respect to the galactic halo in summer and winter, is due to a simple variation of the incoming flux and should be essentially the same for all nuclei. Thus here also the observation of the same percentage variation on the different nuclei of a target material would help greatly in establishing the effect.

2.4. *Fast neutrons*

A particular difficulty would be a fast neutron background. Fast neutrons would induce a recoil spectrum like that from the nuclear form factor. And if an annual variation effect is due to fast neutrons created by cosmic muons, these would also show the same relative variation for different nuclei. However–and here we again see the advantage of being able to distinguish which nucleus is recoiling–fast neutrons undergo diffraction scattering and diffraction scattering shows a characteristic variation from one nucleus to another, different from that expected for WIMPs. This is discussed in ref [4] and in table 1 we illustrate the point by showing for different nuclei the rate, relative to that on oxygen, for coherently scattering WIMPs and for diffractively scattering neutrons. One notes different patterns in the variation with nucleus. This could be used to distinguish WIMPs from neutrons.

3. The Light Channel

In CRESST, information about the nature of the recoil is obtained via the scintillation light, the "light channel". Fig 6 shows a module with the target crystal and light detector, all surrounded by a reflecting and scintillating foil.

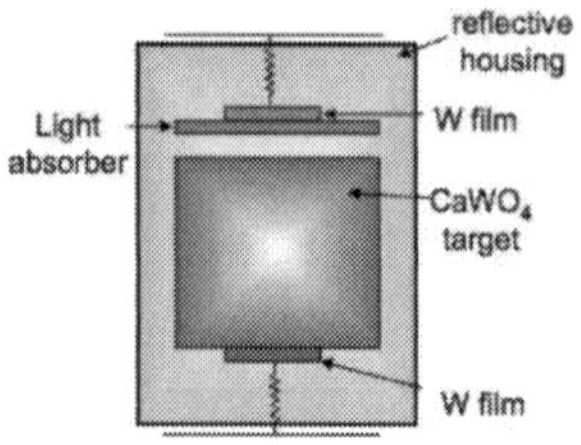

Figure 6. A module with main detector and light detector, surrounded by a reflecting and scintillating housing.

The distinction as to the type of recoil particle is based on the fact that the ratio (light output/energy of the event) is high for fast, light, particles and low for slow, heavy, particles. This distinction is quantified in term of the "quenching factor", QF, which is defined as the ratio of the light output from a given nucleus to that for an electron of the same energy. The QF has been extensively studied by the CRESST collaboration, with the values $QF_\alpha \approx 0.22$, $QF_O \approx 0.10$, $QF_{Ca} \approx 0.064$, $QF_W \approx 0.040$, in $CaWO_4$. There are in general large fluctuations around these average values, as one sees in the plots below. Events from a run are plotted in the (energy, light yield) plane, where 'light yield' is defined as the light output relative to that for 122 keV photons from a ^{57}Co source used for calibration.

First of all, the light channel is used to separate the very large electron-photon background from nuclear recoils. This seems to work quite well, as we illustrate by comparing a run with a neutron test source Fig 7, and a run without the test source Fig 8, from ref [3]. One sees that in the latter the neutron-induced nuclear recoils have disappeared. WIMP candidates will thus appear on such plots as events with low energy and low light yield.

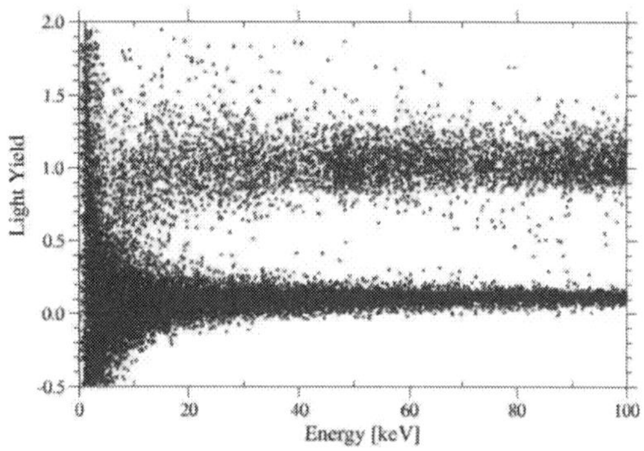

Figure 7. Events in the light yield-energy plane with a neutron source present. Two bands are seen, one for e/γ events, the other for neutron-induced nuclear recoils.

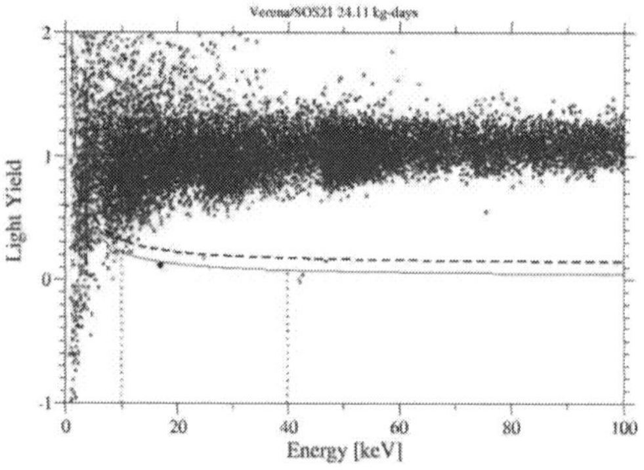

Figure 8. The same with the neutron source removed. The lower band is now absent. The event marked with a larger dot would be a WIMP candidate. 'Light yield' is defined as the light output relative to that for a 122 keV photon in the same detector.

With the present detectors, the separation of the recoils of the different nuclei from each other is not as clear as it is for the $e - \gamma$ band. Since the QF get small, there is little light and the separation becomes difficult at low energy.

4. Results from 730 kg-days

In the data analysis of a dark matter run an acceptance region at low energy and light yield is defined. This is shown for one module in Fig 9. The lower boundary in energy is set so that

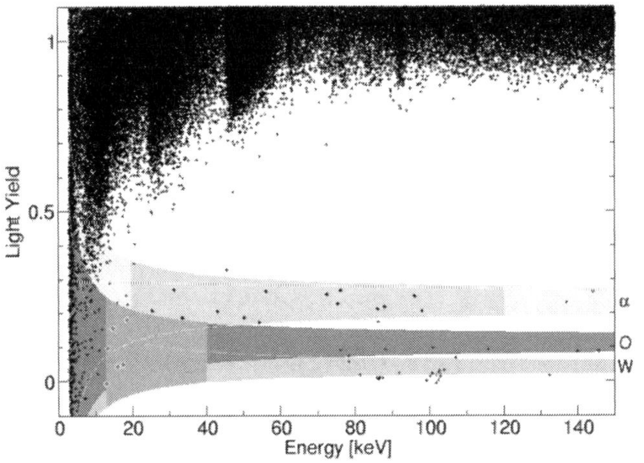

Figure 9. Events in the light yield-energy plane from one of the modules from a long dark matter run, with a total of 730 kg-days for all 8 modules. The acceptance region for WIMP scattering candidates is colored orange and for this module contains 6 events (red dots). Also shown are the expected (90%) bands for α's, and oxygen and tungsten nuclei. The calcium band is not shown but is included in the analysis.

only one "leakage" event from the $e - \gamma$ band is expected; the upper boundary is set at 40 keV, where according to calculations as in Fig 1, neglible WIMP rates are expected. The boundaries in light yield are chosen to include oxygen, calcium, and tungsten recoils. In all, from eight modules, the acceptance regions are found to contain 67 events.

5. Background

The question then is: can all the events in the acceptance regions can be explained as background? Three forms of background are evident in Fig 9. First there is the "leakage" from the $e - \gamma$ band, whose expected value has been used to set the lower boundary of the acceptance region. Secondly and thirdly there are $\alpha's$ and Pb ions from the decay $^{210}Po \rightarrow ^{206} Pb\,(103\,keV) + \alpha\,(5.3\,MeV)$. We believe these originate in the clamps holding the crystals. The α's and Pb ions can lose energy on their way out of the clamps and so possibly 'leak' down into the acceptance region. The probability of this occuring can be estimated by taking the observed events outside the acceptance region and extrapolating into the acceptance region. For example the blue region in Fig 9 is used to determine the α spectrum.

Finally there is a fourth and perhaps most difficult background, one not seen on the plot: neutrons. The scattering of neutrons on nuclei can resemble that expected from WIMPs [4]. But there is one important difference. Neutrons, being strongly interacting, are expected to multiple scatter, while WIMPs, of course, should not. The neutron possibility was examined in two ways. One was with a neutron test source, as in Fig 7. The other was using the muon veto. CRESST has a muon veto used to exclude events which are in coincidence with an incoming muon. However one may also use some of these coincidences to see the effects of neutrons induced by muons. From both of these methods, the neutron source or the muon coincidences, one may determine the ratio of multiple hits in different detector modules to single hits. Given this information one may use the number of multiple hits found in the dark matter run to scale up to the number of WIMP-like single hits to be expected from neutrons. In the total dark

Journal of Physics: Conference Series **384** (2012) 012013 doi:10.1088/1742-6596/384/1/012013

Table 2. Backgrounds and possible WIMP signal resulting from maximum likelihood fits.

	M1	M2
e/γ-events	8.00 ± 0.05	8.00 ± 0.05
α-events	$11.5^{+2.6}_{-2.3}$	$11.2^{+2.5}_{-2.3}$
neutron events	$7.5^{+6.3}_{-5.5}$	$9.7^{+6.1}_{-5.1}$
Pb recoils	$14.8^{+5.3}_{-5.2}$	$18.7^{+4.9}_{-4.7}$
signal events	$29.4^{+8.5}_{-7.7}$	$24.2^{+8.1}_{-7.2}$
m_χ [GeV]	25.3	11.6
σ_{WIMP} [pb]	$1.6 \cdot 10^{-6}$	$3.7 \cdot 10^{-5}$

matter run there were three events with multiple hits, which when scaled up, imply relatively few neutrons.

The sum of the estimates for the backgrounds does not appear to be able to explain all the events. In the analysis we thus also include a possible WIMP signal, assuming coherent scattering for the WIMP, as throughout. The results of an elaborate maximum likelihood analysis is shown in table 2.

The fitting procedure finds two minima M1 and M2 for the likelihood function, both of about the same strength (4.7 and 4.2 σ), with WIMP masses of 25 GeV and 12 GeV respectively. The WIMP cross section for M1 can be appreciably smaller than for M2 since the heavier mass allows the scattering on the tungsten to be above threshold and the coherence A^2 factor has a large effect.

It must be said that maximum likelihood analyses should be appreciated with care. The analysis takes place within the context of an assumed model, and the resulting σ's characterize how sharply the parameters used in the model are determined. It is thus perfectly possible to have a bad model with a good sigma. For this reason some kind of additional 'goodness of fit' characterization is needed. We have examined the "p-value" [8] which estimates how probable it is that the model parameters found from the maximum likelihood fit would give rise to the actually observed data. The value of p turns out to be the not very small 0.35, which appears acceptable.

6. Future Plans

Further analysis using other procedures is underway, and these may yield somewhat different results. But it is intriguing that the obvious backgrounds do not seem to entirely explain the data. To clarify the situation, a further CRESST run, in which it is hoped there will be substantially reduced backgrounds, is in preparation. Since it is believed that the ^{210}Po background mentioned above originates in the clamps holding the crystals, a special effort is devoted to producing clamps of highly pure material.

Also the installation of a further layer of polyethlyne in the inner region around the detectors is being studied in order to provide additional shielding against neutrons originating in the inner parts of the setup.

Finally, efforts aimed at improvment of the light detection, which as discussed above, could play an important role in the identification of the recoil nucleus, continue to be a focus of detector development.

References

[1] "Detectability of Certain Dark Matter Candidates," M. W. Goodman and E. Witten, Phys. Rev. D **31**, 3059 (1985).

[2] For a general introduction to the problems and ideas of detecting low energy interactions and using low temperatures see: L. Stodolsky, *Questions in the Detection of Very Low Energy Intertactions*, in the Workshop on Theoretical and Phenomenological Aspects of Underground Physics, L'Aquila, Italy *TAUP 89*, A. Bottino and P. Monacelli eds. Editions Frontières, Gif-sur-Yvette (1989); and *Neutrino and Dark Matter Detection at Low Temperature* in *Physics Today*, August, 1991.

[3] From "Commissioning Run of the CRESST-II Dark Matter Search," G. Angloher, M. Bauer, I. Bavykina, A. Bento, A. Brown, C. Bucci, C. Ciemniak and C. Coppi *et al.*, Astropartphys. **31**, 270 (2009). arXiv:0809.1829 [astro-ph].

[4] L. Stodolsky, "Features of Fast Neutrons in Dark Matter Searches," Astropart. Phys. **35**, 114 (2011) [arXiv:1009.3791 [astro-ph.IM]].

[5] See Fig 24 of "Light neutralinos as dark matter in the unconstrained minimal supersymmetric standard model," A. Gabutti, M. Olechowski, S. Cooper, S. Pokorski and L. Stodolsky, Astropart. Phys. **6**, 1 (1996), [hep-ph/9602432].

[6] "Detecting Cold Dark Matter Candidates," A. K. Drukier, K. Freese and D. N. Spergel, Phys. Rev. D **33**, 3495 (1986).

[7] "Results from 730 kg days of the CRESST-II Dark Matter Search," G. Angloher, M. Bauer, I. Bavykina, A. Bento, C. Bucci, C. Ciemniak, G. Deuter and F. von Feilitzsch *et al.*, arXiv:1109.0702 [astro-ph.CO].

[8] "p-values for model evaluation," F. Beaujean, A. Caldwell, D. Kollar and K. Kroninger, Phys. Rev. D **83**, 012004 (2011).

Light Higgs and neutralino dark matter in the NMSSM

Daniel E. Lopez-Fogliani

Laboratoire de Physique Théorique, Université Paris-Sud, F-91405 Orsay, France

Abstract.
The next-to-minimal supersymmetric standard model (NMSSM) is a R-parity conserving model that solves the μ-problem of the minimal supersymmetric standard model (MSSM) by adding a singlet superfield. Here we study different aspects of this model. Firstly, using Bayesian statistics, we discuss the constrained next-to-minimal supersymmetric standard model (CNMSSM) scenario. We place special emphasis on analysing the neutralino as a dark matter candidate. Additionally, using the nested sampling (NS) algorithm, we focus our analysis on a particular region of the parameter space. Results obtained scanning the low energy parameter space of the NMSSM, searching for low mass neutralinos, are discussed.

1. Introduction

In the next-to-minimal supersymmetric standard model (NMSSM) the singlet field mixes with the minimal supersymmetric standard model (MSSM) Higgses, and the singlino field mixes with the MSSM neutralinos, offering a richer phenomenology. In the general case, it is possible to have light neutralinos, and in addition it is also possible to have a very light Higgs, which is experimentally viable for a sufficiently high singlet composition [1]. In particular, the exchange of very light Higgses can lead to large direct detection cross sections, within the reach of the present generation of dark matter detectors [2, 3].
The NMSSM superpotential is,

$$W = \epsilon_{ij} \left(Y_u \, H_2^j \, Q^i \, u + Y_d \, H_1^i \, Q^j \, d + Y_e \, H_1^i \, L^j \, e \right) - \epsilon_{ij} \lambda \, S \, H_1^i H_2^j + \frac{1}{3} \kappa S^3 \, . \tag{1}$$

In the context of supergravity, SUGRA, the soft breaking terms are giving by:

$$\begin{aligned}
-\mathcal{L}_{\rm soft} =& m_{\tilde{Q}}^2\,\tilde{Q}^*\,\tilde{Q} + m_{\tilde{U}}^2\,\tilde{u}^*\,\tilde{u} + m_{\tilde{D}}^2\,\tilde{d}^*\,\tilde{d} + m_{\tilde{L}}^2\,\tilde{L}^*\,\tilde{L} + m_{\tilde{E}}^2\,\tilde{e}^*\,\tilde{e} \\
&+ m_{H_d}^2\,H_d^*\,H_d + m_{H_u}^2\,H_u^* H_u + m_S^2\,S^* S \\
&+ \epsilon_{ij}\left(A_u\,Y_u\,H_u^j\,\tilde{Q}^i\,\tilde{u} + A_d\,Y_d\,H_d^i\,\tilde{Q}^j\,\tilde{d} + A_e\,Y_e\,H_d^i\,\tilde{L}^j\,\tilde{e} + \text{H.c.}\right) \\
&+ \left(-\epsilon_{ij}\lambda\,A_\lambda S H_d^i H_u^j + \frac{1}{3}\kappa\,A_\kappa\,S^3 + \text{H.c.}\right) \\
&- \frac{1}{2}\left(M_3\,\lambda_3\,\lambda_3 + M_2\,\lambda_2\,\lambda_2 + M_1\,\lambda_1\,\lambda_1 + \text{H.c.}\right) . \qquad (2)
\end{aligned}$$

In addition to terms from $\mathcal{L}_{\rm soft}$, the tree-level scalar Higgs potential receives the usual D and F term contributions:

$$\begin{aligned}
V_D =& \frac{g_1^2+g_2^2}{8}\left(|H_d|^2 - |H_u|^2\right)^2 + \frac{g_2^2}{2}|H_d^\dagger H_u|^2 , \\
V_F =& |\lambda|^2\left(|H_d|^2|S|^2 + |H_u|^2|S|^2 + |\epsilon_{ij}H_d^iH_u^j|^2\right) + |\kappa|^2|S|^4 \\
&- \left(\epsilon_{ij}\lambda\kappa^* H_d^i H_u^j S^{*2} + \text{H.c.}\right) . \qquad (3)
\end{aligned}$$

When the scalar component of S acquires a VEV, an effective interaction $\mu H_1 H_2$ is generated, with $\mu \equiv \lambda\langle S\rangle$. We refer to [2] for a detailled analysis of the Higgs scalar potential and minimization conditions.

Using the minimization equations we can re-express the soft breaking Higgs masses in terms of λ, κ, A_λ, A_κ, $v_d = \langle H_d^0\rangle$, $v_u = \langle H_u^0\rangle$ (with $\tan\beta = v_u/v_d$), and $s =< S >$:

$$m_{H_d}^2 = -\lambda^2\left(s^2 + v^2\sin^2\beta\right) - \frac{1}{2}M_Z^2\cos 2\beta + \lambda s\tan\beta\left(\kappa s + A_\lambda\right) , \qquad (4)$$

$$m_{H_u}^2 = -\lambda^2\left(s^2 + v^2\cos^2\beta\right) + \frac{1}{2}M_Z^2\cos 2\beta + \lambda s\cot\beta\left(\kappa s + A_\lambda\right) , \qquad (5)$$

$$m_S^2 = -\lambda^2 v^2 - 2\kappa^2 s^2 + \lambda\kappa v^2\sin 2\beta + \frac{\lambda A_\lambda v^2}{2s}\sin 2\beta - \kappa A_\kappa s , \qquad (6)$$

In the next sections, to conduct the investigation of the parameter space of the model we use the nested sampling (NS) algorithm [4], implemented in the Bayesian interface tool `Multinest` [5], that is incorporated into the `SuperBayeS` [6] and NMSPEC [7] packages. Details of the analysis are given in the following next sections.

In section 2 we are going to discuss the results obtained in [8], regarding the investigation of the constrained NMSSM, CNMSSM, using Bayesian statistics.

The NS algorithm is very powerful, allowing us to scan over the parameter space of the NMSSM. Particular regions of the parameter space can be studied using this technique. As for example, the analysis of the ideal Higgs scenario carried out in [9]. Nevertheless to say, statistical interpretation must be taken with great care when particular regions of the parameter space are scanned. However the power of this technique can help to have a better

CNMSSM parameters θ	SM (nuisance) parameters ψ
$50 < m_{1/2} < 4$ TeV	$160 < M_t < 190$ GeV
$50 < m_0 < 4$ TeV	$4 < m_b(m_b)^{\overline{MS}} < 5$ GeV
$\|A_0\| < 7$ TeV	$0.10 < \alpha_s(M_Z)^{\overline{MS}} < 0.13$
$2 < \tan\beta < 65$	
$10^{-3} < \lambda < 0.7$	

Table 1. Initial ranges for our basis parameters $m = (\theta, \psi)$.

understanding of a particular chosen region of the parameter space of the model. For an analysis of the ideal Higgs scenario using this scan technique we refer the interested reader to [9].

In section 3, based in a scan of the low mass parameters space using the power of the NS algorithm, we focus our investigation in the low mass neutralinos region [10].

2. The CNMSSM under the light of a Bayesian analysis

In this section we want to discuss a satistical Bayesian analysis of the CNMSSM. In this case the boundary conditions at the grand unification scale $M_{\rm GUT} \simeq 2 \times 10^{16}$ GeV are analogous to those of the CMSSM, with the exception of m_S. We took m_S indenpendent from m_0, and κ, m_S and s are fixed by the minimization equations (4)-(6) which leads to five continuous free parameters of the CNMSSM: $m_{1/2}$, m_0, A_0, $\tan\beta$ and λ. In addition we fix sgn(μ), chosen $\lambda > 0$ (without loss of generality [2]) and $s > 0$.

We adopt flat prior in $\log m_{1/2}$, $\log m_0$, A_0, $\tan\beta$ and λ, with the last two quantities taken at the scale $M_{\rm SUSY}$. In Table 1 we give the range where our priors vary.

For the nuisance parameters flat priors (although this is not important as they are directly constrained by measurements) are used as well as Gaussian likelihoods representing the experimental observations (see [8] for details).

Mass spectra and observables quantities are computed as was explained in [8], using NMSPEC Ref. [7]. The scanning technique adopted is a "nested sampling" (NS) method [4] as implemented in the MultiNest [11] algorithm, which computes the Bayesian evidence primarily but produces posterior pdfs in the process. MultiNest provides an extremely efficient sampler even for likelihood functions defined over a parameter space of large dimensionality with a very complex structure. (See, *e.g.*, Refs. [12, 13].) We list the observables that were included in the likelihood function in Table 2, for details see [8].

To start the discussion of the results, we show Fig. 1. Focusing on the left panel in the plane spanned by $m_{1/2}$ and m_0, we can see some prominent features: a rather strong preference for the stau coannihilation region of $m_{1/2} \gtrsim m_0 \lesssim 0.5$ TeV, although the 68% total probability region extends to larger m_0 because of the pseudoscalar funnel effect contribution

Observable	Mean value	Uncertainties		Ref.
	η	σ (exper.)	τ (theor.)	
$\delta(g-2)_\mu \times 10^{10}$	29.5	8.8	1	[14]
$BR(\overline{B} \to X_s\gamma) \times 10^4$	3.55	0.26	0.21	[14]
$BR(\overline{B}_u \to \tau\nu) \times 10^4$	1.32	0.49	0.38	[15]
$\Omega_\chi h^2$	0.1099	0.0062	$0.1\,\Omega_\chi h^2$	[16]
	Limit (95% CL)		τ (theor.)	Ref.
$BR(\overline{B}_s \to \mu^+\mu^-)$	$< 5.8 \times 10^{-8}$		14%	[17]
m_h	As implemented in NMSSMTools.			[7]
sparticle masses	As implemented in NMSSMTools.			[7]

Table 2. Summary of the observables used in the analysis. Upper part: Observables for which a positive measurement has been made. $\delta(g-2)_\mu$ denotes the discrepancy between the experimental value and the SM prediction of the anomalous magnetic moment of the muon $(g-2)_\mu$. For central values of the SM input parameters used here, the SM value of $BR(\overline{B} \to X_s\gamma)$ is 3.11×10^{-4}, while the theoretical error of 0.21×10^{-4} includes uncertainties other than the parametric dependence on the SM nuisance parameters, especially on M_t and $\alpha_s(M_Z)^{\overline{MS}}$. For each quantity we use a likelihood function with mean η and standard deviation $s = \sqrt{\sigma^2 + \tau^2}$, where σ is the experimental uncertainty and τ represents our estimate of the theoretical uncertainty (see Ref. [18] for details). Lower part: Observables for which only limits currently exist. The likelihood function is given in Ref. [18], including in particular a smearing out of experimental errors and limits to include an appropriate theoretical uncertainty in the observables in $BR(\overline{B}_s \to \mu^+\mu^-)$. The limit on the light Higgs mass m_h is applied in a simplified way, see text for details.

to $\Omega_\chi h^2$. The wedge of $m_{1/2} \gg m_0$ is disallowed because of charged LSP (normally the stau). On the right panel we can see that big and low values of $\tan\beta$ are preferred.

On the left panel of Fig. 2 we can see that in general λ prefers to be small, which leads to a statistical preference for a very CMSSM-like behavior. Large values of λ, bigger than around 0.6 are disfavored due to a Landau pole in the running of the parameter. At "intermediate" values, $0.1 \lesssim \lambda \lesssim 0.6$, the constraints become weaker but there remain problems with tachyons, seen most clearly in the (λ, κ) plane, presented in Fig. 2, which also shows how both parameters are rather closely correlated, $\kappa \propto \lambda$, and favor small values, towards the decoupling limit. The region with $\kappa \gg \lambda$ is disfavored by the presence of tachyonic CP-odd scalars, and similarly with CP-even scalars for $\kappa \ll \lambda$. The preference for low λ could be due to the fact that there are fewer tachyonic directions in the potential close to the decoupling limit $\lambda \to 0$ (for this to happen $\lambda \lesssim 0.1$ is sufficient). For a more detailed discussion, see Ref. [2].

On the right panel of Fig. 2 we present the 2D posterior pdfs in the usual plane spanned by the spin-independent cross section σ_p^{SI} and the neutralino mass m_χ. For comparison, some of the most stringent 90% CL experimental upper limits are also marked [19, 20, 21, 22, 23, 24], although they have not been imposed in the likelihood because of substantial astrophysical uncertainties, especially in the value of the local density.

Several key features can be seen on the right panel of Fig. 2. The banana shaped high-probability region corresponds to the Higgs funnel and the stau coannihilation regions. The horizontal branch of $\sigma_p^{SI} \simeq 7 \times 10^{-8}$ pb corresponds to the focus point region of large m_0. Next, the overall shape rather closely resembles the case of the CMSSM, see, *e.g.*, Fig. 18 of Ref. [12] or Fig. 13 of Ref. [18].

Basically the whole 68% and 95% total probability regions are likely to be within the planned reach of 10^{-10} pb of future 1-tonne detectors. Some of the currently operating detectors are already probing some of the high probability regions, and with a "modest" improvement down to $\sim 10^{-8}$ pb, they will be testing some of the most likely cross sections.

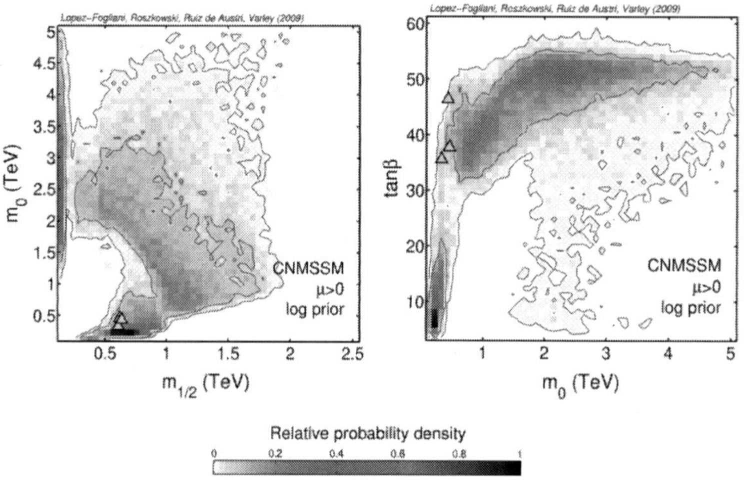

Figure 1. The 2D relative probability density functions in the planes spanned by the CNMSSM parameters $m_{1/2}$, m_0, $\tan\beta$ and A_0. The pdfs are normalized to unity at their peak. The inner (outer) blue solid contours delimit regions encompassing 68% and 95% of the total probability, respectively. All other basis parameters, both CNMSSM and SM ones, in each plane have been marginalized over (i.e., integrated out). Green triangles indicate some of the best fit points.

3. Low mass neutralinos

In this section we are going to use the power of the NS algorithm to analyse a particular region of the parameter space, searching for low mass neutralinos. Statistical interpretations in this case must be taken with great care.

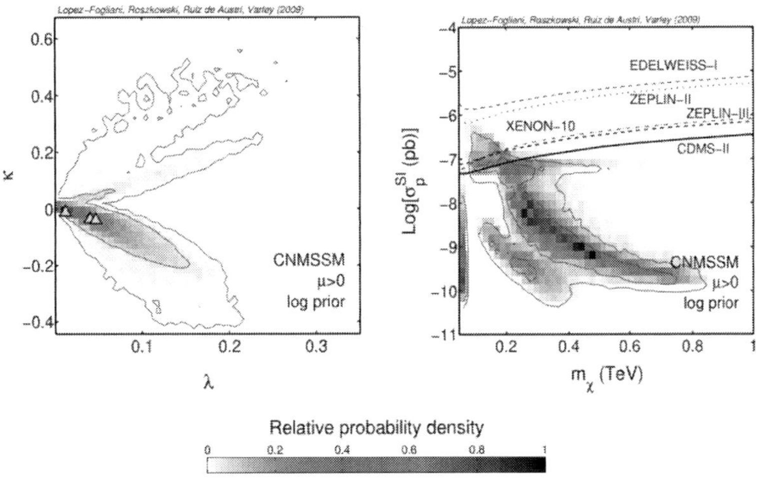

Figure 2. Left: The 2D relative pdfs in the plane of (λ, κ). Right: For the dark matter spin-independent cross section σ_p^{SI} vs the neutralino mass m_χ we show the 2D relative probability density.

Instead of scan the parameters of the model at the GUT scale, as in the previous section, we are going to make a scan of the parameters at low scale.

We are going to assume minimal flavour violation, and perform a scan as described in [10], over the following NMSSM parameters:

$$M_1, M_2, \mu, m_A, m_{\tilde{l}_L}, m_{\tilde{l}_R}, A_\kappa, \tilde{A}, \tan\beta, \lambda, \kappa. \tag{7}$$

where m_A denotes the CP-odd Higgs that is mainly doublet-like, as defined in the NMSPEC code [7]. For simplicity, we take all the trilinear couplings degenerated at EW scale: $A_t = A_b = A_\tau = \widetilde{A}$, with the exception of A_κ, which freedom is going to be crucial in order to find low mass neutralinos. We take all left-handed slepton soft mass parameters to be family degenerate (i.e., $m_{\tilde{e}_L} = m_{\tilde{\mu}_L} = m_{\tilde{\tau}_L} = m_{\tilde{l}_L}$), and likewise for their right-handed partners. On the other hand, gluino and squark masses will not play any significant role in determining either the relic density $\Omega_\chi h^2$ or the spin-independent elastic scattering cross section σ_p^{SI}, and hence we fix them at 1 TeV. In order to focus our scan towards the low mass region of a few GeV, we adopt a log prior distribution of the bino mass M_1, which primarily determines the mass, m_χ, of the lightest neutralino.

The observables we include in the likelihood are the same as in the previous section, listed in Table 2, for details see [10]. Also for this analysis, unlike the analysis done in the previous section, we impose (2σ) hard cuts, in all the available experimental constraints, on the resulting scan data. In Table 3 we give the range where our priors vary.

In Fig. 3 we display our results in the m_χ - σ_p^{SI} plane. We can see in the plot the regions

Parameter	Range	Parameter	Range
bino mass	$0.1 < M_1 < 30$	CP-odd neutral scalar Higgs mass	$85 < m_A < 600$
wino mass	$90 < M_2 < 500$	slepton-left mass	$70 < m_{\tilde{l}_L} < 3000$
μ parameter	$90 < \mu < 500$	slepton-right mass	$70 < m_{\tilde{l}_R} < 3000$
ratio of Higgs doublet s	$2 < \tan\beta < 65$	trilinear terms	$\|A_\kappa\| < 100$, $\|\tilde{A}\| < 4000$
Higgs sector coupling	$10^{-4} < \lambda < 0.5$	Higgs sector coupling	$10^{-4} < \kappa < 0.5$

Table 3. The prior ranges of input parameters over which we perform our scan of the NMSSM. All displayed mass ranges are given in GeV. We adopt log priors for all input parameters except $\tan\beta$, for which we use a flat prior.

favoured by DAMA/LIBRA [25] and CoGeNT [26] experiments. On the contrary, XENON-100 recent data seems to exclude this regions [27]. Likewise, the results from CDMS-II [28] seems to partially exclude the region favoured by CoGENT. At this point we want to recall that we are not including neither of this experiments in the likelihood function. We can clearly see that there are only a minority of our selected points that possess a significant singlino component that survive our cuts and generate a value of $\sigma_p^{SI} > 10^{-7}$ pb. The consequences of imposing the constraints from flavour physics are severe.

Finally, in Fig. 4 we present our results in the λ - κ plane with third axes displaying the singlet composition of the lightest Higgs scalar (left panel) and singlino composition of the lightest neutralino (right panel). We clearly observe that our scan points are segregated into several different regions in the λ - κ plane, each with a distinctive minimum value of $\log_{10}\lambda$, containing points that we can ascribe to one of approximately four different categories that we will now discuss.

Starting at small values of λ we observe the following categories of points: (i) At $\lambda \sim 10^{-4}$ we can see the onset of the MSSM-like region, consisting of bino-dominated LSPs with $m_\chi \gtrsim 7$ GeV, with the lightest Higgs scalar being doublet-like with $m_{h_1} \gtrsim 114.4$ GeV in order to evade LEP bounds. In this case the neutralino annihilates mainly through a t-channel slepton exchange. (ii) For $\lambda \sim 10^{-3}$ and $\kappa \sim 10^{-4}$ we observe the onset of a region where $m_\chi \gtrsim 15$ GeV and, as with the MSSM-like region, the LSP is bino-dominated and still annihilates mainly via t-channel slepton exchange, but in this case the lightest Higgs is singlet-like. (iii) For $\lambda \sim 10^{-2}$ we find the onset of those points involving bino-dominated neutralino LSPs that annihilate through a s-channel resonance involving a singlet-like Higgs, possessing a mass following the relation: $m_\chi \approx 2m_{h_1}$. iv) Lastly, for $\lambda \gtrsim 10^{-1}$ we find neutralinos with $m_\chi \lesssim 15$ GeV which are now a mixture of both singlino and bino and, as with (iii), annihilate via a resonance involving a singlet-like Higgs. A degree of fine-tunning is necessary in order to generate very light-singlet Higgses, pure enough to escape colliders constraints, far from the limit $\lambda \to 0$. Such solutions can potentially generate large values of the SI elastic scattering cross section that are several orders of magnitude larger than

Journal of Physics: Conference Series **384** (2012) 012014 doi:10.1088/1742-6596/384/1/012014

that associated with the bino-dominated case, for $\lambda \sim 10^{-2}$, depending on the degree of fine-tuning involved. We must notice that we are using a log prior in λ and we are not taken into account dark matter direct detection experiments in the likelihood function. As a result, solution in the $\lambda \sim 10^{-2}$ region are much more frequently sampled than the extremely finely-tuned solutions in $\lambda \gtrsim 0.1$ and within the DAMA/LIBRA region. We can see that these extremely finely-tuned solutions are not surviving our (2σ) hard cuts, however such solutions have been found by several authors, see for example [30, 31].

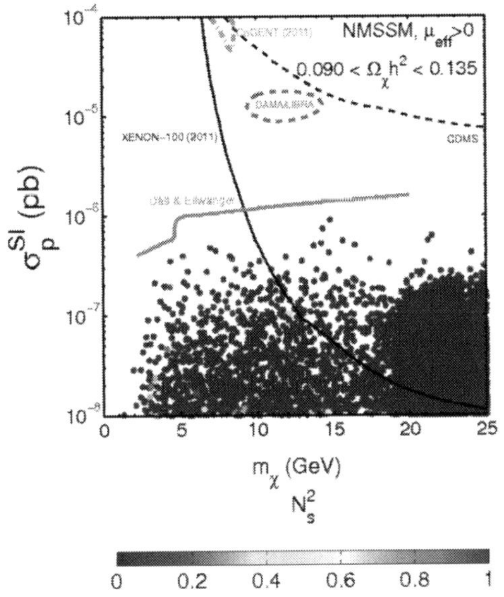

Figure 3. m_χ vs σ_p^{SI} in the NMSSM, for the points from our initial NS scan that survive subsequent 2σ hard cuts on $\Omega_\chi h^2$ and when using constraints from colliders (including flavour physics). The colour scale indicates the singlino fraction of each respective neutralino LSP. We also illustrate the regions of parameter space currently favoured by CoGENT (outlined by the magenta dash-dotted curve), DAMA/LIBRA (with ion channelling, outlined by the dashed grey curve), and illustrate the current limits from CDMS-II (dashed black curve) and XENON-100 (solid black curve). We also display the estimated upper bound on σ_p^{SI} in the NMSSM according to Das & Ellwanger [29] when using default values of strange quark content in nucleons (orange curve).

4. Summary

A global exploration of wide ranges of the CNMSSM parameters and a Bayesian analysis, show that from a statistical point of view, it is difficult to distinguish the model from the CMSSM, i.e., small values of λ are preferred (decoupled limit). As a result the neutralino dark matter direct detection cross section resembles the one obtained for the CMSSM.

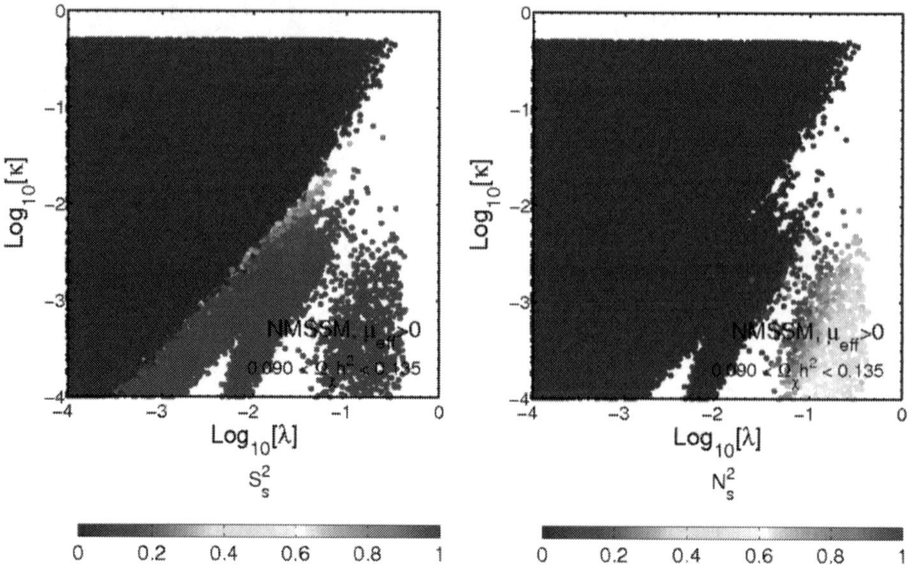

Figure 4. *Left panel:* $\log_{10}\lambda$ vs $\log_{10}\kappa$ in the NMSSM, for the points from our initial NS scan that survive subsequent 2σ hard cuts on $\Omega_\chi h^2$ and when using constraints from colliders (including flavour physics). The colour scale illustrates the singlet fraction S_S^2 of the lightest Higgs. *Right panel:* Identical to the left panel except now the colour scale indicates the singlino fraction of each respective neutralino LSP.

In addition, using the power of the nested sampling (NS) algorithm, we investigate low mass neutralinos in the case of the NMSSM, making a scan of the low energy parameter space. For $m_\chi \lesssim 15$ GeV, we found a favoured region of Bino-dominated neutralinos, for $\lambda \sim 10^{-2}$, accessible to dark matter direct detection experiments. After invoking experimental constraints as 2σ hard cuts, particularly those coming from $BR(\overline{B}_s \to \mu^+\mu^-)$ and $\Omega_\chi h^2$, we find results very similar to those obtained in [29] where $\sigma_p^{SI} \lesssim 10^{-6}$ pb.

Acknowledgements

The work reported here was based in collaborations with D. T. Cumberbatch, J. F. Gunion, L. Roszkowski, R. Ruiz de Austri, Y-L S. Tsai and T. A. Varley. I gratefully acknowledge them. I wishes to thank the organizers of the DSU 2011, at Beijing, for their hospitality. This work was supported by the French ANR TAPDMS ANR-09-JCJC-0146, and in part by the Project of Knowledge Innovation Program (PKIP) of Chinese Academy of Sciences, Grant No. KJCX2.YW.W10.

References

[1] For a review see, U. Ellwanger, C. Hugonie and A. M. Teixeira, Phys. Rept. **496**, 1 (2010) [arXiv:0910.1785 [hep-ph]].

[2] D. G. Cerdeño, C. Hugonie, D. E. López-Fogliani, C. Muñoz and A. M. Teixeira, JHEP **0412** (2004) 048 [arXiv:hep-ph/0408102].
[3] D. G. Cerdeno, E. Gabrielli, D. E. Lopez-Fogliani, C. Munoz and A. M. Teixeira, JCAP **0706** (2007) 008 [arXiv: hep-ph/0701271]; D. E. Lopez-Fogliani, J. Phys. AA **40** (2007) 6889 [arXiv:hep-ph/0703181].
[4] J. Skilling, *Nested sampling*, AIP Conference Proceedings, 735 (2004), pp. 395–405.
[5] http://projects.hepforge.org/multinest/
[6] `http://superbayes.org/`
[7] U. Ellwanger and C. Hugonie, Comput. Phys. Commun. **177** (2007) 399 [arXiv:hep-ph/0612134]; http://www.th.u-psud.fr/NMHDECAY/nmssmtools.html
[8] D. E. Lopez-Fogliani, L. Roszkowski, R. R. de Austri and T. A. Varley, Phys. Rev. D **80** (2009) 095013 [arXiv:0906.4911 [hep-ph]].
[9] J. F. Gunion, D. E. Lopez-Fogliani, L. Roszkowski, R. Ruiz de Austri and T. A. Varley, Phys. Rev. D **84** (2011) 055026 [arXiv:1105.1195 [hep-ph]].
[10] D. T. Cumberbatch, D. E. Lopez-Fogliani, L. Roszkowski, R. R. de Austri and Y. -L. S. Tsai, arXiv:1107.1604 [astro-ph.CO].
[11] F. Feroz and M. P. Hobson, Mon. Not. Roy. Astron. Soc. **384** (2008) 449 [astro-ph/0704.3704].
[12] R. Trotta, F. Feroz, M. P. Hobson, L. Roszkowski and R. Ruiz de Austri, JHEP **0812** (2008) 024 [arXiv:hep-ph/0809.3792].
[13] L. Roszkowski, R. Ruiz de Austri, R. Trotta, Y.-L. S. Tsai, and T. A. Varley, Phys. Rev. D **83** (2011) 015014, [hep-ph/0903.1279].
[14] J. P. Miller, E. de Rafael and B. L. Roberts, Rept. Prog. Phys. **70** (2007) 795 [arXiv:hep-ph/0703049].
[15] The CDF Collaboration, Phys. Rev. Lett. **97** (2006) 062003 [hep-ex/0606027]; Phys. Rev. Lett. **97** (2006) 242003 [hep-ex/0609040].
[16] J. Dunkley et al. [The WMAP Collaboration], Astrophys. J. Suppl. **180** (2009) 306, arXiv:0803.0586 [astro-ph].
[17] The CDF Collaboration, *Search for* $B_s \to \mu^+\mu^-$ *and* $B_d \to \mu^+\mu^-$ *decays in* $p\bar{p}$ *collisions with CDF-II*, CDF note 8956 (August 2007).
[18] R. Ruiz de Austri, R. Trotta and L. Roszkowski, JHEP **0605** (2006)002 [arXiv:hep-ph/0602028]; see also R. Trotta, R. Ruiz de Austri and L. Roszkowski, New Astron. Rev. **51** (2007) 316 [astro-ph/0609126].
[19] The CDMS Collaboration, *Phys. Rev. Lett.* **96** (2006) 011302 [astro-ph/0509259].
[20] V. Sanglard *et al.*[EDELWEISS Collaboration], Phys. Rev. D **71** (2005) 122002 [astro-ph/0503265].
[21] G. J. Alner *et al.*[UK Dark Matter Collaboration], *Astropart. Phys.* **23** (2005)444.
[22] G. J. Alner *et al.*, Astropart. Phys. **28** (2007) 287 [astro-ph/0701858].
[23] V. N. Lebedenko *et al.*, Phys. Rev. D **80** (2009) 052010 arXiv:0812.1150 [astro-ph].
[24] J. Angle *et al.* [XENON Collaboration], Phys. Rev. Lett. **100** (2008) 021303 [astro-ph/0706.0039].
[25] R. Bernabei *et al.*, Eur. Phys. J. C **67** (2010) 39 [arXiv:1002.1028 [astro-ph.GA]].
[26] C. E. Aalseth *et al.* [CoGeNT collaboration], Phys. Rev. Lett. **106** (2011) 131301 [arXiv:1002.4703 [astro-ph.CO]]; C. E. Aalseth *et al.*, Phys. Rev. Lett. **107** (2011) 141301 [arXiv:1106.0650 [astro-ph.CO]].
[27] E. Aprile *et al.* [XENON100 Collaboration], Phys. Rev. Lett. **107** (2011) 131302 [arXiv:1104.2549 [astro-ph.CO]].
[28] Z. Ahmed *et al.* [The CDMS-II Collaboration], Science **327** (2010) 1619 [arXiv:0912.3592 [astro-ph.CO]].
[29] D. Das and U. Ellwanger, JHEP **1009** (2010) 085, [arXiv:1007.1151 [hep-ph]].
[30] G. Belanger, F. Boudjema, C. Hugonie, A. Pukhov and A. Semenov, JCAP **0509**, 001 (2005) [arXiv:hep-ph/0505142].
[31] http://higgs2.ucdavis.edu/gunion/blois_6_2_11.pdf

Journal of Physics: Conference Series **384** (2012) 012015 doi:10.1088/1742-6596/384/1/012015

Signatures of long-lived exotic strongly interacting massive particles on light element abundances through reactions triggered by the particles

M. Kusakabe[1] **and M. Kawasaki**[1,2]

[1] Institute for Cosmic Ray Research, University of Tokyo, Kashiwa, Chiba 277-8582, Japan
[2] Institute for the Physics and Mathematics of the Universe, University of Tokyo, Kashiwa, Chiba 277-8582, Japan

E-mail: `kusakabe@icrr.u-tokyo.ac.jp`

Abstract. A primordial ^{7}Li abundance inferred from observations of metal-poor halo stars is a factor of about three smaller than predictions of standard big bang nucleosynthesis (BBN) model. Some particle models beyond the standard model include heavy long-lived colored particles with mass much larger than 1 GeV. They would be confined inside exotic heavy hadrons, i.e., strongly interacting massive particles. We have found possible reactions which destroy ^{7}Be and ^{7}Li in the scenario of BBN including a long-lived sub-strongly interacting massive particle (sub-SIMP, X). The destruction is associated with non radiative X captures of the nuclei, and it can be realized only if the interaction strength between an X and a nucleon is properly weaker than that between two nucleons. Binding energies of nuclei to an X and nuclear reaction rates associated with the X are estimated. We perform a network BBN calculation using the estimated rates, and suggest that the ^{7}Li problem can be solved if long-lived sub-SIMPs have existed.

1. Introduction

The standard big bang nucleosynthesis (BBN) model is characterized by one parameter, the baryon-to-photon number ratio η. The ratio is precisely determined for a specified cosmological model from the observation of the cosmic microwave background (CMB) radiation with Wilkinson Microwave Anisotropy Probe (WMAP) [1]. Theoretical values of primordial abundances of light element is rather consistent with observations. However, the ^{7}Li abundances of the standard BBN (SBBN) model are different from those inferred from observations of metal-poor halo stars (MPHSs) [2, 3]. Observed abundance levels are roughly constant as a function of metallicity [4, 5, 2, 3, 6, 7, 8] at ^{7}Li/H$=(1-2)\times10^{-10}$. The SBBN prediction is a factor of 2–4 higher, e.g., ^{7}Li/H$=(5.24^{+0.71}_{-0.67})\times10^{-10}$ [9], when the baryon-to-photon ratio η from WMAP determination [1] is adopted. In SBBN, ^{7}Li nuclei are produced almost as ^{7}Be in BBN epoch, and they are converted to ^{7}Li via electron capture at a later epoch of cosmological recombination. The ^{7}Li problem may be caused by an astrophysical process reducing ^{7}Li abundances in surface layers of observed stars or by a cosmological process reducing a ^{7}Be abundance in the early universe.

A possible primordial abundance of ^{6}Li also has been suggested from spectroscopic observations of MPHSs [3, 10, 11]. The observed level is about thousand times as high as

Journal of Physics: Conference Series **384** (2012) 012015 doi:10.1088/1742-6596/384/1/012015

theoretical predictions of SBBN. Primordial abundances of nuclides heavier than Li isotopes, i.e., Be [12, 13, 14, 15], B [16, 17], and C [18] are not yet detected by observations of MPHSs.

Long-lived heavy ($m \gg 1$ GeV) colored particles appear in some particle models beyond the standard model, e.g. split supersymmetry [19, 20]. The long-lived colored particles would be confined at temperatures $T_C \sim 180$ MeV inside exotic heavy hadrons [21], i.e., strongly interacting massive particles (SIMPs) [22, 23, 24] which we call X particles.

If the SIMPs annihilate with their cross sections given by a typical value for strong interaction, i.e., $\sigma \sim \mathcal{O}(\mathrm{GeV}^{-1})^2$, the relic abundance of X particles in the early universe is naively estimated [21] to be $N_X/n_b \sim 0.5 \times 10^{-8}$, where N_X is the number density of the X particle and n_b is the number density of baryons. A real relic abundance should, however, reflects microscopic details on a hadronic annihilation of X [25]. Thermal relic abundances depend on the annihilation cross section which is not known precisely. We then take the freezeout X abundance as a free parameter in this study.

Effects of exotic neutral stable hadrons on BBN were studied in Refs. [26, 27]. In the two studies, binding energies of an exotic hadron X to nuclei are taken to be those of a Λ hyperon. The strong force between an N and an X was, therefore, assumed to be similar to that between an nucleon (N) and a Λ. It has also been assumed that the mass of X is about that of Λ, i.e., $m_X \sim 1.116$ GeV [28] implicitly.

Recently effects on BBN of long-lived SIMPs of $m \gg 1$ GeV, such as a weak-scale mass [$m = \mathcal{O}(0.1$–1 TeV)] have been studied [29]. The interaction strength between an X and an N is assumed to be similar to that between two nucleons. Rates of many reactions involving the X particle were estimated, and a network calculation of the nucleosynthesis including effects of the X was performed. The constraint on the decay lifetime of such particles, i.e., $\tau_X \lesssim 200$ s was derived from a comparison of calculated light element abundances with observational abundance constraints.

This model predicts two interesting observable signatures of the X. One is that ^{9}Be and B can be produced in amounts more than predicted in the SBBN. Another is the possibility of high isotopic ratio, i.e., ^{10}B/^{11}B. The high ratio is never reproduced by boron productions from the cosmic ray nucleosynthesis [30, 31, 32] and the supernova neutrino process [33, 34].

The strength of interaction between an X and an N is unknown. We then investigate effects of a long-lived X in multiple cases of interaction strengths, and report a new possibility that reactions associated with the X particle can reduce ^{7}Be abundance. The reduction of ^{7}Be (and ^{7}Li) can be a solution to the ^{7}Li problem.

In this paper the destruction mechanism of ^{7}Be is reported. We perform a network calculation of BBN assuming the existence and an interaction of a long-lived SIMP. In Sec. 2 assumptions on the X^0 particle and estimations for binding energies between nuclei and an X^0 are described. In Sec. 3 the destruction processes of ^{7}Be and ^{7}Li are described. Based on results of the network calculations of BBN, we delineate the parameter region in which the ^{7}Be and ^{7}Li destructions possibly operate. In Sec 4 we summarize this work. See Ref. [35] for details of this work.

We use the notation, i.e., 1(2,3)4, for a reaction $1 + 2 \rightarrow 3 + 4$.

2. Model

A strongly interacting massive particle (SIMP), i.e., X, is assumed to exist during the BBN epoch. The spin is zero, and charge is also zero. The mass is one parameter. Two types of nuclear potentials between an X^0 and a nucleon (N) are considered in Ref. [35]. In this paper, however, the result for only the Gaussian potential between an X^0 and an N (XN) is shown. The potential is given by $v(r) = v_0 \delta \exp\left[-(r/r_0)^2\right]$, where $v_0 = -72.15$ MeV and $r_0 = 1.484$ fm [36]. The interaction strength is varied by changing δ, the second parameter. When the δ equals unity then the binding energy of deuteron, i.e., 2.224 MeV, is derived.

The potential between an X^0 and a nuclide A (XA) is given by $V(r) = \int v(\boldsymbol{x}) \rho(\boldsymbol{r}')\, d\boldsymbol{r}'$, where

$\boldsymbol{r}$ is the radius from an X to the center of mass of A, $\boldsymbol{r}'$ is the distance between the center of mass of A and a nucleon inside the nuclide A, $\boldsymbol{x} = \boldsymbol{r} + \boldsymbol{r}'$ is the distance between the X and the nucleon, and $\rho(\boldsymbol{r}')$ is the nucleon density of the nucleus which is generally distorted by potential of an X^0 from the density of normal nucleus.

The nucleon density $\rho(\boldsymbol{r})$ is approximately given by the undistorted one for normal nucleus. The nucleon density of nuclei with mass number $A_{\rm m} \geq 2$ is assumed to be spherically symmetric and of Gaussian shape, i.e,

$$\rho(r) = \rho(0) \exp\left[-(r/b)^2\right], \tag{1}$$

where $\rho(0) = A_{\rm m}\pi^{-3/2}b^{-3}$ is the nucleon density at $r = 0$ and satisfies the normalization $\int \rho(r)d\boldsymbol{r} = A_{\rm m}$, with $A_{\rm m}$ the mass number. The parameter for the width of density, i.e., b, is related to the root mean square nuclear matter radius determined from experiments, i.e., $b = \sqrt{2/3}r_{\rm m}^{\rm RMS}$.

2.1. Nuclear Binding Energies

Binding energies of nuclides to an X are estimated. We solve the two-body Schrödinger equation by a variational calculation with the Gaussian expansion method [37]. Binding energies and eigenstate wave functions of the bound states of nuclei A and an X, i.e., X-nuclei or A_X, are obtained.

If the mass of the X^0 particle, i.e., m_X, is much heavier than light nuclides, the reduced mass μ approaches asymptotically to the mass of the nuclide. The binding energies, therefore, also approach to asymptotic values in the limit of large m_X.

The interaction strength δ and the mass m_X of the X particle are taken as parameters. Binding energies of ground state X-nuclei are calculated. Using the obtained binding energies, Q-values of various reactions are calculated. We also calculate binding energies of nuclear excited states of $^4{\rm He}^*_X$ and $^8{\rm Be}^*_X$ with relative angular momentum of $L = 1$.

2.2. Reaction rates

It is hard to calculate rates for nonradiative nuclear reactions precisely. We then approximately take rates of different reactions which has been measured experimentally. For example, the rate of $X(^7{\rm Be},^3{\rm He})^4{\rm He}_X$ is taken from the nonresonant component in the rate of $^6{\rm Li}(n,\alpha)^3{\rm H}$, and we corrected the rate for a reduced mass dependence of cross sections.

Calculations of rates for radiative reactions are relatively easy. Rates of reactions, i.e., $X(\alpha,\gamma)^4{\rm He}_X$, $^4{\rm He}_X(d,\gamma)^6{\rm Li}_X$ and $^4{\rm He}_X(\alpha,\gamma)^8{\rm Be}_X$ are calculated with the code RADCAP [38].

Our estimation for rates of nonradiative and radiative nuclear reactions and β-decay as well as a nuclear reaction network are explained in Ref. [35]. There are large uncertainties in the estimation so that realistic calculations with quantum many-body models are needed in future.

3. Results

Figure 1 shows calculated results of the abundances of normal and X-nuclei as a function of temperature for $\delta = 0.1$ (left panel) and $\delta = 0.2$ (right), respectively. The initial abundance of the X is $N_X/n_b = 1.7 \times 10^{-4}$ ($Y_X \equiv N_X/s = 1.5 \times 10^{-14}$), where N_X and n_b are the number densities of the X^0 particles and baryons, respectively (s is the entropy density). The X^0 particle is assumed to be long-lived compared to the BBN time scale, and have been long extinct by now, i.e., 200 s $\ll \tau_X \ll$ 10 Gyr.

- Case 1 ($\delta = 0.1$)
  At high temperatures $T_9 \gtrsim 1$, the X^0 particles are in the free state since efficient photodisintegrations of X-nuclei destroy bound states A_X. At $T_9 \sim 1$ the ^{4}He is produced as in SBBN, and about one third of X^0 particles are captured by ^{4}He. $^4{\rm He}_X$ nuclei

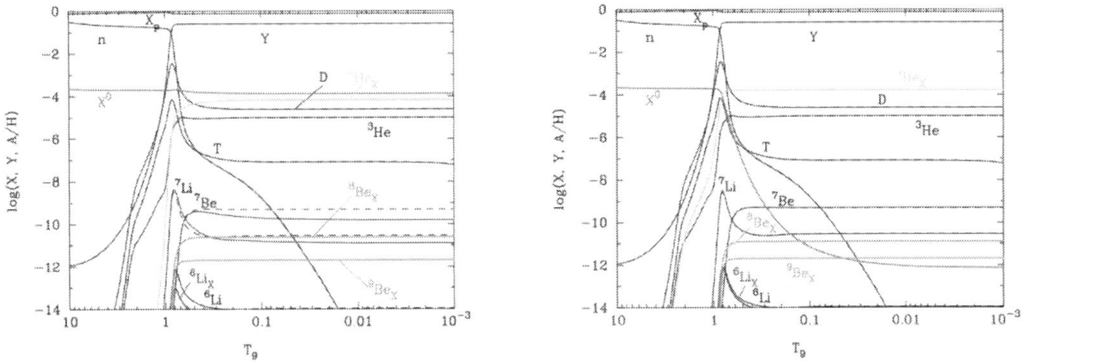

Figure 1. Calculated abundances of normal and X-nuclei (solid lines) as a function of $T_9 \equiv T/(10^9$ K). The mass of X^0 particle was assumed to be $m_X = 100$ GeV, and the interaction strength of XN force is 0.1 and 0.2 times that of NN force, i.e., $\delta = 0.1$ (left panel) and $\delta = 0.2$ (right), respectively. The X^0 abundance is taken to be $N_X/n_b = 1.7 \times 10^{-4}$ ($Y_X \equiv N_X/s = 1.5 \times 10^{-14}$), and its lifetime is assumed to be much longer than BBN time scale, i.e., $\tau_X \gg 200$ s. The X^0 reaction rates are given in Ref. [35] for cases of the two δ values. The dashed lines correspond to the abundances of normal nuclei in SBBN model. This figure is reprinted from Ref. [35]

react with normal nuclei, and affect abundances of ^{7}Li [by ^{4}He$_X(t,^7$Li$)X$], ^{6}Li$_X$ [by ^{4}He$_X(d,\gamma)^6$Li$_X$], and ^{8}Be$_X$ [by ^{4}He$_X(\alpha,\gamma)^8$Be$_X$]. Note that ^{6}Li$_X$ nuclei produced at $T_9 \sim 1$ experience a strong destruction process, i.e., ^{6}Li$_X(p,^3$He$\alpha)X$. ^{9}Be$_X$ is produced by ^{8}Be$_X(d,p)^9$Be$_X$. Finally, free X^0 particles which survived the capture by ^{4}He react with ^{7}Be [by $X(^7$Be$,^3$He$)^4$He$_X$] and ^{7}Li [by $X(^7$Li$,t)^4$He$_X$]. Abundances of ^{7}Be and ^{7}Li thus decrease.

- Case 2 ($\delta = 0.2$)
  The abundance evolution is similar to that in Case 1 at high temperatures $T_9 \gtrsim 1$. However, almost all of X^0 particles are captured by ^{4}He following the ^{4}He synthesis at $T_9 \sim 1$. The abundance of free X^0 particle then significantly reduces at this time. The destruction of ^{7}Be and ^{7}Li, therefore, does not occur as seen in Fig 1.

The reason for the difference between Cases 1 and 2 is as follows: When the interaction strength (δ) is large and the excited state ^{4}He$^*_X$ ($L = 1$) exists, the reaction rate of $X(^4$He$,\gamma)^4$He$_X$ increases. The ^{7}Be and ^{7}Li destruction reactions are then hindered by the depletion of free X^0 particles [35]. In Case 1, there is no excited state of ^{4}He$^*_X$. The reaction $X(^4$He$,\gamma)^4$He$_X$ then proceeds mainly from the initial scattering p-wave to the final bound S-state. In Case 2, on the other hand, an excited state of ^{4}He$^*_X$ exists. The reaction $X(^4$He$,\gamma)^4$He$_X$ can then proceed mainly from the initial scattering s-wave to the final bound P-state.

Figure 2 shows abundances (solid curves) of ^{4}He (mass fraction), D, ^{3}He, ^{7}Li and ^{6}Li (by number relative to H) as a function of the baryon-to-photon ratio η or the baryon energy density $\Omega_B h^2$ of the universe. The solid curves are the calculated result in this BBN affected by X for (m_X, δ, Y_X, τ_X)=(100 GeV, 0.1, 1.5×10^{-14}, ∞). The dashed curves are those in the SBBN. The boxes correspond to the adopted constraints on primordial abundances [35]. The vertical stripe is the 2σ limits on $\Omega_B h^2 = 0.02258^{+0.00057}_{-0.00056}$ provided by WMAP [1] for the ΛCDM+SZ+lens model.

A decrease in ^{7}Li abundance is found, while other nuclear abundances are not changed. A solution to the ^{7}Li problem is thus found in this model.

Journal of Physics: Conference Series **384** (2012) 012015 doi:10.1088/1742-6596/384/1/012015

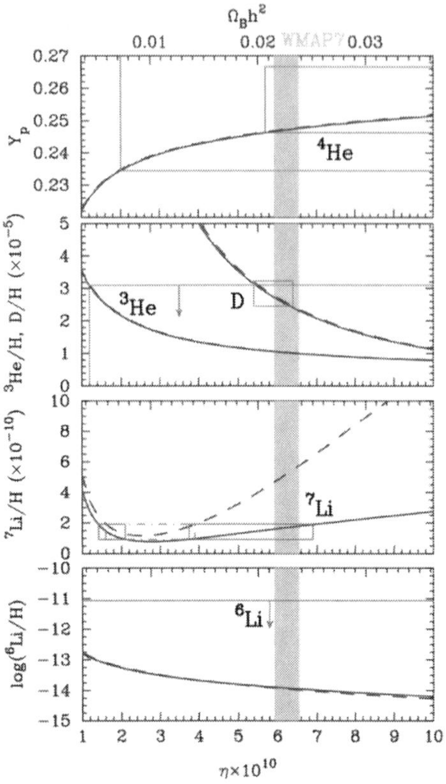

Figure 2. Abundances of ^{4}He (mass fraction), D, ^{3}He, ^{7}Li and ^{6}Li (by number relative to H) as a function of the baryon-to-photon ratio η or the baryon energy density $\Omega_B h^2$ of the universe. The solid curves are the calculated results in the BBN model including X for the parameter set of (m_X, δ, Y_X, τ_X)=(100 GeV, 0.1, 1.5×10^{-14}, ∞), while the dashed curves are those in SBBN. The boxes represent the adopted abundance constraints from Refs. [39, 40] for ^{4}He, [41] for D, [42] for ^{3}He, [5] for ^{7}Li, and [3] for ^{6}Li, respectively. The vertical stripe corresponds to the 2 σ $\Omega_B h^2$ limits provided by WMAP [1]. This figure is reprinted from Ref. [35]

If the strength of XN interaction is relatively weak, most strong reactions for the X^0 particle to get bound to nuclei are non-radiative X^0-capture reactions which are found important in the present model. $X(^6\text{Li}, d)^4\text{He}_X$, $X(^7\text{Li}, t)^4\text{He}_X$, and $X(^7\text{Be},^3\text{He})^4\text{He}_X$ are important reactions in this BBN scenario including X effects.

Figure 3 shows contours in the parameter space (m_X, δ) for critical binding energies of X-nuclei (thin and thick smooth curves). Above the respective contours, Q-values for respective reactions [35] are positive. Zigzag curves show boundaries above which an excited state of $^4\text{He}^*_X$ with $L = 1$ (upper line) and $^8\text{Be}^*_X$ with $L = 1$ (lower) exist, respectively.

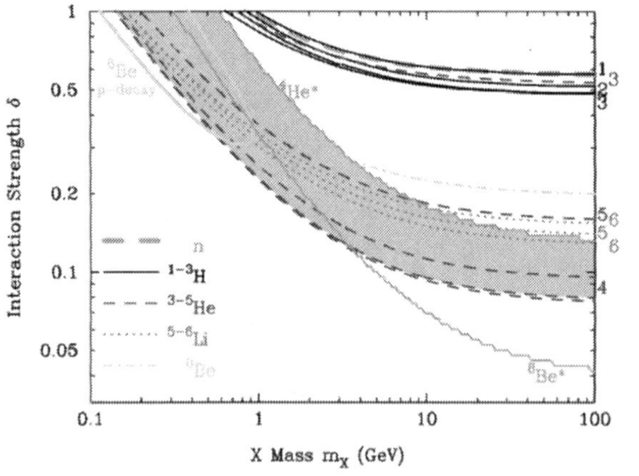

Figure 3. Contours of binding energies between nuclei and an X^0 particle corresponding to $Q = 0$ of reactions [35] (thin and thick smooth curves). Numbers attached to the contours indicate mass numbers of nuclei. Above the contours, reaction Q-values are positive. Zigzag curves correspond to boundaries above which an excited state of $^4\text{He}^*_X$ (upper line) and $^8\text{Be}^*_X$ (lower) exist, respectively. In the shaded region, the ^{7}Li problem can be resolved. This figure is reprinted from Ref. [35]

In Fig. 3, it is found that the contour of the boundary for existence of ${}^4\mathrm{He}^*_X$ are above the contours of the reaction $X({}^7\mathrm{Be},{}^3\mathrm{He}){}^4\mathrm{He}_X$ (the second lowest thin dashed lines). In the parameter region in right upper side from the curve of ${}^4\mathrm{He}^*_X$, free X^0 particles are captured onto ^{4}He before they can react with ^{7}Be to reduce its abundance.

In the shaded region below that curve and above the curve of $X({}^7\mathrm{Be},{}^3\mathrm{He}){}^4\mathrm{He}_X$, free X^0s would possibly remain, and they can reduce the ^{7}Be abundance. This shaded region is, therefore, a possible parameter region solving the ^{7}Li problem.

4. Summary

We study effects of a long-lived strongly interacting massive particle (SIMP) X^0 on BBN. The property of the SIMP is roughly described by the mass m_X and the strength of XN interaction, i.e., δ. Binding energies of X-nuclei (bound states of nuclei and an X^0 particle) are calculated for two types of XN potentials, i.e., Gaussian and well types [35].

We estimate Q-values and rates for possibly important reactions which exist in the presence of X^0 particle. Evolutions of elemental abundances are then calculated under the assumption of existence of a SIMP which interacts with a nucleon by a strength about 0.1 times as large as that of a nucleon. Such a SIMP with relatively weak strong interaction is called sub-SIMP. It is found that if there is no excited bound state of ${}^4\mathrm{He}_X$ with a relative angular momentum $L = 1$, a significant fraction of the X can escape the capture by ^{4}He. In such a situation, ^{7}Be and ^{7}Li can be destroyed by the nuclear capture reactions by free X^0 particles, i.e., $X({}^7\mathrm{Be},{}^3\mathrm{He}){}^4\mathrm{He}_X$ and $X({}^7\mathrm{Li},t){}^4\mathrm{He}_X$. We suggest that the ^{7}Li problem could be solved based on a network calculation of BBN, and constrain the parameter region in the (m_X, δ) plane where the ^{7}Li problem can be solved.

Acknowledgments

This work is supported by Grant-in-Aid for JSPS Fellows No.21.6817 (Kusakabe) and Grant-in-Aid for Scientific Research from the Ministry of Education, Science, Sports, and Culture (MEXT), Japan, No.22540267 and No.21111006 (Kawasaki) and also by World Premier International Research Center Initiative (WPI Initiative), MEXT, Japan.

References

[1] Larson D *et al.* 2011 *Astrophys. J. Suppl.* **192** 16 (*Preprint* 1001.4635)
[2] Melendez J and Ramirez I 2004 *Astrophys. J.* **615** L33 (*Preprint* astro-ph/0409383)
[3] Asplund M, Lambert D L, Nissen P E, Primas F and Smith V V 2006 *Astrophys. J.* **644** 229–259 (*Preprint* astro-ph/0510636)
[4] Spite F and Spite M 1982 *Astron. Astrophys.* **115** 357–366
[5] Ryan S G, Beers T C, Olive K A, Fields B D and Norris J E 2000 *Astrophys. J.* **530** L57–L60 (*Preprint* arXiv:astro-ph/9905211)
[6] Bonifacio P, Molaro P, Sivarani T, Cayrel R, Spite M, Spite F, Plez B, Andersen J, Barbuy B, Beers T C, Depagne E, Hill V, François P, Nordström B and Primas F 2007 *Astron. Astrophys.* **462** 851–864 (*Preprint* arXiv:astro-ph/0610245)
[7] Shi J R, Gehren T, Zhang H W, Zeng J L and Zhao G 2007 *Astron. Astrophys.* **465** 587–591
[8] Aoki W *et al.* 2009 *Astrophys. J.* **698** 1803–1812 (*Preprint* 0904.1448)
[9] Cyburt R H, Fields B D and Olive K A 2008 *JCAP* **0811** 012 (*Preprint* 0808.2818)
[10] Asplund M and Meléndez J 2008 Primordial and Pre-Galactic Origins of the Lithium Isotopes *First Stars III* (*American Institute of Physics Conference Series* vol 990) ed B W O'Shea & A Heger pp 342–346
[11] Steffen M, Cayrel R, Bonifacio P, Ludwig H G and Caffau E 2010 ^{6}Li in metal-poor halo stars: real or spurious? *IAU Symposium* (*IAU Symposium* vol 265) ed K Cunha, M Spite, & B Barbuy pp 23–26 (*Preprint* 0910.5917)
[12] Boesgaard A M, Deliyannis C P, King J R, Ryan S G, Vogt S S and Beers T C 1999 *Astron. J.* **117** 1549–1562
[13] Primas F, Asplund M, Nissen P E and Hill V 2000 *Astron. Astrophys.* **364** L42–L46 (*Preprint* arXiv:astro-ph/0009482)

[14] Primas F, Molaro P, Bonifacio P and Hill V 2000 *Astron. Astrophys.* **362** 666–672 (*Preprint* arXiv:astro-ph/0008402)
[15] Ito H, Aoki W, Honda S and Beers T C 2009 *Astrophys. J.* **698** L37–L41 (*Preprint* 0905.0950)
[16] Duncan D K, Primas F, Rebull L M, Boesgaard A M, Deliyannis C P, Hobbs L M, King J R and Ryan S G 1997 *Astrophys. J.* **488** 338
[17] Garcia Lopez R J *et al.* 1998 *Astrophys. J.* **500** 241 (*Preprint* astro-ph/9801167)
[18] Suda T, Katsuta Y, Yamada S, Suwa T, Ishizuka C, Komiya Y, Sorai K, Aikawa M and Fujimoto M Y 2008 *Publ. Astron. Soc. Jpn.* **60** 1159– (*Preprint* 0806.3697)
[19] Arkani-Hamed N and Dimopoulos S 2005 *JHEP* **06** 073 (*Preprint* hep-th/0405159)
[20] Arkani-Hamed N, Dimopoulos S, Giudice G F and Romanino A 2005 *Nucl. Phys.* **B709** 3–46 (*Preprint* hep-ph/0409232)
[21] Kang J, Luty M A and Nasri S 2008 *JHEP* **09** 086 (*Preprint* hep-ph/0611322)
[22] Wolfram S 1979 *Phys. Lett.* **B82** 65
[23] Dover C B, Gaisser T K and Steigman G 1979 *Phys. Rev. Lett.* **42** 1117
[24] Starkman G D, Gould A, Esmailzadeh R and Dimopoulos S 1990 *Phys. Rev.* **D41** 3594
[25] Kusakabe M and Takesako T 2011 (*Preprint* 1112.0860)
[26] Dicus D A and Teplitz V L 1980 *Phys. Rev. Lett.* **44** 218–221
[27] Mohapatra R N and Teplitz V L 1998 *Phys. Rev. Lett.* **81** 3079–3082 (*Preprint* hep-ph/9804420)
[28] Nakamura K and Particle Data Group 2010 *Journal of Physics G Nuclear Physics* **37** 075021–+
[29] Kusakabe M, Kajino T, Yoshida T and Mathews G J 2009 *Phys. Rev.* **D80** 103501 (*Preprint* 0906.3516)
[30] Prantzos N, Casse M and Vangioni-Flam E 1993 *Astrophys. J.* **403** 630–643
[31] Ramaty R, Kozlovsky B, Lingenfelter R E and Reeves H 1997 *Astrophys. J.* **488** 730–748 (*Preprint* arXiv:astro-ph/9610255)
[32] Kusakabe M 2008 *Astrophys. J.* **681** 18–26 (*Preprint* 0803.3401)
[33] Woosley S E and Weaver T A 1995 *Astrophys. J. Suppl.* **101** 181–235
[34] Yoshida T, Kajino T and Hartmann D H 2005 *Phys. Rev. Lett.* **94** 231101 (*Preprint* astro-ph/0505043)
[35] Kawasaki M and Kusakabe M 2011 *Phys. Rev.* **D83** 055011 (*Preprint* 1012.0435)
[36] Yahiro M, Iseri Y, Kameyama H, Kamimura M and Kawai M 1986 *Progress of Theoretical Physics Supplement* **89** 32–83
[37] Hiyama E, Kino Y and Kamimura M 2003 *Prog. Part. Nucl. Phys.* **51** 223–307
[38] Bertulani C A 2003 *Comput. Phys. Commun.* **156** 123–141 (*Preprint* nucl-th/0305099)
[39] Izotov Y I and Thuan T X 2010 *Astrophys. J.* **710** L67–L71 (*Preprint* 1001.4440)
[40] Aver E, Olive K A and Skillman E D 2010 *JCAP* **1005** 003 (*Preprint* 1001.5218)
[41] Pettini M, Zych B J, Murphy M T, Lewis A and Steidel C C 2008 *Mon. Not. R. Astron. Soc.* **391** 1499–1510 (*Preprint* 0805.0594)
[42] Bania T M, Rood R T and Balser D S 2002 *Nature* **415** 54–57

Journal of Physics: Conference Series **384** (2012) 012016 doi:10.1088/1742-6596/384/1/012016

The status of the $e^{\pm}$ cosmic-ray anomaly

Katie Auchettl and Csaba Balázs

Monash Centre for Astrophysics and
ARC Centre of Excellence for Particle Physics at the Tera-scale
School of Physics, Monash University, Victoria 3800 Australia

E-mail: Katie.Auchettl@monash.edu; Csaba.Balazs@monash.edu

Abstract. Via a Bayesian likelihood analysis using 219 cosmic ray data points we extract the anomalous part of the cosmic $e^{\pm}$ flux. First we show that a serious tension exists between the $e^{\pm}$ fluxes and the rest of the data. Interpreting this tension as effect of an anomalous component on the $e^{\pm}$ related data, we then infer the values of selected cosmic ray propagation parameters excluding the anomalous data sample from the analysis. Based on these values we calculate background predictions with theoretical uncertainties for PAMELA and Fermi-LAT. We find a statistically significant deviation between the Fermi-LAT $e^- + e^+$ data and the predicted background even when (systematic) uncertainties are taken into account. Identifying this deviation as an anomalous $e^{\pm}$ contribution, we make an attempt to distinguish between various sources that may be responsible for the anomalous $e^{\pm}$ flux.

1. Introduction

Cosmic ray observations provided significant and puzzling deviations from theoretical predictions over the last decades. Experiments such as TS [1], AMS [2], CAPRICE [3], MASS [4], and HEAT [5, 6] established a hint of an excess of high energy electrons and/or positrons. Measurements of the PAMELA satellite confirmed these results by finding an excess over the theoretical predictions in the $e^-/(e^- + e^+)$ flux for $E > 10$ GeV [7]. The PAMELA data seem to deviate from the theoretical predictions even when theoretical and experimental uncertainties are taken into account [8]. An excess in the $e^- + e^+$ flux was also found by AMS [9], PPB-BETS [10], and HESS [11, 12]. Recently the Fermi-LAT satellite confirmed the $e^- + e^+$ excess above 100 GeV [13]. The deviation between the Fermi-LAT data and the theoretical $e^- + e^+$ prediction is significant. This deviation was reinforced by PAMELA which found the e^- flux to be consistent with the Fermi-LAT data.

Many attempts were to explain the deviation between the data and theory. New physics was invoked ranging from modification of the cosmic ray propagation, through supernova remnants, to dark matter annihilation. Ref. [14] summarizes these speculations. Whether the $e^{\pm}$ anomaly exists depends on the the cosmic ray background prediction. The theoretical prediction is challenging because of the lack of precise knowledge of the cosmic ray sources, and because the cosmic ray propagation model has numerous free parameters, such as convection velocities, spatial diffusion coefficients and momentum loss rates.

Motivated by traces of possible new physics in the Fermi-LAT data, we attempt to determine the size of the anomalous component in the $e^{\pm}$ flux. Our method involves the following steps. First we find the parameters of the cosmic ray propagation that influence the $e^{\pm}$ flux measured by Fermi-LAT and PAMELA the most. Then we subject the cosmic ray data, other than the

Journal of Physics: Conference Series **384** (2012) 012016 doi:10.1088/1742-6596/384/1/012016

Fermi-LAT and PAMELA $e^{\pm}$ measurements to a Bayesian likelihood analysis, to determine the preferred values and the 68 % (1-σ) credibility regions of the relevant propagation parameters. Based on the central values and 1-σ credibility regions of these propagation parameters we then predict the background flux, with uncertainties, for Fermi-LAT and PAMELA. Finally, we extract the anomalous part of the spectrum by subtracting the background prediction from the Fermi-LAT and PAMELA measurement.

2. Theory of cosmic ray propagation

Galactic cosmic ray propagation is modeled by the diffusion-convection theory [15]. This model assumes the homogeneous propagation of charged particles within the Galactic disk and it also includes energy loss effects. The phase-space density $\psi_a(\vec{r},p,t)$ of a cosmic ray species, labelled by a, at a Galactic radius of $\vec{r}$ can be calculated solving the transport equation which has the general form [16]

$$\begin{aligned}\frac{\partial\psi_a(\vec{r},p,t)}{\partial t} &= Q_a(\vec{r},p,t) + \nabla\cdot(D_{xx}\nabla\psi_a - \vec{V}\psi_a) - (\frac{1}{\tau_f}+\frac{1}{\tau_r})\psi_a \\ &+ \frac{\partial}{\partial p}\left(p^2 D_{pp}\frac{\partial}{\partial p}\frac{1}{p^2}\psi_a\right) - \frac{\partial}{\partial p}\left(\dot{p}\psi_a - \frac{p}{3}(\nabla\cdot\vec{V})\psi_a\right). \quad (1)\end{aligned}$$

Here $q(\vec{r},p,t)$ is the source term of primary and secondary cosmic ray contributions. The spatial diffusion coefficient D_{xx} has the form

$$D_{xx} = D_{0xx}\beta\left(\frac{R}{\text{GeV}}\right)^{\delta}, \quad (2)$$

where $\beta = v/c$, and $R = pc/eZ$ is the (magnetic) rigidity of the particles which describes a particle's resistance to deflection by a magnetic field. Above Z is the effective nuclear charge of the particle, e is its charge, p is its momentum, v is its velocity, and c is the speed of light. The exponent δ indicates the power law dependence of the spatial diffusion coefficient D_{xx}.

Diffusion in momentum space (diffusive re-acceleration) is described by the coefficient D_{pp}

$$D_{pp}D_{xx} = \frac{4p^2{v_A}^2}{3\delta(4-\delta^2)(4-\delta)w}. \quad (3)$$

Here v_A is the Alfven speed, the parameter w characterises the level of hydromagnetic turbulence experienced by the cosmic rays in the interstellar medium [17]. In Eq.(1), $\vec{V}$ is the convection velocity and the parameter τ_f is the time-scale of the fragmentation loss, and τ_r is the radioactive decay time-scale.

The GalProp numerical package solves the propagation equation numerically for $Z \geq 1$ nuclei, as well as for electrons and positrons on a two dimensional spatial grid with cylindrical symmetry in the Galaxy [16]. The input parameter file for GalProp has a number of free parameters which are available for the author to define. These can be classified into a number of subsets: the diffusion of cosmic ray, the primary cosmic ray sources and radiative energy losses of these primary cosmic rays. The diffusion parameters, defined above, are:

$$D_{0xx}, \delta, L, v_A, \partial\vec{V}/\partial z. \quad (4)$$

Parameters that are the most relevant for the primary cosmic ray sources are:

$$R^{e^-}_{ref}, \gamma_1^{e^-}, \gamma_2^{e^-}, R^{nucleus}_{ref}, \gamma_1^{nucleus}, \gamma_2^{nucleus}. \quad (5)$$

Here $\gamma_1^{e^-}$ and $\gamma_2^{e^-}$ are primary source electron injection indices. They specifying the steepness of the electron injection spectrum, $dq(p)/dp \propto p^{\gamma_i^{e^-}}$, below ($i=1$) and above ($i=2$) of a reference rigidity ($R^{e^-}_{ref}$). There are injection indices for nuclei defined by $\gamma_1^{nucleus}$ and $\gamma_2^{nucleus}$ below and above $R^{nucleus}_{ref}$. For further details we refer the reader to [16].

3. Parameter space, uncertainties and experimental input

We tested the robustness of the $e^\pm$ flux against the variation of nearly all propagation parameters individually. We found that the $e^\pm$ flux is mostly sensitive to the following parameters:

$$P = \{\gamma_1^{e^-}, \gamma_2^{nucleus}, \delta_1, \delta_2, D_{0xx}\}. \tag{6}$$

Here δ_1 and δ_2 are spatial diffusion coefficients below and above a reference rigidity ρ_0, $\gamma_1^{e^-}$ and $\gamma_2^{nucleus}$ are the primary electron and nucleus injection indices, and D_{0xx} determines the normalization of the spatial diffusion coefficient.

Our calculations confirmed the findings of the study by [18] that the $e^\pm$ flux is sensitive to the value of the Galactic plane height L. Indeed [17] have shown that there is a connection between L and D_{0xx}:

$$D_{0xx} = \frac{2c(1-\delta)L^{1-\delta}}{3\pi w\delta(\delta+2)}. \tag{7}$$

Thus, varying the cylinder height is the same as the redefinition of D_{0xx} [19]. Realizing this we use D_{0xx} as free parameter and fix L to 4 kpc.

We treat the normalizations of the e^-, e^+, $\bar{p}$/p, B/C, (SC+Ti+V)/Fe and Be-10/Be-9 fluxes as nuisances parameters.

$$P_{nuisance} = \{\Phi^0_{e^-}, \Phi^0_{e^+}, \Phi^0_{\bar{p}/p}, \Phi^0_{B/C}, \Phi^0_{(SC+Ti+V)/Fe}, \Phi^0_{Be-10/Be-9}\}. \tag{8}$$

When evaluating uncertainties, following [20], we ignore theory uncertainties and combine statistical and systematic experimental uncertainties as

$$\sigma_i^2 = \sigma^2_{i,statistical} + \sigma^2_{i,systematic}. \tag{9}$$

This can be done for Fermi-LAT and the latest PAMELA e^- flux. Unfortunately, systematic uncertainties are not available for the rest of the cosmic ray measurements. When this is the case, to parametrize the systematics, we rescale the statistical uncertainty to define σ_i

$$\sigma_i^2 = \sigma^2_{i,statistical}/\tau_i. \tag{10}$$

For simplicity, we use the same scale factor for all data points where systematic uncertainty is not available. To remain consistent with Ref. [20], we set this common scale factor to a value that they use ($\tau_i = 0.2$). We checked that our conclusions are robust against this choice. Further details about our Bayesian parameter inference can be found in [21].

We included 219 of the most recent experimental data points in our statistical analysis. These contained 114 $e^\pm$ related, and 105 $\bar{p}/p$, B/C, (Sc+Ti+V)/Fe and Be-10/Be-9 cosmic ray flux measurements. These data are summarized in Table 1.

4. The size of the $e^\pm$ anomaly

We begin by investigating whether the present cosmic ray data justify the existence of an anomaly in the $e^\pm$ spectrum. To this end we divide the cosmic ray data into two groups: 114 measurements containing observations of $e^\pm$ fluxes (AMS, Fermi, HESS, and PAMELA) and the rest of 105 data points ($\bar{p}/p$, B/C, (Sc+Ti+V)/Fe, Be-10/Be-9). We perform a Bayesian analysis independently on these two sets of data extracting the preferred values of the propagation parameters.

Fig. 1 shows that the two subsets of cosmic ray data are not consistent with the sources implemented in GalProp, or with the cosmic ray propagation model altogether. Our interpretation of this tension between the $e^\pm$ data and the rest of the cosmic ray fluxes is the

Table 1. Cosmic ray experiments and their energy ranges over which we have chosen the data points for our analysis. We split the data into two groups: $e^{\pm}$ flux related (first five lines in the table), and the rest. We do two Bayesian analyses in parallel to show the significant tension between the two data sets.

Measured flux	Experiment	Energy (GeV)	Data points
	AMS [9]	0.60 - 0.91	3
$e^+ + e^-$	Fermi-LAT [13]	7.05 - 886	47
	HESS [11, 12]	918 - 3480	9
$e^+/(e^+ + e^-)$	PAMELA [22]	1.65 - 82.40	16
e^-	PAMELA [23]	1.11 - 491.4	39
$\bar{p}$/p	PAMELA [24]	0.28 - 129	23
	IMP8 [25]	0.03 - 0.11	7
	ISEE3 [26]	0.12 - 0.18	6
B/C	Lezniak et al. [27]	0.30 - 0.50	2
	HEAO3 [28]	0.62 - 0.99	3
	PAMELA [29]	1.24 - 72.36	8
	CREAM [30]	91 - 1433	3
(Sc+Ti+V)/Fe	ACE [31]	0.14 - 35	20
	SANRIKU [32]	46 - 460	6
	Wiedenbeck et al. [33]	0.003 - 0.029	3
	Garcia-Munoz et al. [34]	0.034 - 0.034	1
	Wiedenbeck et al. [33]	0.06 - 0.06	1
Be-10/Be-9	ISOMAX98 [35]	0.08 - 0.08	1
	ACE-CRIS [36]	0.11 - 0.11	1
	ACE [37]	0.13 - 0.13	1
	AMS-02 [38]	0.15 - 9.03	15

following. The measurements of PAMELA and Fermi-LAT may be affected by new physics. This new physics is unaccounted for by the cosmic ray sources included in our calculation or by the propagation model.

We use the central values and credibility regions of the parameters determined using the non-$e^{\pm}$ related data to calculate a background prediction for the $e^{\pm}$ fluxes. Fig. 2 shows the the measured $e^{\pm}$ fluxes and the calculated background. Statistical and systematic uncertainties combined in quadrature are shown for Fermi-LAT, while ($\tau = 0.2$) scaled statistical uncertainties are shown for PAMELA $e^+/(e^+ + e^-)$ as gray bands. Our background prediction is overlaid as magenta bands. The central value and the 1-σ uncertainty of the calculated anomaly is displayed as green dashed lines and bands. As the first frame shows the Fermi-LAT measurements deviate from the predicted background both below 10 GeV and above 100 GeV.

In our interpretation the deviation is a statistically significant signal of the presence of new physics in the $e^+ + e^-$ flux. Based on the difference between the central values of the background and the data a similar conclusion can be drawn from PAMELA. Unfortunately, the large PAMELA uncertainties prevent us to claim a significant deviation. After having determined the background for the $e^{\pm}$ fluxes, we subtract it from the measured flux to obtain the size of the new physics signal. Results for the $e^{\pm}$ anomaly are also shown in Fig. 2. As expected based on the background predictions a non-vanishing anomaly can be established for the Fermi-LAT $e^+ + e^-$ flux, while no anomaly with statistical significance can be claimed for

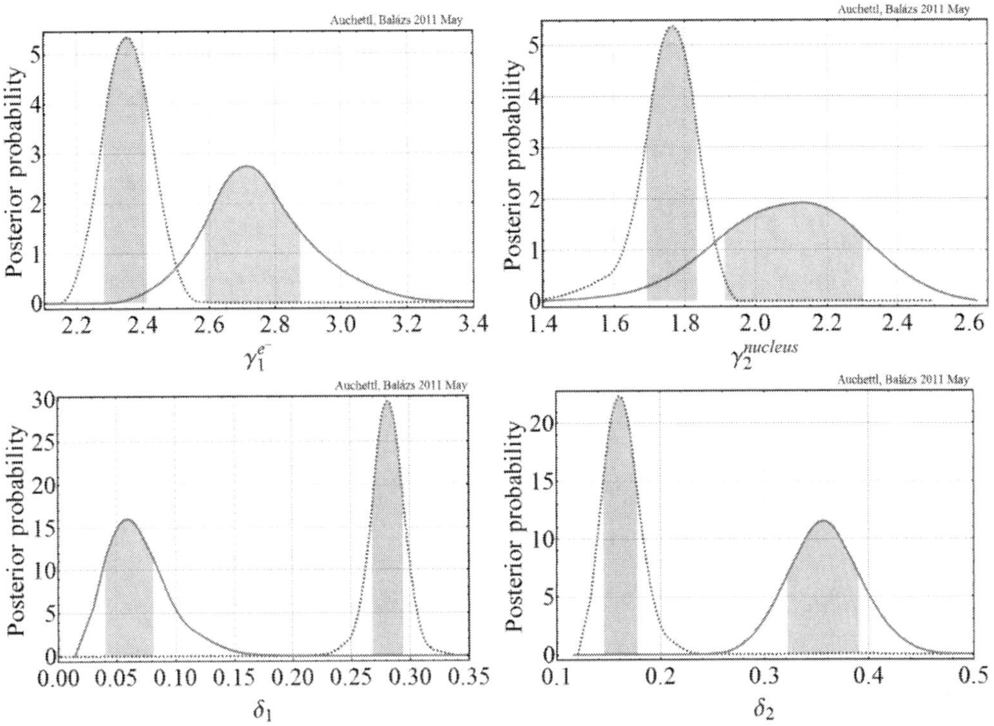

Figure 1. Marginalized posterior probability distributions of propagation parameters listed in Eq.(6). The dashed blue curves show results with likelihood functions containing $e^{\pm}$ flux data while the likelihood functions for the solid red curves contain only the rest of the comic ray data. Shaded areas show the 68 % credibility regions. A statistically significant tension between the $e^{\pm}$ and the rest of the data is evident in the lower frames.

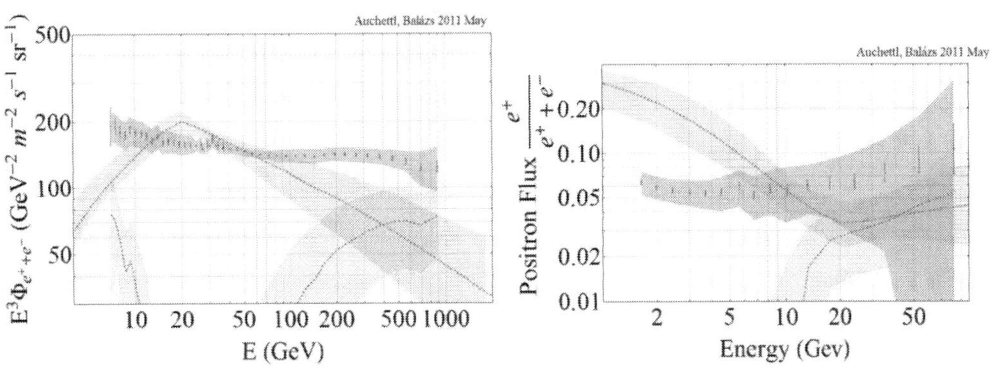

Figure 2. Electron-positron fluxes measured by Fermi-LAT and PAMELA (gray bands) with the extracted size of the $e^{\pm}$ anomaly (green bands). Combined statistical and systematic uncertainties are shown for Fermi-LAT and PAMELA e^-, while ($\tau = 0.2$) scaled statistical uncertainties are shown for PAMELA $e^+/(e^+ + e^-)$. Our background predictions (magenta bands) are also overlaid.

PAMELA due to the large uncertainties.

At this stage we can only speculate about the origin of the difference between the data and background predictions of the cosmic $e^{\pm}$ spectra. One possibility is that some aspect of the propagation model is insufficient for the proper description of the electron-positron fluxes arriving at Earth [39]. In this case there exists no anomaly in the data. Assuming that the propagation model satisfactorily describes physics over the Galaxy the next reasonable thing is to suspect local effects modifying the electron-positron distribution [40]. Further suspicion falls on the lack of sources included in the calculation [41].

Two major categories of possible new sources that could account for the anomaly have been proposed. The first category is standard astrophysical objects such as supernova remnants, pulsars, various objects in the Galactic centre, etc. Finally, more exotic explanations call for new astronomical and/or particle physics phenomena, such as dark matter. In Ref. [21] we compared our extracted signal to recent predictions of anomalous sources. We considered predictions from supernova remnants, nearby pulsars and dark matter annihilation. We concluded that presently uncertainties are too large and prevent us from judging the validity of these as explanations of the anomaly. With more data and more precise calculations the various suggestions of the cosmic $e^- + e^+$ anomaly can be confirmed or ruled out.

Acknowledgments

We are indebted to R. Cotta, A. Donea, Y. Levin and T. Porter for invaluable discussions on various aspects of cosmic ray physics and likelihood analysis. C.B. wishes to thank the organizers and participants of the 2011 DSU workshop for stimulating discussions. K.A. is thankful to T. Porter and to A. Vladimirov for help with GalProp, to J. Lazendic-Galloway for helpful discussions, to M. Jasperse for assistance in Mathematica programming, and to P. Chan for help with issues of parallel computing. This research was funded in part by the Australian Research Council under Project ID DP0877916 and in part by the Project of Knowledge Innovation Program (PKIP) of Chinese Academy of Sciences, Grant No. KJCX2.YW.W10. The use of Monash University Sun Grid, a high-performance computing facility, is also gratefully acknowledged.

References

[1] Golden R L et al. (1994) *Astrophys. J.* **436**, 769–775.
[2] Alcaraz J et al. (2000) *Phys. Lett.* **B484**, 10–22.
[3] Boezio M et al. (2001) *Astrophys. J.* **561**, 787–799.
[4] Grimani C et al. (2002) *Astron. Astrophys.* **392**, 287–294.
[5] Barwick S W et al. (1997) *Astrophys. J.* **482**, L191–L194.
[6] Beatty J J et al. (2004) *Phys. Rev. Lett.* **93**, 241102.
[7] Adriani O, et al. (2009) *Nature* **458**, 607–609.
[8] Delahaye T, Armand F, Pohl M and Salati P (2011) *Preprint* 1102.0744 [astro-ph].
[9] Aguilar M et al. (2002) *Phys. Rept.* **366**, 331–405.
[10] Torii S et al. (2008) *Preprint* 0809.0760 [astro-ph].
[11] Aharonian F et al. (2008) *Phys. Rev. Lett.* **101**, 261104.
[12] Aharonian F et al. (2009) *Astron. Astrophys.* **508**, 561.
[13] Ackermann M et al. (2010) *Phys. Rev.* **D82**, 092004.
[14] Serpico P D (2011) *Preprint* 1108.4827 [astro-ph.HE].
[15] Ginzburg V et al. (1990) *Astrophysics of cosmic rays* (Amsterdam) pp 534.
[16] Strong A W, Moskalenko I V and Ptuskin V S (2007) *Ann. Rev. Nucl. Part. Sci.* **57**, 285–327 .
[17] Seo E S and Ptuskin V S (1994) *Astrophys. J.* **431**, 705–714.
[18] Cotta R C et al (2011) *JHEP* **01**, 064.
[19] Di Bernardo G et al (2011)*Astropart.Phys.* **34**, 528–538.
[20] Trotta R et al (2010) *The Astrophysical Journal* **729**, 106.
[21] Auchettl K and Balazs C (2011) *Preprint* 1106.4138 [astro-ph.HE].
[22] Adriani O et al (2010) *Astropart.Phys.* **34**, 1–11.

Journal of Physics: Conference Series **384** (2012) 012016 doi:10.1088/1742-6596/384/1/012016

[23] Adriani O et al (2011) *Phys. Rev. Lett.* **106**, 201101.
[24] Adriani O et al (2010) *Phys. Rev. Lett.* **105**, 121101.
[25] Moskalenko I V et al (2002) *Astrophys.J.* **565**, 280–296.
[26] Krombel K E and Wiedenbeck M E (1988) *Astrophys. J.* **328**, 940–953.
[27] Lezniak J A and Webber W R (1978) *Astrophys. J.* **223**, 676–696.
[28] Engelmann J J et al (1990) *Astronomy and Astrophysics* **233**, 96–111.
[29] Mocchiutti E et al (2008) *Proceeding of 21st European Cosmic Ray Symposium* 396–401.
[30] Ahn H S et al (2008) *Astropart. Phys.* **30**, 133–141.
[31] Davis A J et al (2000) *American Institute of Physics Conference Series* vol. 528, ed R A Mewaldt, J R Jokipii, M A Lee, E Möbius and T H Zurbuchen 421–424.
[32] Hareyama M (1999) *International Cosmic Ray Conference*, vol. 3 105–.
[33] Wiedenbeck M E and Greiner D E (1980 *The Astrophysical Journal Letters* **239**, L139–L142.
[34] Garcia-Munoz M et al (1981) *International Cosmic Ray Conference*, vol. 9 195–.
[35] Hams T et al (2001) *International Cosmic Ray Conference*, vol. 5 1655–.
[36] Davis A J et al (2000) *AIP Conference Proceedings* **528**, 421–424.
[37] Yanasak N E et al (2001) *Advances in Space Research* **27**, 727–736.
[38] Burger J (2004) *European Physical Journal C* **33**, 941–943.
[39] Tawfik A and Saleh A (2010) *Preprint* 1010.5390 [astro-ph.HE].
[40] Pesce-Rollins M and Fermi-LAT collaboration (2009) *Preprint* 0907.0387 [astro-ph.HE].
[41] Frandsen M T, Masina I and Sannino F (2010) *Preprint* 1011.0013 [astro-ph].

Journal of Physics: Conference Series **384** (2012) 012017 doi:10.1088/1742-6596/384/1/012017

Lepton Flavor Violation at a Future LC

Mario E. Gómez

Department of Applied Physics, University of Huelva,
Avenida 3 de Marzo, 21071 Huelva, Spain

E-mail: mario.gomez@dfa.uhu.es

Abstract.
We investigate the prospects for detection of lepton flavour violation (LFV) in sparticle production and decays at a Linear Collider (LC). We study the Constrained Minimal Supersymmetric extension of the Standard Model (CMSSM), focusing on the subset of the supersymmetric parameter space that also leads to cosmologically interesting values of the relic neutralino LSP density. Emphasis is given to the complementarity between the LC and the LHC signals.

1. Introduction

Data from both atmospheric [1], solar [2] and long-baseline reactor [3] and accelerator [4, 5] neutrino experiments have by now confirmed the existence of neutrino oscillations with near-maximal $\nu_\mu - \nu_\tau$ mixing and large $\nu_e \to \nu_\mu$ one. These observations would also imply violation of the corresponding charged-lepton numbers, which in supersymmetric theories might be significant and observable in low-energy experiments [6, 7]. This may be enhanced sufficiently to become observable in a class of theories predicting new physics at the TeV scale accessible to colliders, particularly in supersymmetric theories. In this presentation, we review the detection channels for lepton flavour violation at a Linear Collider [8] and compare with a previous work [9] where we found that a signal for τ flavour-violating χ_2 decays may be observable at the LHC. In our sampling of the parameter space, we focus on regions that satisfy not only all phenomenological constraints, but also the cosmological relic density considerations.

2. LFV in Slepton Production

2.1. CMSSM with LFV

In the unrotated charged-lepton flavour basis $\tilde{\ell}_i = (\tilde{e}_L, \tilde{\mu}_L, \tilde{\tau}_L, \tilde{e}^*_R, \tilde{\mu}^*_R, \tilde{\tau}^*_R)$, the charged slepton mass matrix is:

$$M^2_{\tilde{\ell}} = \begin{pmatrix} M^2_{LL} & M^2_{LR} \\ M^2_{RL} & M^2_{RR} \end{pmatrix}, \tag{1}$$

where

$$\begin{aligned} M^2_{LL} &= m^\dagger_\ell m_\ell + M^2_L - \frac{1}{2}(2m^2_W - m^2_Z)\cos 2\beta\ I, \\ M^2_{RR} &= m^\dagger_\ell m_\ell + M^2_R - (m^2_Z - m^2_W)\cos 2\beta\ I, \\ M^2_{LR} &= (A^e - \mu \tan\beta)\ m_\ell, \\ M^2_{RL} &= (M^2_{LR})^\dagger. \end{aligned} \tag{2}$$

Here we parametrize trilinear soft supersymmetry-breaking terms as $A^e_{ij}(\lambda_e)_{ij}$, where the λ_e are the respective Yukawa couplings. For universal soft terms at some high input scale, one has

$$M_L^2 = M_R^2 = m_0^2 \, I, \quad A^e_{ij} = A_0 \delta_{ij}, \tag{3}$$

whereas flavour-mixing entries may be parametrized by:

$$\delta^{ij}_{XX} = (M^2_{XX})^{ij}/(M^2_{XX})^{ii} \qquad (X = L, R). \tag{4}$$

The evaluation of the LFV observables is done by performing the diagonalization of the slepton mass matrices (see [10], for instance), inserting the full rotation matrices in the lepton-slepton-gaugino vertices and summing over all the mass eigenstates of the exchanged particles.

2.2. LFV at he LHC.

At the LHC sleptons may appear in cascade decays of heavy squarks or gluinos, we pay particular attention to regions that lead to large values of $\Gamma(\chi_2 \rightarrow \chi + \tau^\pm + \mu^\mp)$ via the on-shell slepton production mechanism: $BR(\chi_2 \rightarrow \chi\tau^\pm\mu^\mp) = \sum_{i=1}^3 \left[BR(\chi_2 \rightarrow \tilde{l}_i\mu)BR(\tilde{l}_i \rightarrow \tau\chi) + BR(\chi_2 \rightarrow \tilde{l}_i\tau)BR(\tilde{l}_i \rightarrow \mu\chi)\right]$. The simulations performed in [9] indicate that this BR must be above 0.1 to allow the identification of LFV events. Furthermore, these ratios can only be achieved if the sleptons mass matrix contain large RR entries in a basis where the charged sleptons are diagonal.

The inclusion of a *see-saw* mechanism to explain neutrino masses may lead to LL mixings with interesting predictions for radiative decays like $\tau \rightarrow \mu\gamma$. However, the required LL mixing values to observe stau flavour oscillation will imply a large violation of this radiative decay. For instance, the cases presented in [9] show that in order to achieve a branching ratio for $\tilde{\chi}_2 \rightarrow \tilde{\chi}_1\tau^\pm\mu^\mp$ that is of interest for the LHC, we need $\delta_{RR} \sim 0.15$ for $\delta_{LL} = 0$ or $\delta_{LL} \sim 0.35$ for $\delta_{RR} = 0$. While the bound on BR($\tau \rightarrow \mu\gamma$) is very restrictive on the size of δ_{LL}, imposing a maximum value ~ 0.03 on it . These large mixings are difficult to generate in models with universal soft terms at hight energy scales, however they may be present in non- minimal GUT models as shown in ref.[11].

2.3. LFV at he LC.

Charged-lepton flavour violation at a LC may occur either directly in slepton pair production or indirectly via slepton production in cascade decays [12]. Processes leading to lepton production in the decays of a pair of sleptons include

$$\begin{aligned} e^+e^- &\rightarrow \tilde{\ell}^-_i\tilde{\ell}^+_j \rightarrow \tau^\pm\mu^\mp\tilde{\chi}^0_1\tilde{\chi}^0_1, \\ e^+e^- &\rightarrow \tilde{\nu}_i\tilde{\nu}^c_j \rightarrow \tau^\pm\mu^\mp\tilde{\chi}^+_1\tilde{\chi}^-_1, \end{aligned} \tag{5}$$

for which representative Feynman diagrams are shown in Fig. 1. Slepton production may also result from the the cascade decays of the heavier gauginos, e.g., via the processes

$$\begin{aligned} e^+e^- &\rightarrow \tilde{\chi}^\pm_2\tilde{\chi}^\mp_1 \rightarrow \tau^\pm\mu^\mp\tilde{\chi}^+_1\tilde{\chi}^-_1, \\ e^+e^- &\rightarrow \tilde{\chi}^0_2\tilde{\chi}^0_1 \rightarrow \tau^\pm\mu^\mp\tilde{\chi}^0_1\tilde{\chi}^0_1. \end{aligned} \tag{6}$$

Complete expressions for the respective cross sections are given in [13], and used in our work. The channel $e^+e^- \rightarrow \tau^\pm\mu^\mp + 2\chi^0$ was studied in [16], where it was found that a signal of $\sim$ 1 fb could be distinguished from the background, we will consider this level as a reference value in our study.

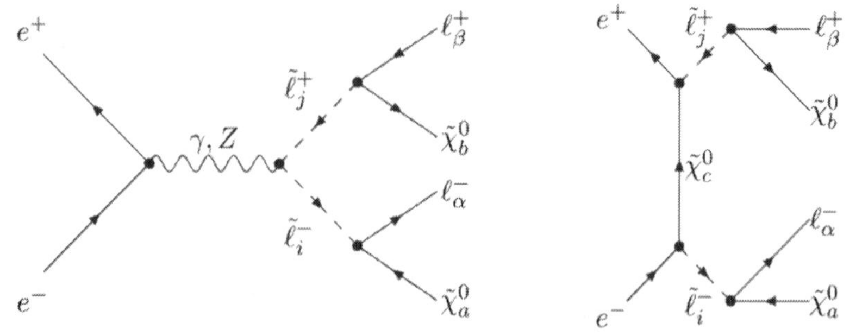

Figure 1. *Feynman diagrams for $e^+e^- \to \tilde{\ell}_j^+\tilde{\ell}_i^- \to \ell_\beta^+\ell_\alpha^-\tilde{\chi}_b^0\tilde{\chi}_a^0$. The arrows on scalar lines indicate the flow of lepton number. Similar diagrams - appropriately modified - exist for charginos.*

3. LFV observation at the LC.

We take regions of SUSY parameter space relevant from the cosmological point of view, in which sizable $\tilde{\tau} - \chi$ coannihilations or direct-channel H/A resonances lead to values of $\Omega_\chi h^2$ consistent with WMAP. As a representative of the coannihilation case, we chose $\tan\beta = 35$, a value that leads to a relatively heavy sparticle spectrum compatible with WMAP. At $\tan\beta = 45$ we found points in the resonance funnel, which have larger scalar masses since m_0 can take higher values than in the coannihilation area. We need a choice of flavour mixing parameters for a more detailed discussion of LFV cross sections. We assume the same values of the mixing parameters for both choices of $\tan\beta$, namely:

$$(\delta_{LL})_{13} = 0.02, \quad (\delta_{LL})_{23} = 0.02, \tag{7}$$

$$(\delta_{RR})_{13} = 0.04, \quad (\delta_{RR})_{23} = 0.15. \tag{8}$$

For completeness, we also introduce a small mixing between the first and second generation: $(\delta_{LL/RR})_{12} = 0.2 \cdot (\delta_{LL/RR})_{13}$. The LFV decays into $\tau - \mu$ pairs are not heavily dependent on this parameter while, with this choice, the bound on $\mathrm{BR}(\mu \to e\gamma)$ does not over-constrain the parameter space.

Cross sections for $e^+e^- \to \tau^\pm\mu^\mp + 2\chi^0$ above 1 fb can be reached in the areas of parameters shown in Fig. 2 for energies below those indicated. The large areas between the thick solid red, black and blue lines, on the one hand, and the shaded regions that are excluded by the indicated present experimental constraints, on the other hand, demonstrate that there are ample opportunities for LFV discovery and measurement at the LC. These opportunities are exemplified by the benchmark points marked by crosses in Fig. 2 and studied with more detail in ref. [8].

In Fig. 3 we present the maximum values of the cross sections in the areas allowed by all the current constraints. The shaded areas show the possible ranges of $\sigma(e^+e^- \to \tau^\pm\mu^\mp + 2\chi^0)$, and we note that points along the WMAP strips (solid lines) generally have high values of the cross sections. The dashed lines show the possible values of $\sigma(e^+e^- \to \tau^\pm e^\mp + 2\chi^0)$ along the WMAP strips, and we see that these cross sections may be of the same order of magnitude as those for $\mu - \tau$ pairs.

4. Conclusions

The LC enhances significantly the prospects of detecting LFV for heavy sparticle spectra, where flavour-violating rare decays and conversions are significantly suppressed. This allows probing an entirely different range of the flavour-violating parameters. Unlike the LHC of ref. [9], at the LC we can have significant LFV within the CMSSM via LL mixing.

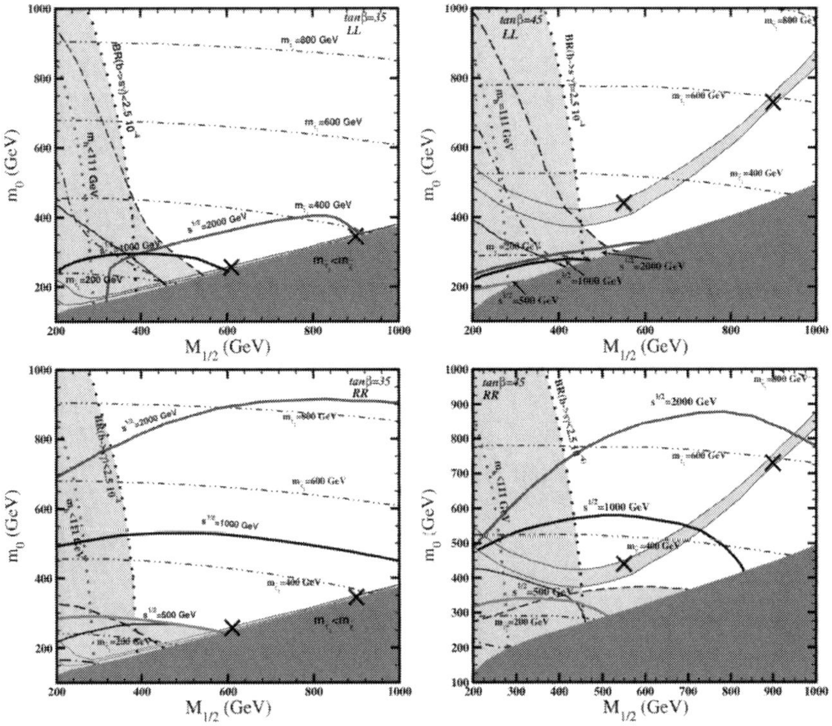

Figure 2. *The solid red (black) (blue) lines are contours where* $\sigma(e^+e^- \to \tau^\pm\mu^\mp + 2\chi^0) = 1$ *fb at* $\sqrt{s} = 500, 1000, 2000$ GeV, *assuming the left mixing parameters (7) in the upper panels and the right mixing (8) in the lower panels. The left panels are for for* $\tan\beta = 35$ *and the right panels for* $\tan\beta = 45$, *assuming* $A_0 = 0$ *in both cases. Each panel also shows the areas excluded by current bounds on* $BR(\mu \to e\gamma)$ *(thin solid line),* $BR(\tau \to e\gamma)$ *(thin dash line),* $BR(\tau \to \mu\gamma)$ *(thin dot-dash line). The areas excluded by* $BR(b \to s\gamma)$ *and the LEP Higgs search are also displayed, and the green area denotes the WMAP favored region. We see that there are ample opportunities for LFV discovery and measurement at the LC, and the benchmark points chosen for further studies are indicated by crosses.*

The expectations for models of massive neutrinos, in which quantum corrections provide a significant source of LFV in the *LL* channel, provide a potential link between LC observables and neutrino mass and mixing parameters. The cross sections expected in non-minimal extensions of the theory might not only enhance channels that in the current scheme are more suppressed, but could also enable a comparison of the allowed range of mixing parameters in different models.

Overall, it seems that the LC provides an optimal environment for the study of LFV, whereas the LHC is limited to specific channels that have significant backgrounds. The fact that the LC opens up additional possibilities may prove significant for making the link between observable cross sections and flavour model building.

Acknowledgments

This work is supported by the Spanish MCINN Consolider-Ingenio 2010 Programme under grant MultiDark CSD2009-00064, the project FPA2006-13825 and the project P07FQM02962 funded by "Junta de Andalucia".

References

[1] Y. Fukuda *et al.* [Super-Kamiokande Collaboration], Phys. Rev. Lett. **81** (1998) 1562.

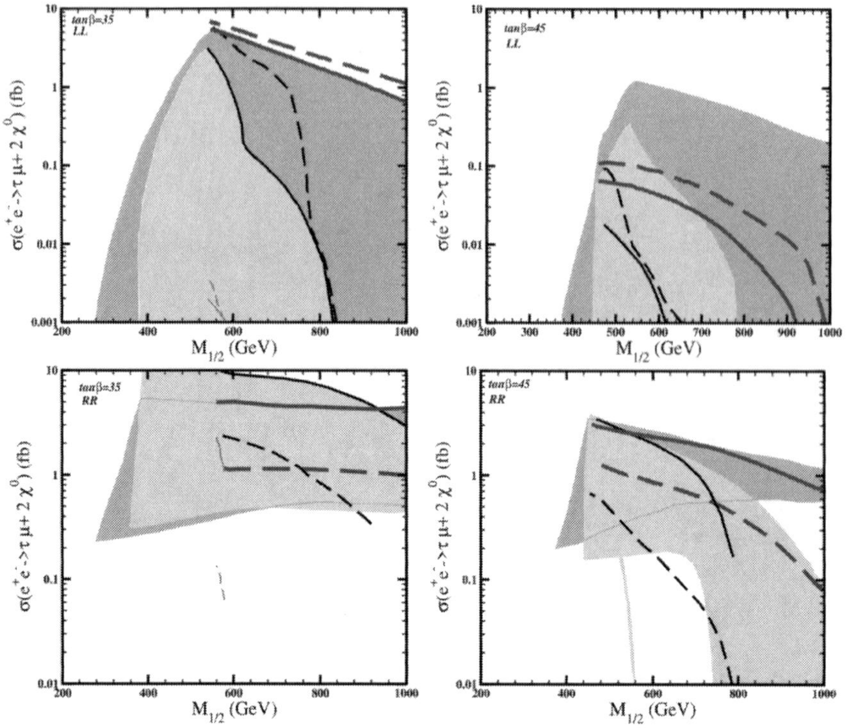

Figure 3. *The shaded areas show the possible ranges of $\sigma(e^+e^- \to \tau^\pm\mu^\mp + 2\chi^0)$ for $m_0 < 1000$ GeV, $\tan\beta = 35$ (left panels) and $\tan\beta = 45$ (right panels), assuming $A_0 = 0$ in both cases, with $\sqrt{s}$ fixed to 500 GeV (grey), 1000 GeV (green) and 2000 GeV (orange). In the upper panels, we assume the left mixing parameters (7), whereas in the lower panels we assume the right mixing (8). The solid lines present the possible values for models along the center of the WMAP strips, with the lines corresponding to $\sqrt{s}$=500, 1000, 2000 GeV being progressively thicker. The corresponding predictions for $\sigma(e^+e^- \to \tau^\pm e^\mp + 2\chi^0)$ are shown by the dashed lines.*

[2] Y. Fukuda *et al.* [Super-Kamiokande Collaboration], Phys. Rev. Lett. **82** (1999) 1810; Phys. Rev. Lett. **82** (1999) 2430; Q. R. Ahmad *et al.* [SNO Collaboration], Phys. Rev. Lett. **87** (2001) 071301.
[3] K. Eguchi et al., KamLAND Collaboration, Phys. Rev. Lett. **90** (2003) 021802; T. Araki et al., KamLAND Collaboration, Phys. Rev. Lett. **94** (2005) 081801.
[4] M.H. Ahn et al., K2K Collaboration, Phys. Rev. Lett. **90** (2003) 041801.
[5] D.G. Michael et al., MINOS colaboration, Phys. Rev. Lett. **97** (2006) 191801.
[6] Y. Kuno and Y. Okada, Rev. Mod. Phys. **73** (2001) 151.
[7] M. E. Gómez, G. K. Leontaris, S. Lola and J. D. Vergados, Phys. Rev. D **59** (1999) 116009; J. R. Ellis, M. E. Gómez, G. K. Leontaris, S. Lola and D. V. Nanopoulos, Eur. Phys. J. C **14** (2000) 319;
[8] E. Carquin, J. Ellis, M. E. Gomez and S. Lolab, JHEP **1111**, 050 (2011).
[9] E. Carquín, J. Ellis, M. E. Gómez, S. Lola and J. Rodríguez-Quintero, JHEP 0905 (2009) 026.
[10] J. Hisano, T. Moroi, K. Tobe and M. Yamaguchi, Phys. Rev. D **53** (1996) 2442.
[11] M. E. Gómez, S. Lola, P. Naranjo and J. Rodriguez-Quintero, JHEP **1006** (2010) 053.
[12] N. Arkani-Hamed, H. Cheng, J. L. Feng and L. J. Hall, Phys. Rev. Lett. **77** (1996) 1937; Nucl. Phys. B **505** (1997) 3.
[13] J. Hisano, M. M. Nojiri, Y. Shimizu and M. Tanaka, Phys. Rev. D **60**, 055008 (1999).
[14] M. Guchait, J. Kalinowski and P. Roy, Eur. Phys. J. C **21**, 163 (2001).
[15] F. Deppisch, J. Kalinowski, H. Pas, A. Redelbach and R. Ruckl, arXiv:hep-ph/0401243, work performed for the LHC study group.
[16] F. Deppisch, H. Pas, A. Redelbach, R. Ruckl, Y. Shimizu, Phys. Rev. D69 (2004) 054014.

Journal of Physics: Conference Series **384** (2012) 012018 doi:10.1088/1742-6596/384/1/012018

Cold Dark Matters in $U(1)'$ Models

Yuji Omura

School of Physics, KIAS, Seoul 130-722, Korea

E-mail: omura@kias.re.kr

Abstract. We consider models with extra $U(1)'$ gauge symmetry that is broken spontaneously. In the models, there are cold dark matter candidates which are charged under $U(1)'$ symmetry. Depending on the charge assignment, we can evade the strong bound from the Drell-Yang process, and consider the scenario that the dark matters strongly interact with the fermions in the Standard Model through the extra gauge boson exchanging. As an illustrative example, we discuss a supersymmetric gauged $U(1)_B \times U(1)_L$ model and see that not only correct relic density but also the DAMA/CoGeNT excess can be achieved by the dark matters.

1. Introduction

Many astrophysical observations support the existence of cold dark matter (CDM) in our universe, but its properties, such as its mass, spin, and interactions, are still largely unknown. Several decades of experimental work have been spent trying to detect dark matter particles coming to Earth (direct detection), or more generally products of dark matter reactions in outer space (indirect detection). In the both cases, very interesting signals, which might be originated by dark matters, were observed, but we could have not yet concluded that they are really the evidences of dark matters.

On the other hand, it is well known that the Standard Model (SM) describes the particle physics, up to the weak scale. Except for the Higgs sector, the SM has been investigated very well and we know that it can realize most of all the experimental results at high accuracy. However, one of the insufficient points of the SM might be dark matter physics. The SM does not include a heavy neutral stable particle which could be a CDM candidate, so that the extension of the SM may be required.

In this letter, we propose the extension to gauged $U(1)'$ symmetric models, motivated by CDM. In our models, the additional gauged symmetry is assigned to the SM fermions and extra chiral fermions are also added to satisfy the anomaly conditions. Some of the extra fermions become neutral and stable after the electro-weak symmetry breaking (EWSB). Furthermore, an extra scalar which is only charged under $U(1)'$ is required to avoid stable extra quarks. It also becomes stable because of the remnant global symmetry after $U(1)'$ symmetry breaking, and interacts with the SM fermions through the $U(1)'$ gauge boson exchanging. When we consider the extension of our models to the supersymmetric, we could realize another CDM candidates such as the superpartner of the additional scalar and neutralino.

This letter is devoted as follows. In the next section, we introduce our $U(1)'$ models, where the SM particles are charged under $U(1)'$ symmetry and extra fermions are added for the anomaly conditions. We also see that some of the extra particles become neutral and good CDM candidates in the section 3. The CDMs interact with the SM particles through the new gauge boson exchanging, and could achieve the correct relic density and predict the signal of direct detections. As an illustrative model, we consider a supersymmetric gauged $U(1)_B \times U(1)_L$ model [1] in the section 4, and estimate the thermal relic density and the cross section of direct detection with proton.

2. Setups of our models

We start with $U(1)'$ gauge models with the following flavor-dependent charge assignments:

	$SU(3)_c$	$SU(2)_L$	$U(1)_Y$	$U(1)'$
Q_i	3	2	1/6	q_i
D_{Ri}	3	1	$-1/3$	d_i
U_{Ri}	3	1	2/3	u_i
L_i	1	2	$-1/2$	l_i
E_{Ri}	1	1	-1	e_i
H	1	2	1/2	q_h
Φ	1	1	0	q_Φ

with $Q_i^T = (U_{Li}, D_{Li})$, $L_i^T = (\nu_{Li}, E_{Li})$ and $i = 1, 2, 3$ are generation indices. Φ is a SM-gauge singlet with nonzero $U(1)'$ charge which is required to break $U(1)'$ spontaneously and generate nonzero masses for Z' and extra fermions.

The flavor-dependent $U(1)'$ models generally become anomalous without extra chiral fields. One of the simplest ways to construct anomaly-free theory is to add one extra generation and two SM gauge vector-like pairs as follows:

	$SU(3)_c$	$SU(2)_L$	$U(1)_Y$	$U(1)'$
Q'_R	3	2	1/6	$(q_1 + q_2 + q_3)$
D'_L	3	1	$-1/3$	$(d_1 + d_2 + d_3)$
U'_L	3	1	2/3	$(u_1 + u_2 + u_3)$
L'_R	1	2	$-1/2$	$(l_1 + l_2 + l_3)$
E'_L	1	1	-1	$(e_1 + e_2 + e_3)$
l_{L1}	1	2	$-1/2$	Q_L
l_{R1}	1	2	$-1/2$	Q_R
l_{L2}	1	2	$-1/2$	$-Q_L$
l_{R2}	1	2	$-1/2$	$-Q_R$

Q'_R, D'_L, U'_L, L'_R, and E'_L are like the extra SM fermions of the fourth generation and their chiralities are opposite. The $U(1)'$ charges are defined to satisfy the conditions for $U(1)'$, $U(1)'SU(2)_L^2$, $U(1)'SU(3)_c^2$ and $U(1)'U(1)_Y^2$. We just assume that they also satisfy the condition for $U(1)'^3$, but there might be not so many simple solutions, if one requires rational numbers for all the $U(1)'$ charges. $Q_{L,R}$ contribute to only $U(1)_Y U(1)'^2$, and we can easily find the solution with $Q_L - Q_R = 1$, as discussed in Ref. [2, 3].

One can replace the $SU(2)_L$ doublets (l_{LI}, l_{RI}) $(I = 1, 2)$ with $SU(3)_c$ triplets, whose $U(1)_Y$ charges are $-1/3$ and $U(1)'$ charges are the same. One can also replace them with fields charged or not charged under $SU(3)_c \times SU(2)_L$. As discused in the section 4, we can also consider the case with $q_i = d_i = u_i = 1/3$ and $l_i = e_i = 0$ where the anomaly condition for $U(1)_Y U(1)'^2$ is satisfied without the SM gauge vector-like pairs.

We also have constraints on the $U(1)'$ charges from the Yukawa couplings which generate masses of the SM fermions and extra fermions. Furthermore, our charge assignment may allow the mixing between the SM fermions and the extra fermions, which causes the flavor-changing neutral currents (FCNCs). For example, in the case that q_h, l_i and e_i are set to 0, we can generally write down the mixing terms between SM leptons and extra leptons, such as $m_i \overline{L'_R} L^i$. We assume that such tree-level mixing terms are controlled to avoid the FCNC problem.

3. Cold dark matter

As we discuss in the above, we require the extra fermions, depending on the charge assignment. In this section, we comment on the possibility that the required chiral fields give rise to CDM candidates, based on Ref. [2, 3].

First, let us focus on models with $SU(2)_L$ doublet pairs, (l_{LI}, l_{RI}). l_{LI} and l_{RI} are chiral fields charged under $SU(2)_L \times U(1)_Y$ like the left-handed lepton, so we can expect one component to be charged and the other to be neutral after EWSB. The Yukawa couplings corresponding to the mass terms are given by the extra scalar, Φ with $U(1)'$ charge, q_Φ. When q_Φ is set to 1, mass terms of(l_{LI}, l_{RI}) (or (q_{LI}, q_{RI})) are linear to the $\langle\Phi\rangle$, because the solution with $Q_L - Q_R = 1$ is found in any cases.

l_{LI} and l_{RI} do not have Yukawa couplings with the SM fermions because of $U(1)'$ symmetry, so that $U(1)$ global symmetry, which is phase rotation of l_{LI} and l_{RI}, could be assigned, and their stabilities are guaranteed by the global symmetry. After EWSB, one component is charged and the other is neutral. Their masses are degenerate before EWSB, but especially the charged particles get enough large radiative corrections and then heavier than the neutral. If the mass split is bigger than the electron mass, the charged can decay to the neutral, e, and ν_e through the weak interaction. We conclude that the light neutral components could be good CDM candidates.

The CDM can annihilate to 2 light SM fermions through Z' and Z boson exchanging, and 2 gauge boson in the t-channel, so that the thermal relic density could be achieved by the gauge interactions. The mass of the CDM is constrained by the search for extra leptons, that is, it should be heavier than 100 GeV, where the direct searches for dark matters shows negative results. Even if we consider the scenario that Z' interaction is negligible in the direct scattering, Z boson exchanging will work at the higher order. Such heavy CDM scenario is not favored by the direct searches, such as XENON100 [4].

Next, we discuss the case with extra $SU(3)_c$ triplets, (q_{LI}, q_{RI}) or only the extra family, such as Q'_R, U'_L, and D'_L. The mass terms of (q_{LI}, q_{RI}) are give by Φ like the masses of l_{LI} and l_{RI}. The masses of the extra family are given by the $SU(2)_L$ doublet Higgs like the SM fermions. Those extra colored particles would be stable because of the $U(1)'$ symmetry and the chirality. In order to allow the decay of the extra colored particles, we introduce mixing term between the extra quarks and the SM quarks according to adding X, where X is a SM gauge-singlet scalar with $U(1)'$ charge. X has the Yukawa couplings, such as $\lambda_i X^\dagger \overline{D_{Ri}} q_{L1}$, and the extra quarks could decay to X and the SM quarks if X is lighter that the extra quarks. If X does not get nonzero vev, we can avoid tree-level FCNC and expect X to be stable because of the remnant global symmetry after $U(1)'$ breaking. This type of dark matter has been well investigated in Ref. [5, 6, 7], and we discuss the dark matter physics in the gauged $U(1)_B (\equiv U(1)')$ model, in the next section.

4. Supersymmetric Gauged $U(1)_B \times U(1)_L$ Model

As an illustrative model, we discuss the supersymmetric gauged $U(1)_B \times U(1)_L$ model, which has been proposed in Ref. [1]. The gauged $U(1)_B$ is anomalous in the $SU(2)_L^2 U(1)_B$ and the $U(1)_Y^2 U(1)_B$ products, so that the extra generation, which was introduced in Sec. 2, is required. Based on the argument in Sec. 3, the extra scalar, X, with nonzero $U(1)_B$ charge should be added and X will be a good CDM candidate [5]. If we consider the extension to the supersymmetric, we can realize several CDM candidates, such as X, the superpartner of X, $\tilde{X}$, and the neutralino, because of the remnant global $U(1)$ symmetry, which is originated by the gauged $U(1)_B$, and the R-parity, which could be realized spontaneously after $U(1)_B \times U(1)_L$ breaking [1]. The CDMs, X and $\widetilde{X}$, could annihilate to the SM particles through the $U(1)_B$ gauge boson (Z_B) exchanging, and the new gauge interaction could be strong, compared with the well-known gauge symmetry, such as $U(1)_{B-L}$. This is because such leptophobic charge assignment can evade the strong bound from the Drell-Yang process, assuming that the kinetic mixing is enough small. In the following, we discuss the dark matter physics and introduce the constraints on the leptophobic gauge interaction.

In this model, X and $\widetilde{X}$ can annihilate to the SM quarks though the Z_B exchanging. If the Yukawa couplings with the extra quarks and the SM quarks are sizable, the CDMs can also annihilate in the t-channel. However, the Yukawa couplings should be controlled to avoid the FCNCs, so that we simply assume that the Yukawa couplings are enough small, compared with the gauge interaction.

Depending on the mass spectrum of X and $\widetilde{X}$, we can describe the each scenario: X or $\widetilde{X}$ is a CDM. The thermal relic density is given by the annihilation cross section of the CDM to the SM particles, and it is well known that the cross section should be $\langle \sigma v \rangle \sim$ 1pb to explain the WMAP data. X and $\widetilde{X}$ annihilate to the SM quarks through the Z_B exchanging: the cross section, σv, could be estimated as $\sigma v \sim v^2 m_X^2 g_B^4 / M_{Z_B}^4$ in the scalar case and $\sigma v \sim m_X^2 g_B^4 / M_{Z_B}^4$ in the Dirac case, if the each CDM mass is far from the resonance, $m_X \sim M_{Z_B}/2$. On the other hand, the experiments of the direct detections succeed in limiting the scattering cross section of the CDM with nuclei. For example, if m_X is $O(100)$GeV, the spin-independent cross section, σ_{SI}, should be smaller than 10^{-44}cm^2 [4]. Now σ_{SI} is also linear to $g_B^4/M_{Z_B}^4$, so that $\langle \sigma v \rangle$ will be less than 1pb, which induce the overproduction of the CDM. If m_X is lower than the energy threshold of the direct detections, the upper bound on g_B/M_{Z_B} would be relaxed. In Ref. [9, 10], we discussed the scenario that $m_X \sim O(1)$GeV and $M_{Z_B} \sim$ 150GeV, motivated by the Wjj excess which was reported by the CDF collaboration [8]. In Fig. 1, the dotted pink line also corresponds to the $M_{Z_B} =$ 150GeV case with $\Omega_X h^2 = 0.1123$. The CRESST experiment [11] gives the strongest bound and the bound is just below the dotted pink line. Furthermore, it has been pointed out that the mono-jet signals, $p\bar{p} \to g, Z_B \to g, X, X^\dagger(\widetilde{X}, \overline{\widetilde{X}})$, strongly constrain the heavy Z_B scenario [10]. In fact, the bound seems to appear at the order, $\sigma_{SI} \sim 10^{-38}\text{cm}^2$ in Fig. 1, which totally cover the dotted-pink line [12].

On the other had, we could consider the scenario with very small M_{Z_B} and small g_B, because of the leptophobic charge assignment. In this case, the mono-jet bound could be relaxed because of the small gauge coupling. As discussed in Ref. [1, 9], the strongest bounds come from the invisible and hadronic decay widths of the Υ meson [13, 14, 15]. Contour lines of $\Omega_X h^2 = 0.1123$ in the m_X-σ_{Xp} plane are shown in Fig. 1 for light Z_B. The parameter α' changes along each line. Below each line, one has $\Omega_X h^2 > 0.1123$. The thick red and purple contours correspond to m_{Z_B} =12 GeV/c^2 and 20 GeV/c^2, respectively. Each contour shows a dip at $m_X = m_{Z_B}/2$ due to the annihilation through the Z_B resonance.

Even in this case, the contour lines are just above the upper bound from the CRESST experiment, except for the resonance. On the other hand, around the resonance, we see that the contour lines sweep the DAMA/CoGeNT region [16] for Z_B masses in the range $\sim$ 1 to $\sim$ 20 GeV/c^2, touching the DAMA/COGeNT region on the left at the lowest m_{Z_B} and on the right at highest m_{Z_B}.

5. Summary

We discussed $U(1)'$ models, especially motivated by CDM. Most of $U(1)'$ models, where the SM fermions are charged under the extra $U(1)'$ symmetry, are anomalous without extra chiral fermions. We introduced one extra generation with the opposite chirality, two SM vector-like pairs and scalars to achieve all anomaly conditions and to avoid conflicts with experiments. We

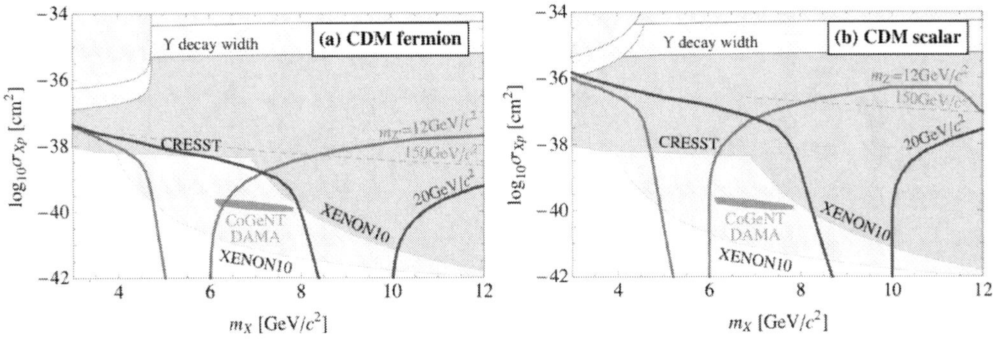

Figure 1. Contour lines of $\Omega_X h^2 = 0.1123$ for several values of the Z_B boson mass m_{Z_B} [9]. On each contour, the cosmic density of particles X (fermions in panel (a), and scalars in panel (b)) equals the cosmic density of CDM. Also shown are the DAMA/CoGeNT region (in orange), direct detection constraints (in blue) [17], and accelerator constraints (in yellow).

found that the neutral particles, which are included in the extra matters, become good CDM candidates because of the remnant global symmetry. As an illustrative model, we discussed the supersymmetric gauged $U(1)_B \times U(1)_L$ model, and investigated the relic density and the direct detection. Thanks to the leptophobic charge assignment, we can evade the strong bounds at LEP and Tevatron. If the Z_B is heavy, the bound from the mono-jet signal may constrain the interaction of CDM strongly. We also discussed the light Z_B scenario, and investigated how we can achieve the DAMA/CoGeNT signal. We have to require the fine tuning, $m_{Z_B} \sim 2m_X$, but we can surely evade the bounds from collider experiments.

Acknowledgments
I am grateful to P. Gondolo, P. Ko, and Chaehyun Yu for collaborations.

References

[1] Ko P and Omura Y 2011 *Phys. Lett.* B **701** 363
[2] Ko P, Omura Y and Yu C 2011 *Preprint*: 1108.4005 [hep-ph] *to be published in JHEP*
[3] Ko P, Omura Y and Yu C *Preprint* 1108.0350 [hep-ph]
[4] Aprile E *et al.* [XENON100 Collaboration] 2011 *Phys. Rev. Lett.* **107** 131302
[5] Perez P F and Wise M B 2010 *Phys. Rev.* D **82** 011901; *Erratum-ibid.* D **82** 079901
[6] Dulaney T R, Perez P F and Wise M B 2011 *Phys. Rev.* D **83**, 023520
[7] Alwall J, Feng J L, Kumar J and Su S 2010 *Phys. Rev.* D **81** 114027
[8] Aaltonen T *et al.* [CDF Collaboration] 2011 *Phys. Rev. Lett.* **106** 171801
[9] Gondolo P, Ko P and Omura Y *Preprint* 1106.0885 [hep-ph]
[10] Ko P, Omura Y and Yu C *Preprint* 1104.4066 [hep-ph]
[11] Angloher G *et al.* 2002 *Astropart. Phys.* **18** 43
[12] Bai Y, Fox P J and Harnik R 2010 *JHEP* **1012** 048
[13] Carone C D and Murayama H 1995 *Phys. Rev. Lett.* **74** 3122
[14] Aranda A and Carone C D 1998 *Phys. Lett.* B **443** 352
[15] Fayet P 2010 *Phys. Rev.* D **81** 054025
[16] Hooper D, Collar J I, Hall J and McKinsey D 2010 *Phys. Rev.* D **82** 123509
[17] Angle J *et al.* [XENON10 Collaboration] 2011 *Phys. Rev. Lett.* **107**, 051301

Flavored Dark Matter and its Laboratory Tests

Steve Blanchet

Instituto de Física Teórica, IFT-UAM/CSIC Nicolas Cabrera 15, UAM Cantoblanco, 28049 Madrid — Spain

Institut de Théorie des Phénomènes Physiques, École Polytechnique Fédérale de Lausanne, CH-1015 Lausanne — Switzerland

E-mail: steve.blanchet@epfl.ch

Abstract.
In this talk, based on the paper 1109.3516, we consider theories where the dark matter particle carries lepton flavor quantum numbers, and has renormalizable contact interactions with the Standard Model fields. We find that the region of parameter space where dark matter has the right abundance to be a thermal relic is in general within reach of current direct detection experiments. In order to evaluate the collider prospects, we focus on a class of models where dark matter carries tau flavor, and show that the collider signals of these models include events with four or more isolated leptons and missing energy. A significant part of parameter space in these theories can be discovered above Standard Model backgrounds at the 14 TeV Large Hadron Collider. We also study the extent to which flavor and charge correlations among the final state leptons allows models of this type to be distinguished from theories where dark matter couples to leptons but does not carry flavor, similarly to a neutralino.

1. Introduction

It is now well established that about 80% of the matter in the universe is dark [1]. However, the masses and interactions of the particles of which dark matter is composed are not known. One simple and well-motivated possibility is that dark matter is made up of particles with masses close to the weak scale that have weak scale annihilation cross section to Standard Model (SM) particles. Dark matter candidates with these properties neatly fit into the 'Weakly Interacting Massive Particle' (WIMP) paradigm, and therefore naturally tend to have the right relic abundance to explain observations.

The matter fields (Q, U^c, D^c, L, E^c) of the SM each come in three copies, or flavors, that differ only in their masses. This reflects the fact that the Lagrangian of the SM possesses an approximate $\mathrm{U}(3)^5$ flavor symmetry acting on the matter fields, which is explicitly broken by the Yukawa couplings that generate the quark and lepton masses. An interesting possibility is that the dark matter field, which we label by χ, also carries flavor quantum numbers, with the physical dark matter particle being the lightest of three copies. Several specific dark matter candidates of this type have been studied extensively in the literature, including sneutrino dark matter [2, 3] in the Minimal Supersymmetric Standard Model (MSSM), and Kaluza-Klein (KK) neutrino dark matter [4] in extra dimensional models.

To incorporate three flavors of the dark matter field, the flavor symmetry of the SM is extended from $\mathrm{U}(3)^5$ to $\mathrm{U}(3)^5\times\, \mathrm{U}(3)_\chi$, if χ is a complex field such as a complex scalar, Dirac fermion or complex vector boson. If instead χ is a real field, such as a real scalar, Majorana fermion or real vector boson the flavor symmetry is extended from $U(3)^5$ to $U(3)^5 \times O(3)_\chi$. The new flavor symmetry $U(3)_\chi$ (or $O(3)_\chi$) may be exact, or it may be explicitly broken as in the SM.

Our focus here will be on theories where dark matter has renormalizable contact interactions with the SM fields, which, by analogy with the SM Yukawa couplings, could represent an explicit breaking of the flavor symmetry. Then, if the SM matter field that χ couples to is a lepton, there is an association between the different dark matter flavors and lepton flavors. Accordingly, we refer to this scenario as 'lepton flavored dark matter'. In this scenario the fact that the SM flavor symmetries are not exact naturally results in a splitting of the states in the dark matter multiplet, the physical dark matter particle being identified with the lightest. A more extensive discussion including the case where dark matter carries quark flavor or an internal flavor can be found in [5].

2. Lepton Flavored Dark Matter

The lepton sector of the SM has a $U(3)_L \times U(3)_E$ flavor symmetry, where $U(3)_L$ acts on the SU(2) doublet leptons and $U(3)_E$ on the singlets. This symmetry is explicitly broken down to $U(1)^3$ by the Yukawa interactions that give the charged leptons their masses. (We neglect the tiny neutrino masses, which also break the symmetry).

The characteristic vertex of lepton flavored dark matter involves contact interactions between χ and the SM leptons. For concreteness, in what follows we take χ to be a Dirac fermion and ϕ to be a complex scalar, and restrict our focus to the case where χ couples to the SU(2) singlet lepton field E^c, such that the corresponding term in the Lagrangian is

$$\lambda_\alpha{}^i \chi^\alpha E_i^c \phi \;\; + \text{h.c.}, \tag{1}$$

where i is a $U(3)_E$ flavor index and α is a $U(3)_\chi$ flavor index.

2.1. Flavor Structure

In general the matrix λ will contain both diagonal and off-diagonal elements, thereby giving rise to lepton flavor violation. The experimental bounds on such processes are satisfied if all the elements in $\lambda < 10^{-3}$ for $m_\phi \sim 200$ GeV, even in the absence of any special flavor structure. Unfortunately, however, if χ is a SM singlet and has no sizable couplings beyond those in Eq. (1), couplings of this size are by themselves too small to generate the correct abundance for χ, if it is to be a thermal relic, except if they are aligned with the lepton Yukawa couplings to avoid flavor bounds.

The matrix λ can naturally be aligned with the SM Yukawa couplings if this interaction is consistent with minimal flavor violation (MFV) [6]. In this scenario, the dark matter flavor symmetry $U(3)_\chi$ is identified with either $U(3)_E$ or $U(3)_L$ of the SM, and the matrix λ respects these symmetries up to effects arising from the SM Yukawa couplings.

If we write the lepton Yukawa couplings of the SM as $y_A{}^i L^A E_i^c H + \text{h.c.}$, then the Yukawa matrix $y_A{}^i$ can be thought of as a spurion transforming as $(3, \bar{3})$ under the $SU(3)_L \times SU(3)_E$ subgroup of $U(3)_L \times U(3)_E$. Consider the case where $U(3)_\chi$ is identified with $U(3)_L$. Then

$$\lambda_\alpha{}^i \chi^\alpha E_i^c \phi \;\; + \text{h.c.} \rightarrow \lambda_A{}^i \chi^A E_i^c \phi \;\; + \text{h.c.} \tag{2}$$

MFV restricts the matrix λ to be of the form $\lambda_A{}^i = \alpha \; y_A{}^i$, where we are working only to the leading non-trivial order in an expansion in the SM Yukawa couplings. The dark matter mass term now takes the form

$$[m_\chi]_A{}^B = \left(m_0 \mathbb{1} + \Delta m \; y y^\dagger \right)_A{}^B . \tag{3}$$

Journal of Physics: Conference Series **384** (2012) 012019 doi:10.1088/1742-6596/384/1/012019

We see that in this case the three dark matter flavors are close in mass, but their couplings to the SM fields, though still flavor-diagonal, are hierarchical. In particular, if the relic abundance is determined by λ, we expect that only the tau flavor can constitute dark matter, since the couplings of the other flavors are relatively small. The other case where $\mathrm{U}(3)_\chi$ is identified with $\mathrm{U}(3)_E$ can be worked out in a similar fashion [5].

2.2. Relic Abundance

If χ is a thermal WIMP, its relic abundance is set by its annihilation rate to SM fields. If χ is a SM singlet, and its only interactions are those of Eq. (1), then the primary annihilation mode is through t-channel ϕ exchange to two leptons. In the relevant parameter space, the matrix λ is constrained by flavor bounds to be very nearly flavor diagonal, so that lightest state in the dark matter multiplet is associated with a specific lepton flavor. We assume that the splittings between the different states in this multiplet are large enough so that only the lightest state is stable on cosmological timescales, and constitutes all of dark mater.

Since the dark matter particle is non-relativistic at freeze-out, annihilation is dominated by the lowest partial wave. In this limit

$$\langle \sigma v \rangle = \frac{\lambda^4 m_\chi^2}{32\pi (m_\chi^2 + m_\phi^2)^2}, \tag{4}$$

where we have assumed that $m_\chi \gg m_\ell$, so that the masses of the final state leptons can be neglected.

2.3. Direct Detection

The direct detection signals of this class of theories depend on whether the dark matter particle χ transforms non-trivially under the SM SU(2) gauge symmetry, or remains a SM singlet. If χ is a SM singlet, the leading contribution to dark matter scattering off a nucleus arises from the (triangle) loop diagrams involving leptons and a photon exchange.

In the region of parameter space of interest to current direct detection experiments, bounds on lepton flavor violating processes constrain the coupling matrix λ to be flavor diagonal. Therefore the dark matter candidate carries the flavor of the lepton it couples to. This coupling gives rise to three distinct types of interactions between dark matter and the nucleus, specifically a charge-charge coupling, a dipole-charge coupling, and a dipole-dipole coupling.

The dipole-charge interaction is sub-dominant to the charge-charge interaction. The dipole-dipole coupling, being spin-dependent, is also sub-dominant. Consequently, we use the charge-charge cross section to derive limits. We find

$$\sigma_{ZZ}^0 = \frac{\mu^2 Z^2}{\pi}\left[\frac{\lambda^2 e^2}{64\pi^2 m_\phi^2}\left[1 + \frac{2}{3}\log\left(\frac{m_\ell^2}{m_\phi^2}\right)\right]\right]^2. \tag{5}$$

Here Z is the total charge of the nucleus, σ_0 is the cross section at zero-momentum transfer, and μ is the reduced mass of the dark matter-nucleus system.

The ratio λ/m_ϕ corresponding to a thermal WIMP is plotted in Fig. 1 as a function of the dark matter mass, for the tau flavored and electron flavored cases. The current limits from the Xenon100 experiment [7] are also shown. It is clear from the figure that the expected improvement in sensitivity of the experiment by an order of magnitude will bring a large part of the parameter space of these models within reach.

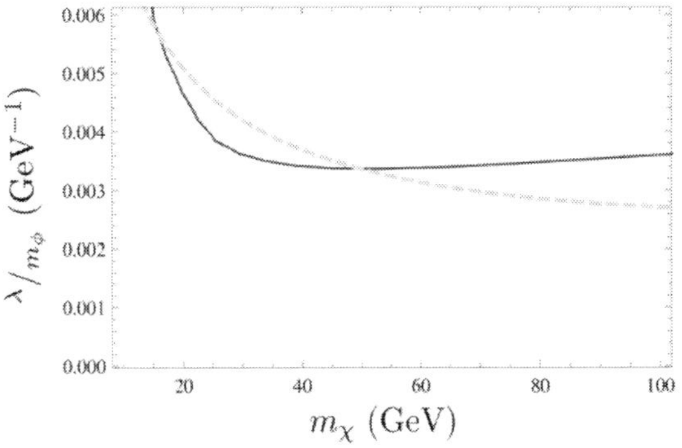

(a)

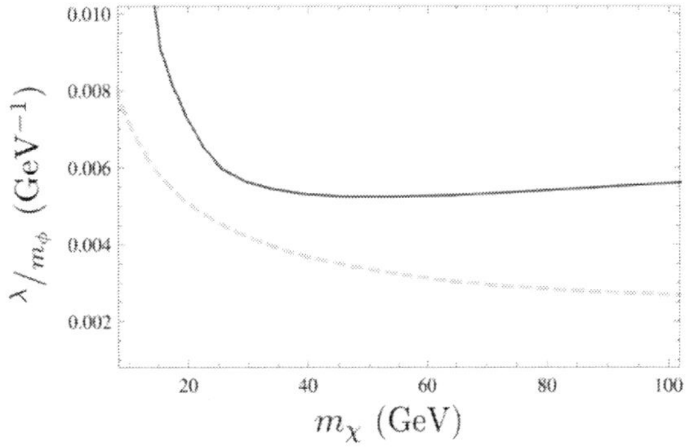

(b)

Figure 1: Direct detection and relic abundance constraints on lepton flavor dark matter for a)χ_e and b)χ_τ, when $m_\phi = 150$ GeV. The area above the solid blue curve is ruled out by the new Xenon100 [7] data. The green dashed curves signify the parameters for which we obtain correct relic abundance.

3. Collider Signals of Tau Flavored Dark Matter

In this section we discuss in detail the collider signals of a specific model where the dark matter particle carries tau flavor. The relevant part of the Lagrangian was given in Eq. (1). This interaction fixes the SM quantum numbers of ϕ, which is charged under the photon and the Z, but does not couple to the W.

For concreteness, we consider two benchmark spectra that are consistent with MFV, with χ transforming under $U(3)_E$. Then χ_e and χ_μ are expected to be nearly degenerate since the corresponding SM Yukawa couplings are very small. We assume that χ_τ is lighter than χ_e or χ_μ, and constitutes dark matter.

We label the first benchmark spectrum τFDM1, where $m_{\chi,e} = m_{\chi,\mu} = 110$ GeV, $m_{\chi,\tau} = 90$ GeV and $m_\phi = 160$ GeV. The second benchmark spectrum we study has a lighter mediator, and therefore leads to a larger production cross section. We label this benchmark spectrum τFDM2, where $m_{\chi,e} = m_{\chi,\mu} = 90$ GeV, $m_{\chi,\tau} = 70$ GeV and $m_\phi = 150$ GeV.

In these simple models, only the mediator ϕ carries SM gauge quantum numbers, so dark matter events at colliders must arise from $\phi^+\phi^-$ production. Since $\phi^+\phi^-$ production proceeds through Drell-Yan and ϕ is a scalar, the ϕ pair comes out in a p-wave, leading to a small cross section, on the order of 10 fb at the 14 TeV LHC run. Therefore, we do not expect the early LHC data to be able to probe this model. In order to obtain reasonable signal over background discrimination, tens of inverse fb of data will be required.

3.1. Signal topologies

Signal events come in three distinct topologies. Each ϕ can decay directly into the dark matter particle and a τ, corresponding to a short chain. Alternatively, it can decay to one of the heavier particles in the dark matter multiplet, which eventually cascades down to the dark matter particle, creating a long chain. Therefore each event can be categorized as comprising of short-short, short-long or long-long chains.

Since τ's are difficult to identify, we implicitly restrict ourselves to $\ell = e, \mu$ final states in this section when we talk about leptons. The events with the long-long decay chain topology have four-lepton final states (not to mention a pair of τ's), which have small SM backgrounds, and therefore it is the most promising channel.

In order to simulate signal and background events we use the usrmod utility of

MadGraph/MadEvent [8, 9], and we use BRIDGE [10] for the $\chi_{e,\mu}$ decays. Pythia [11] is used to simulate parton showers and hadronic physics, and PGS [12] with the default CMS parameter set is used to simulate detector effects.

3.2. Backgrounds

While four-lepton final states are rare in the SM, the signal cross section is also small so we carefully consider the three leading sources of backgrounds and devise cuts to reduce them as much as possible.

$(Z/\gamma)^{(*)}(Z/\gamma)^{(*)}$: One of the dominant backgrounds is the production of two opposite-sign, same-flavor lepton pairs from either on-shell or off-shell Z's and photons. Any missing energy in this background arises from mis-measurement of lepton momenta, which is small.

$t\bar{t}(Z/\gamma)^{(*)}$: This background process, while it has a three-body final state, has a cross section comparable to the above process which is purely electroweak. When both tops decay leptonically and the $(Z/\gamma)^{(*)}$ goes to leptons, the final state is 4ℓ+jets+MET.

$WW(Z/\gamma)^{(*)}$: This process is qualitatively similar to the above process, but has a much smaller production cross section because it is purely electroweak.

Backgrounds with fakes : There are also backgrounds arising from jets that are misidentified as leptons. We find that provided the fake rates are of order 10^{-3} or less, the irreducible backgrounds described above are the dominant ones.

3.3. Cuts

We use the following cut flow in order to maximize signal over background:

- Lepton cuts - We demand events with at least four leptons each with $pT > 7$ GeV. At least two of these leptons are further required to have $E > 50$ GeV.
- Dijet veto - We discard events with two or more jets of $pT > 30$ GeV each.
- Z veto - We veto events if the invariant mass of any Z-candidate (a pair of same-flavor and opposite-charge leptons) falls within 7 GeV of the Z mass. This is a tighter Z-veto than is usually used, but we find that the loss in signal efficiency is more than compensated for by the background reduction.
- Missing energy - We require at least 20 GeV of missing energy in each signal event. Since most backgrounds with high MET have already been eliminated by the previous cuts in the cut flow, we find that a mild threshold such as 20 GeV is sufficient.

3.4. Results

The signal and background events of each type that survive these cuts are listed in Table 1. These results show that it is possible to discover the τFDM2 benchmark above SM backgrounds at 5σ significance with about 20 fb^{-1} of data at the 14 TeV LHC run. A higher luminosity (~ 40 fb^{-1}) would be needed in order to distinguish the τFDM1 benchmark from the SM background.

While ATLAS and CMS have already performed searches in multilepton final states, considering the low cross section of the FDM benchmark model, they are not yet expected to have exclusion level sensitivity to this scenario.

4. Distinguishing τFDM

Multi-lepton events with large missing energy are fairly common signals in theories with neutral stable particles and partners to the SM leptons, which include a variety of dark matter models. We would to like to understand whether it is possible to distinguish τFDM from a specific 'strawman' model, where the dark matter does not carry flavor.

Journal of Physics: Conference Series **384** (2012) 012019 doi:10.1088/1742-6596/384/1/012019

Dataset	Event rate after cuts at 100 fb^{-1}			
	Lepton cuts	Jet cuts	Z veto	MET
τFDM1	46.73	42.83	38.41	35.01
τFDM2	75.39	69.30	63.26	57.04
$\ell^+\ell^-\ell^+\ell^-$	1617.94	1582.42	140.30	13.32
$t\bar{t}\,\ell^+\ell^-$	89.57	19.45	4.92	4.70
$WW\ell^+\ell^-$	14.70	13.98	2.51	2.51

Table 1: Signal and SM background event rates for processes yielding 4-lepton final states after each set of cuts is progressively applied (note that $\ell = e, \mu$). All numbers are reported for the 14 TeV LHC run and include detector effects.

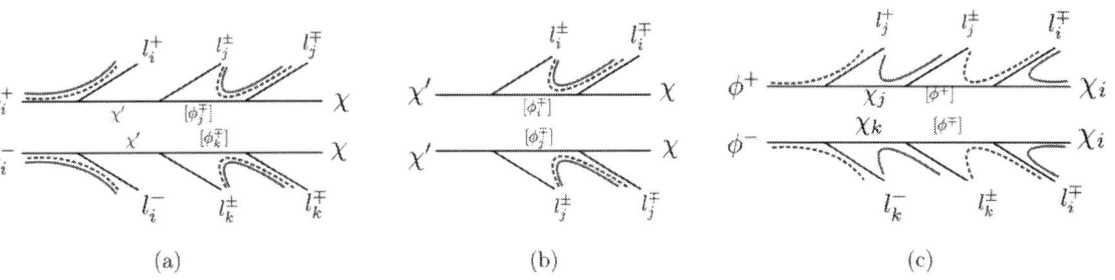

Figure 2: Flavor (red solid) and charge (blue dashed) correlations are shown for topologies in strawman models [(a) and (b)], and for τFDM (c).

The strawman model we choose is related to supersymmetric theories where the bino constitutes dark matter. The form of the lepton-slepton-bino vertex is very similar to the defining vertex of a theory of lepton FDM, except that in the supersymmetric case it is the slepton that carries flavor, not the bino. The strawman model we choose therefore consists of the bino, which we label by χ, along with the three right-handed sleptons, $\widetilde{E}^c_i$. The bino constitutes dark matter. To mimic the collider signals of τFDM, we add to the strawman model an additional 'neutralino' χ', which is heavier than the bino. χ' is an admixture of a SM SU(2) doublet and singlet, so that it can be pair-produced through the Z, and is chosen to couple to leptons and sleptons in a flavor-blind way. This interaction takes the schematic form $\lambda' E^c_i \chi' \widetilde{E}^{c\,i} + \text{h.c.}$

For simplicity, in what follows we assume that the three sleptons are degenerate in mass. In general, χ' could either be lighter than or heavier than the sleptons, while the bino is the lightest of the new states. Both χ' and χ are taken to be Majorana fermions as is the case in the MSSM.

Signal events in the τFDM model involve four or more isolated leptons and missing energy. How does the strawman model generate similar events? The sleptons can be pair-produced in colliders. If they are heavier than χ', this leads to events of the form shown in Fig. 2(a), which involve six leptons, any or all of which could be taus. We label this possibility topology (a). Two χ' particles can also be pair-produced, leading to events of the form shown in Fig. 2(b), which we label topology (b). These events involve four leptons, any or all of which could be taus.

How can we distinguish between signal events in the two classes of models? We focus here on charge and flavor correlations among the final state leptons in the event. In Fig. 2, we exhibit the correlation of flavor and charge among the final state leptons in signal events for the τFDM model and the strawman model.

We generated fake signal events for the strawman model for three benchmark spectra.

Spectrum 1

We assume the masses of the sleptons to be the same as that of the mediator ϕ in τFDM. The masses of χ' and χ are also chosen equal to the $\chi_{e,\mu}$ and χ_τ mass respectively. More specifically, we choose $m_{\chi'} = 110$ GeV, $m_\chi = 90$ GeV and $m_{\tilde{e},\tilde{\mu},\tilde{\tau}} = 160$ GeV. Topology (a) in Fig. 2 dominates the phenomenology of this benchmark. In this topology, the two most upstream leptons are also the hardest, *and* are flavor-correlated. The τFDM leptons, as noted above, have no flavor correlation. Therefore, we expect that the flavor correlation of the two hardest leptons is a good discriminant in this case.

Spectrum 2

If the mass of the sleptons is less than the mass of χ', then only the topology (b) is allowed. The decay of χ' is on-shell in this case. The representative spectrum we study is $m_{\chi'} = 160$ GeV, $m_\chi = 90$ GeV and $m_{\tilde{e},\tilde{\mu},\tilde{\tau}} = 110$ GeV. In this case, the hardest leptons should exhibit neither charge nor flavor correlations, allowing us to distinguish it from τFDM.

Spectrum 3

In the intermediate case, however, the result is a mixture of the two topologies. In order to investigate this we study a third spectrum, $m_{\chi'} = 140$ GeV, $m_\chi = 90$ GeV and $m_{\tilde{e},\tilde{\mu},\tilde{\tau}} = 160$ GeV. In the next section we study the extent to which each of these spectra can be distinguished from τFDM.

4.1. Comparison

The correlations we obtain are listed in the left panel of Fig. 3. The results are in agreement with our expectations. Events with topology (a) in Spectrum 1 clearly exhibit flavor correlation between the two hardest leptons, as expected for the upstream leptons created from (flavor-carrying) sleptons. τFDM, on the other hand exhibits no flavor correlation in the hardest two leptons.

In all the fake spectra with topology (b), the two hardest leptons show no preferential charge assignment beyond the ratio of 1 : 2 for same to opposite charge, as expected from random charge assignment. Consequently these cases have a weaker charge anti-correlation than the τFDM.

Events from topology (a) in Spectrum 3 fall in the middle, with somewhat significant charge anti-correlation, and a weak flavor correlation. While the correlation between charge and flavor is different from the τFDM case, higher statistics might be needed in this case to make a precise distinction.

In the right panel of Fig. 3, we plot the flavor and charge asymmetries of the two hardest leptons, along with statistical error contours for the τFDM1 model at the 14 TeV LHC run, with 100 fb^{-1} of integrated luminosity. The asymmetries for flavor and charge are defined as,

$$a_F, a_C = \frac{n_{\text{same}} - n_{\text{diff.}}}{n_{\text{same}} + n_{\text{diff.}}}. \tag{6}$$

We see that with this amount of data, τFDM can be distinguished from the strawman models at 2σ.

5. Conclusions

In conclusion, we have studied the direct detection and collider prospects of theories where the dark matter particle carries lepton flavor quantum numbers, and has renormalizable contact interactions with the Standard Model fields. Assuming a coupling consistent with relic

Dataset	Frac. events with same	
	Flavor	Charge
τFDM1	0.52	0.14
τFDM2	0.49	0.14
Spectrum 1(a)	0.87	0.13
Spectrum 1(b)	0.61	0.39
Spectrum 2	0.55	0.41
Spectrum 3(a)	0.66	0.33
Spectrum 3(b)	0.60	0.38

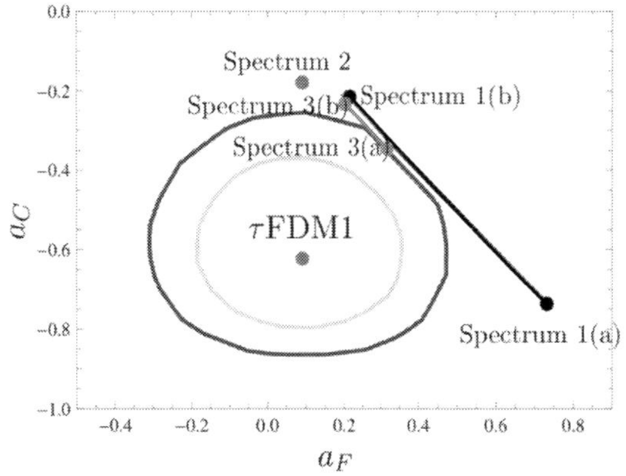

Figure 3: Left panel: Flavor and charge correlations for the two highest pT leptons in events passing cuts for different data samples. Right panel: Flavor and charge asymmetry for different models and event topologies. We show 1σ (green) and 2σ (blue) contours for the τFDM1 model at 100 fb^{-1}. The straight lines interpolate between points which correspond to different event topologies for each fake spectrum, in order to account for cases where both topologies contribute.

abundance considerations, we have shown that this scenario could be probed in the near future by upcoming direct detection experiments

We have then studied in detail a class of models where dark matter carries tau flavor, and the collider signals include events with four or more isolated leptons and missing energy. We have performed a full simulation of the signal and SM backgrounds, including detector effects, and shown that in a significant part of the parameter space favored by MFV, these theories can be discovered above SM backgrounds at the 14 TeV LHC run. We have also shown that flavor and charge correlations among the final state leptons may allow models of this type to be distinguished from simple theories where the dark matter particle couples to leptons but does not carry flavor.

Acknowledgments

I would like to thank Prateek Agrawal, Zackaria Chacko and Can Kilic for a very pleasant collaboration.

References

[1] E. Komatsu *et al.* [WMAP Collaboration], Astrophys. J. Suppl. **192**, 18 (2011) [arXiv:1001.4538 [astro-ph.CO]].
[2] L. E. Ibanez, Phys. Lett. B **137**, 160 (1984).
[3] J. R. Ellis, J. S. Hagelin, D. V. Nanopoulos, K. A. Olive and M. Srednicki, Nucl. Phys. B **238**, 453 (1984).
[4] G. Servant and T. M. P. Tait, Nucl. Phys. B **650**, 391 (2003) [hep-ph/0206071].
[5] P. Agrawal, S. Blanchet, Z. Chacko and C. Kilic, arXiv:1109.3516 [hep-ph].
[6] G. D'Ambrosio, G. F. Giudice, G. Isidori and A. Strumia, Nucl. Phys. B **645**, 155 (2002) [hep-ph/0207036].
[7] E. Aprile *et al.* [XENON100 Collaboration], Phys. Rev. Lett. **107**, 131302 (2011) [arXiv:1104.2549 [astro-ph.CO]].
[8] F. Maltoni and T. Stelzer, JHEP **0302**, 027 (2003) [hep-ph/0208156].
[9] J. Alwall, P. Demin, S. de Visscher, R. Frederix, M. Herquet, F. Maltoni, T. Plehn and D. L. Rainwater *et al.*, JHEP **0709**, 028 (2007) [arXiv:0706.2334 [hep-ph]].
[10] P. Meade and M. Reece, hep-ph/0703031.
[11] T. Sjostrand, S. Mrenna and P. Z. Skands, JHEP **0605**, 026 (2006) [hep-ph/0603175].
[12] J. Conway *et al.*, PGS 4: Pretty Good Simulation of high energy collisions (2006), `physics.ucdavis.edu/~conway/research/software/ pgs/pgs4-general.htm`.

7th International Workshop on the Dark Side of The Universe (DSU 2011) IOP Publishing
Journal of Physics: Conference Series **384** (2012) 012020 doi:10.1088/1742-6596/384/1/012020

The recent results from KIMS experiment

SeungCheon Kim for the KIMS collaboration
Department of Physics and Astronomy, Seoul National University, Seoul, 151-747, Korea
E-mail: sckim@hep1.snu.ac.kr

Abstract. In this paper, the recent results from KIMS (Korea invisible mass search) experiment are presented. KIMS has searched for WIMPs (Weakly Interactig Massive Particles) scattering off the nucleus by using the CsI(Tℓ) scintillator. The detector is an array of 12 CsI(Tℓ) scintilltors, whose total mass is 103.4 kg. The results reported here used the exposure of 24524.3 kg·days. With pulse shape discrimination (PSD) analysis, we estimated the nuclear recoil (NR) event rate and no meaningful excess of NR events rate were found. From this, we derived the improved cross section limit for WIMP–nucleon interaction.

1. Introduction

Since the existence of the dark matter as the majority of the matter in the universe is widely supported by the astronomical observations [1], the explorations to discover the direct or indirect signal from the dark matter have been attempted in various ways. Since what the dark matter is not found yet, many candidates for it have been discussed. WIMP (Weakly Interacting Massive Particle) is one of good candidates, having two important merits. The first is its natural occurrence in the theories of particle physics which is independent on the dark matter problem. The next important thing is that it is supposed to leave tens of keV in the detector when it recoils the nucleus in the detector. The existence of this testable candidate has been pursued through several experiments and KIMS experiment is also searching for the direct signal from the WIMPs.

KIMS uses CsI(Tℓ) scintillator for WIMP search. CsI(Tℓ) sintillator is a very widely used, well-known detector. The ingredient elemets, Cesium and Iodine have large atomic number (A), so that it gives large A^2 scaling effect when WIMP recoils the nucleus coherently. If WIMP nucleus interaction depends on the spin of the proton, because of the high spin expectation value for proton for both elements, this detector also has a good sensitivity. Furthermore, CsI(Tℓ) scintillator enables the pulse shape discrimination (PSD) so that the nuclear recoil (NR) event rate can be estimated statistically. Using PSD analysis, the background like electron recoil (ER) events can be subtracted. However, the weak point of CsI(Tℓ) is its inherent background such as Rb^{87}, Cs^{137}, Cs^{134} [2]. We had made an effort to reduce the internal backgrounds [3] and now we reached the background level of 2-3 count/day/kg/keV around 10 keV.

2. The experimental description

KIMS detector is an array of 3×4 CsI(Tℓ) scintillators. One module is composed of one crystal and two, 3 inch RbCs photo-cathode (green extended) photomultiplier tubes (PMTs) attached at both ends. The photoelectron yield is around 5 photoelectrons per keV. Figure. 1. shows the

Journal of Physics: Conference Series **384** (2012) 012020 doi:10.1088/1742-6596/384/1/012020

result of ^{241}Am calibration for one detector module. The data is well fitted with various X-rays and gamma rays that are expected from ^{241}Am and X-ray escape peaks.

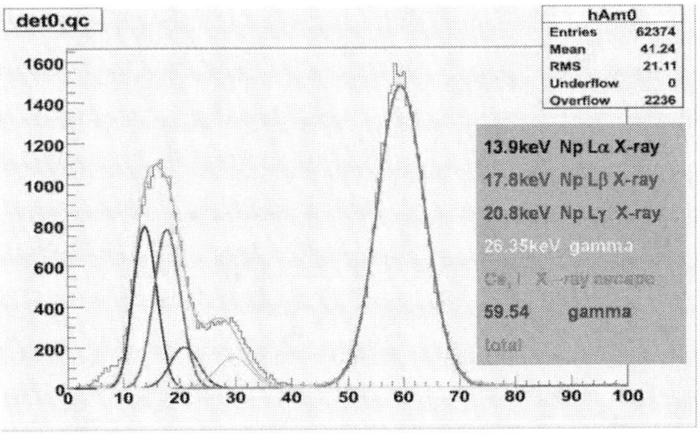

Figure 1. The energy spectrum with ^{241}Am source for one detector module

The size of one crystal is 8×8×30 cm^3 and its weight is around 8.6 kg. The mass of the whole detector array is 103.4 kg. The detector array is shown in Fig. 2 (a). The detector array enables vetoing multiple hit events, which is defined as the event triggering two or more detector modules like Compton scattering events. Because of the extremly low WIMP interaction rate, all multiple hit events are rejected in the analysis for WIMP. Figure 2 (b) shows the energy of one detector versus the energy detected by the others. Here, one can clearly see the structure from various decay modes of ^{134}Cs , one of the dominant backgrounds.

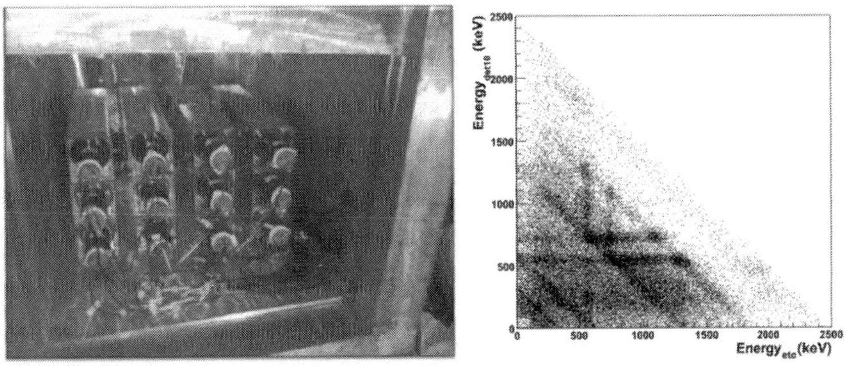

(a) (b)

Figure 2. The detector array (a) The detector array of 12 CsI(Tℓ) scintillators. (b) The two dimensional plot of the energy of one detector and the sum of the energy of the others.

The array is mounted in the shield structure which consists of 10 cm of copper, 5 cm of polyethylene, 15 cm of lead and 30 cm of liquid scintillator based on mineral oil. The mineral oil layer not only moderates the external neutron flux but also serves as the muon veto. The experimental site is Yanyang underground laboratory (Y2L) located east of Gangwon province, South Korea. The experimental hall utilizes the space of the Yangyang pumped storage power plant. The muon flux at the experimental hall is 2.7 × 10^{-7} /cm^2/s.

3. The data analysis

Recently, as the exposure of the data increased, we noticed the new component of the background, surface alpha events. The ^{222}Rn gas in the air readily sticks to any surface. ^{222}Rn attached to the surface of the detector decays to ^{210}Pb via several decay steps, whose half-life is 22 years and produces ^{210}Po which emits a 5304 keV alpha particle with a half-life of 138 days. Since this alpha decays happen at the detector surface, the alpha particle can escape from the detector leaving only small portion of the full energy of 5 MeV. Therefore, this surface alpha (SA) events can show up even in the small energy region below 10 keV. Figure 3 shows the scatter plot of the energy deposits versus the PSD parameter–the characteristics to distinguish the type of particles, LMT10 (the logarithm of the mean time of each event estimated in 10 μs aperture) seen in the KIMS detector. Here, one can see the continuum of the SA events which appear down to the very low energy. In order to understand the characteristics of SA events,

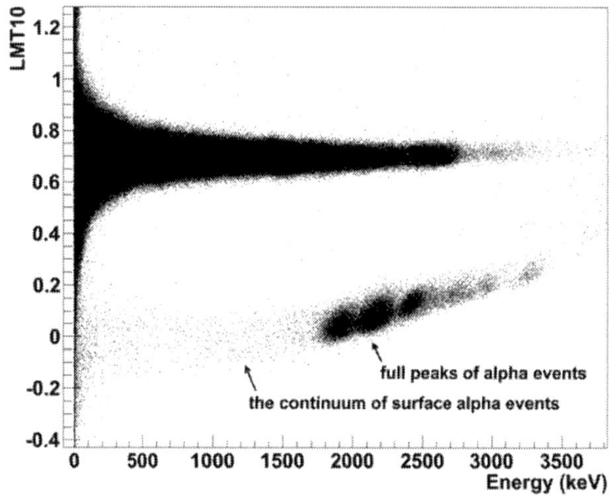

Figure 3. The energy versus the LMT10 (the logarithm of the mean time of each events estimated in 10 μs).

we obtained a small piece of CsI(Tℓ) crystal (3×3×1.5 cm^3, test crystal 1) cut from the large crystal, from which the crystals used in KIMS experiment were also cut and contaminated it with ^{222}Rn progenies by putting it in a special chamber filled with ^{222}Rn gases at the concentration of 4.33MBq/m^3. By tagging the alpha particle going out of the crystal, we have collected the SA events [4]. The characteristic of SA events is seen in Fig. 4, which shows the LMT10 distribution of SA events as well as NR and ER events. To obtain the reference data sample for NR events, we used another piece of crystal (test crystal 2) and exposed it to Am-Be neutron source. This figure shows the SA events are distinctively fast compared to other type of events. As the LMT10 distributions of ER events obtained with various crystals including the KIMS detector show good agreements, we adopted the distribution for SA (and NR) obtained from the test crystal for the PSD analysis of the KIMS detector. The crystal-to-crystal variation is a few percent and it is taken into account as the systematic effect.

Since the energy range of the interest is below 10 keV and the signal event happens rarely, the PMT noise also become a serious background. To investigate the PMT noise background, we developed the PMT Dummy Detector (PDD), which is the same with one CsI(Tℓ) detecotor module except that the crystal is replaced with the clean, hollow acrylic box. We installed this detector besides the detector array and have taken the data at the same time. In the light of the

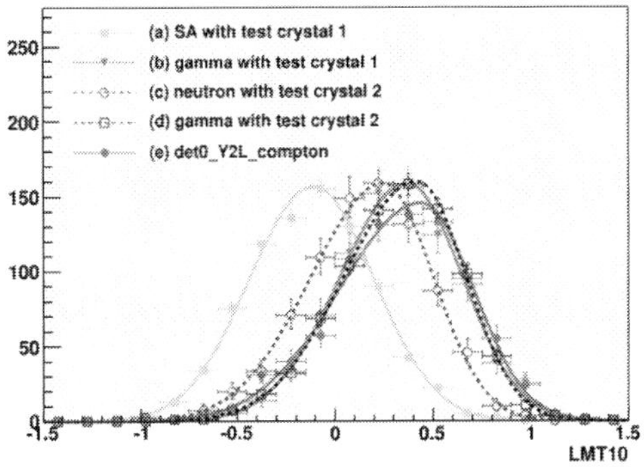

Figure 4. LMT10 distributions at 3 keV (a) SA with test crystal 1,(b) gammas with test crystal 1, (c) neutrons with test crystal 2, (d) gammas with test crystal 2 (e) Compton scattering events in detector 0 used in the WIMP search.

PDD data, we developed the event selection criteria to reject the PMT noise. In general, PMT noise event is the random coincidence between two coupled PMTs. Therefore, the signal size and its start time in two PMTs is asymmetric compared to the real signal. And, PMT noise event have the big spike signal in it. A typical event consists of a series of the photo-electrons, which we named clusters. The PMT noise event usually accompanies the abnormaly big cluster. From the signal asymmetry along the two PMTs and the presence of an abnormal big cluster signal, we can reject the PMT noise events effectively. The major event selection criteria are shown in Fig. 5. In Fig. 5 (a), qc asymmetry means the signal size asymmetry between two PMTs. The x-axis label, fit quality is the value related the goodness of fitting each event with the two exponential functions. As the tail events following the very high energy events such as the muon event show the bad fit quality, we used this variable to reject the tails of the high energy events. The vertical blue line and horizontal green line shows the event selection criteria. Figure 5 (b) shows the two dimensional distribution of the ratio of the whole signal to the biggest cluster by area for two PMTs. As shown in the figure, the PMT noise events are populated around the origin and we rejected the events whose x–value or y–value is smaller than the value marked by the green lines. The event selection criteria for this is marked with two green lines. The event selection efficiency ranges from 20~40 % at 3 keV to 40~70 % at 10 keV.

With the reference distribution of LMT10 for SA, NR and ER events, we have tried to fit the LMT10 distribution of the selected events for the WIMP search. To estimate the fraction of SA and NR events in the data, we used the bayesian analysis method [5]. The most probable value and the 90 % confidence limit (C.L.) value can be estimated from the posterior PDF for NR (SA) components. Figure 6 shows the estimated NR event rate for twelve detectors. We also estimated SA event rate for each detector and found the SA background level of detector 0, 8 and 11 is about three times as high as the average of the remaining detectors. For this reason, in drawing the overall NR event rate, we excluded the detector 0 ,8 and 11. Figure. 7 shows the NR event rate combined from the remaining nine detectors. As one can see, there is no clear excess NR signal. This NR event rate is incompatible with the WIMP-Iodine interaction interpretation for DAMA signal [7]. The annual modulation amplitude claimed by DAMA is 0.0183 ± 0.0022 counts/day/kg/keV in 2-4 keV. Considering the quenching factor of NaI(Tℓ) and CsI(Tℓ) [2, 8], 2-4 keV energy range in NaI(Tℓ) corresponds to the 3.6-5.8 keV in CsI(Tℓ) . In this energy range, we calculated the 90 % C.L. limit of NR event rate and the result is 0.0098 counts/day/kg/keV, which is well below the DAMA signal amplitude. Therefore, any scenario involving Iodine target such as iDM model are incosistent with our results. As an example, we

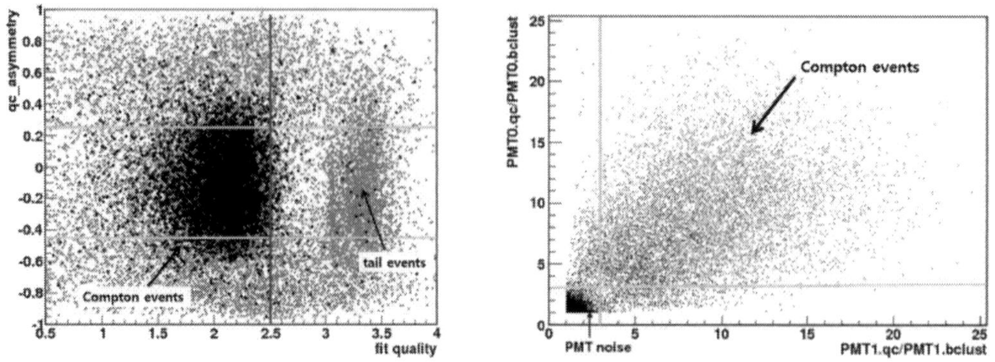

(a) (b)

Figure 5. The event selection criteria (a) The qc asymmetry versus fit quality. (b) The two dimensional plot for the ratio of the whole signal to the biggest cluster by area for the two PMTs.

present the allowed paramter space for DAMA in iDM model and our exclusion limit for 70 GeV mass of WIMP in Fig. 8.

From the total 90 % C.L. limit of NR rate, we can derive cross section limits for the WIMP-nucleon spin independent (SI) and spin dependent (SD) interaction based on the standard halo model [6]. Our limits and other experimental results are presented in Fig. 9.

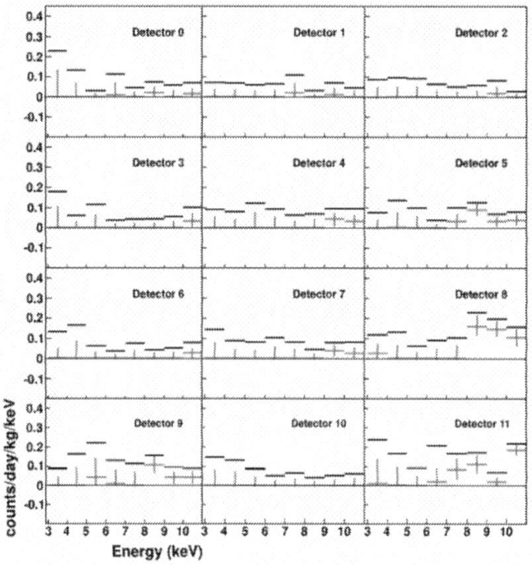

Figure 6. Nuclear recoil event rates for twelve detectors. The black horizontal bar indicate 90% C.L. upper limits, the red vertical lines denote the 68% C.L. interval, and the red horizontal bars the most probable values.

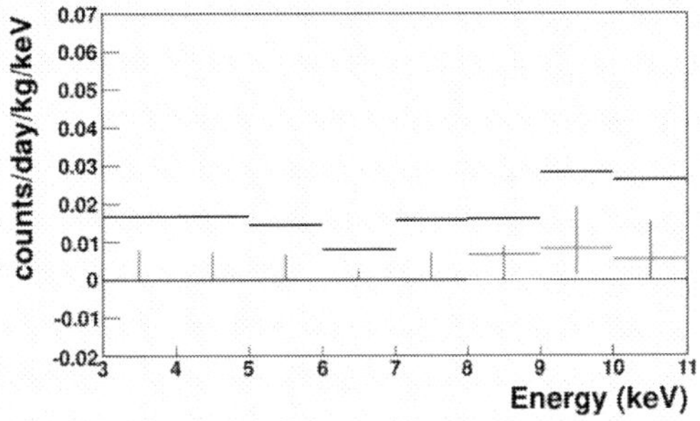

Figure 7. Total nuclear recoil event rates from the combined results from nine detectors (without detector 0, 8 and 11). The details are same with Fig. 6.

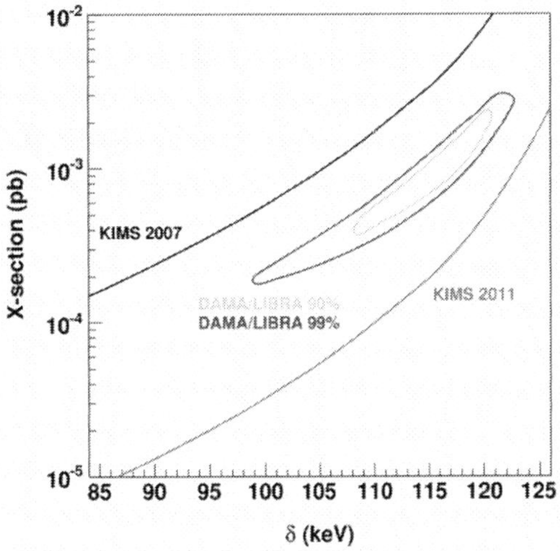

Figure 8. The allowed parameter space for DAMA [9] and the limits reported here for a 70 GeV WIMP mass in iDM model. δ is the mass split between the ground and excited states of the WIMP. The astronomical parameters from Ref. [9] are used.

4. Conclusion

KIMS experiment has been carried out with an array of twelve CsI(Tℓ) scintillators whose total mass is 103.4 kg. From the exposure of 24524.3 kg·days, we estimated the NR event rate in the data with PSD analysis method. We identified and characterized the background from the alpha decay that occurs at the surface of the detector and incorporated it in the PSD analysis. As no excess signal of NR events are seen in the data, we set 90 % C.L. limit for NR event rate. This result is inconsistent with the DAMA annual modulation amplitude and disfavors any WIMP-Iodine interaction scenario including iDM. We derived the improved cross section limit for WIMP-nucleon scattering. Especially, for the WIMP-proton SD scattering, we obtained very competitive result.

References

[1] E. Komatsu *et al.*, 2011 *ApJS* **192** 18
[2] S. K. Kim *et al.*, 2010 *New J. Phys.* **12** 075003

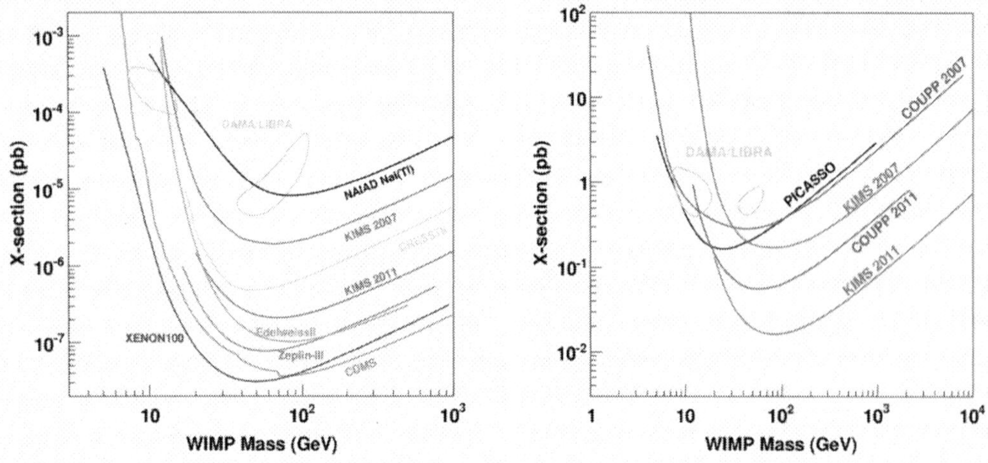

Figure 9. The 90 % exclusion limits on **(Left)** SI WIMP-nucleon and **(Right)** SD WIMP-proton cross sections. In both plots DAMA results interpreted by Savage *et al.* [18] are used (3σ contours are drawn). The SI plot includes NAIAD [10], CRESST [11], Edelweiss [12], Zeplin [13], XENON100 [14] and CDMS [15] limits. The SD plot includes PICASSO [16] and COUPP [17] limits.

[3] H. S. Lee *et al.*, 2007 *Nucl. Instrum. Methods* **571** 644
[4] S. C. Kim *et al.*, arXiv:1108.4353 (2011)
[5] A. Caldwell *et al.*, 2009 *Comp. Phys. Comm* **180** 2197
[6] J.D. Lewin and P.F. Smith, 1996 *Astropart. Phys.* **6** 87
[7] R. Bernabei, *et al.*2010 *Eur. Phys. J. C* **67** 39
[8] H. S. Park *et al.*, 2002 *Nucl. Instrum. Methods* **491** 460
[9] S. Chang *et al.*, 2009 *Phys. Rev. D* **79** 043513
[10] G. J. Alner *et al.*,2005 *Phys. Lett. B* **624** 186
[11] G. Angloher *et al.*, 2009 *Astropart. Phys.* **31** 270
[12] E. Armengaud *et al.*, 2010 *Phys. Lett. B* **687** 294
[13] V. N. Lebedenko *et al.*, 2009 *Phys. Rev. D* **80** 052010
[14] E. Aprile *et al.*, 2010 *Phys. Rev. Lett.* **105** 131302
[15] Z. Ahmed *et al.*, 2010 *science* **327** 1619
[16] S. Archambault *et al.*, 2009 *Phys. Lett. B* **682** 185
[17] E. Behnke *et al.*, 2011 *Phys. Rev. Lett.* **106** 021303
[18] C. Savage *et al.*, 2009 *JCAP* **04** 010

7th International Workshop on the Dark Side of The Universe (DSU 2011) IOP Publishing
Journal of Physics: Conference Series **384** (2012) 012021 doi:10.1088/1742-6596/384/1/012021

Low Mass Dark Matter: Some Perspectives

Shao-Long Chen

Institute of Particle Physics and Key Laboratory of Quark & Lepton Physics, Central China Normal University, Wuhan 430079, China

E-mail: chensl@phy.ccnu.edu.cn

Abstract. The low mass (10 GeV scale) dark matter is indicted and favored by several recent dark matter direct detection experimental results, such as DAMA and CoGeNT. In this talk, we discuss some aspects of the low mass dark matter. We study the indirect detection of dark matter through neutrino flux from their annihilation in the center of the Sun, in a class of models where the dark matter-nucleon spin-independent interactions break the isospin symmetry. The indirect detection using neutrino telescopes can impose a relatively stronger constraint and brings tension to such explanation, if the dark matter self-annihilation is dominated by heavy quarks or τ-lepton final states. The asymmetric dark matter doesn't suffer the constraints from the indirect detection results. We propose a model of asymmetric dark matter where the matter and dark matter share the common origin, the asymmetries in both the matter and dark matter sectors are simultaneously generated through leptogenesis, and we explore how this model can be tested in direct search experiments.

1. Introduction

It is now well established that Dark Matter (DM) dominates the matter in the universe and plays an essential role in the formation of large scale structure in it. From the current observational constraints, we know that dark matter is essentially non-baryonic, cold or warm, neutral and stable or long-lived, but the identity of DM remains unclear.

The most popular class of candidates for dark matter are the stable weakly interacting massive particles (WIMPs), which arise in many well-motivated TeV scale extensions of the standard model. Their stability is guaranteed by some symmetry, e.g., R-parity in supersymmetry or KK-parity in the case of extra dimension models. While the WIMP dark matter does not decay due to the symmetry, pairs of them can annihilate and their relic density is determined by the freeze-out of the annihilation from equilibrium to the SM particles. The fact that their observed relic density can be naturally explained by the weak scale annihilation cross section makes these models quite appealing. This coincidence is usually called as the "WIMP miracle".

The direct detection experiments aim to decode the DM non-gravitational interactions by observing the scattering of DM off detector materials [1]. Many efforts have been made to search for such events for decades. Two collaborations, DAMA [2] and CoGeNT [3], have claimed the evidences for annual modulation in the differential event rate, which is a characteristic property due to the motion of the Earth around the Sun [4]. The simplest explanation points to a low mass $\mathcal{O}(10)$GeV DM spin-independently (SI) elastic scattering off nucleon with cross sections around $(2-5)\times10^{-4}$ pb. In this talk, we will discuss some aspects of the low mass dark matter based on the works [7–9].

Journal of Physics: Conference Series **384** (2012) 012021 doi:10.1088/1742-6596/384/1/012021

2. The IVDM and Neutrinos From the Sun

It turns out that DAMA tends to favor a relatively larger cross section than CoGeNT does. Moreover, they contradict with the null experiments CDMS [5] and XENON [6] results, which put most stringent constraints on the SI DM-nucleon scattering cross sections.

In order to alleviate the tension between the CoGeNT, DAMA results and the constraints of CDMS and XENON, various theoretical attempts and solutions have been put forward [10,11]. Among them, isospin-violating dark matter (IVDM) [11–15] draws a lot of interests. It was proposed that the DM particles might couple differently to the protons and neutrons. Under this generic assumption, one therefore gains an additional degree of freedom, f_n/f_p, the ratio between the two couplings. If it satisfies $f_n/f_p = -Z/(A-Z)$ for a given nuclear isotope (A, Z), the scattering amplitudes will interfere destructively and cancel each other. Therefore, the constraints from the corresponding isotope could be completely evaded .

For ground-based direct detection experimental target containing a certain element i with nucleon and proton numbers (A_i, Z_i), the ratio of the isospin-violating (IV) cross section to isospin-conservative (IC) cross section is

$$\frac{\sigma_i^{\mathrm{IV}}}{\sigma_i^{\mathrm{IC}}} \sim \frac{[Z_i + (A_i - Z_i) f_n/f_p]^2}{A_i^2} . \tag{1}$$

The phenomenologically favored ratio is found to be $f_n/f_p \approx -0.7$. Due to destructive interference in the amplitude, the direct detection rate gets reduced significantly. The suppression factor turns out to be about 10^{-4} for Xenon and 10^{-3} for Germanium, while it is about 10^{-2} for Sodium. This feature acts as the key factor to reconcile the results of DAMA, CoGeNT and XENON experiments.

On the other hand, the capture of DM in the Sun is dominated by light elements for low mass DM favored by CoGeNT and DAMA results, namely Helium for isospin-conserving case and Hydrogen for isospin-violating case. The contributions of heavier elements are suppressed by their small chemical abundance.

Taking into account of the presence of different isotopes, we list the reduction factors in direct detection and solar capture rates in Table. 1. The key observation from Table. 1 is the hierarchy in the suppression factors, amongst which solar capture rate receives the weakest suppression from isospin violation. For DM mass around 10 GeV, the capture rate is reduced only by a factor of 0.04.

Element	Xe	Ge	Na	Solar capture
Suppression	1.3×10^{-4}	2.6×10^{-3}	1.3×10^{-2}	4.0×10^{-2}

Table 1. The suppression factors in the direct detection experiments and solar capture process, with $f_n/f_p = -0.7$ and $m_\chi = 10\,\mathrm{GeV}$.

Therefore, the indirect detection using the neutrino flux can give relatively stronger bounds on the DM-nucleon SI interactions, if the interactions are isospin violating. Weakly interacting DM can be captured in astrophysical bodies like the Sun. The capture process usually happens due to the scattering between DM and the nuclei. As DM particles are accumulated near the core region of the Sun, there can be significant annihilation process whose rate is proportional to its squared number density. There is in principle a competition between the capture and annihilation processes happening around the center of the Sun. It has been shown that for the DAMA and CoGeNT favored region, the capture-annihilation equilibrium has been reached [17]. In fact, for fixed spin-independent interaction $\sigma_{\chi N}^{\mathrm{SI}}$ and annihilate rate $\langle \sigma v \rangle$, in isospin-violating scenario the processes reach equilibrium more quickly due to a smaller capture rate. After the capture and the annihilation processes become balanced, the flux of the annihilation process will be completely controlled by the capture rate.

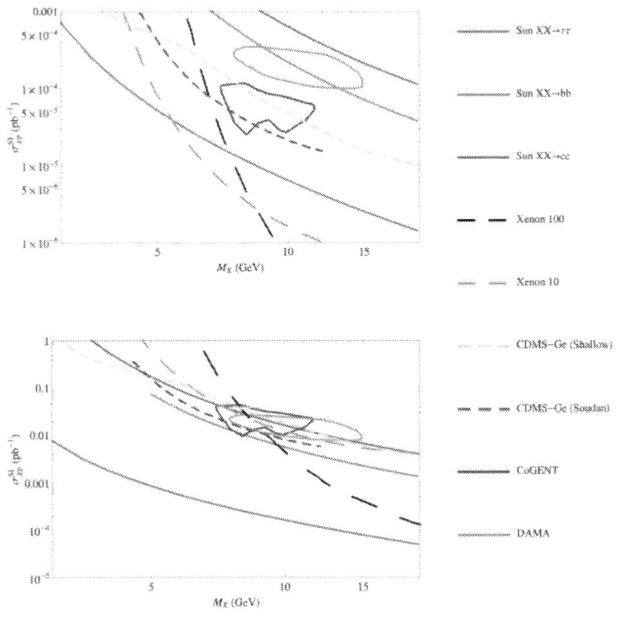

Figure 1. Positive signals from DAMA (orange circle) and CoGeNT (purple circle) in view of other direct detection experiments (dashed line) and indirect detection of neutrino flux (solid curves) from DM solar capture and annihilation, in isospin conserving (upper panel) and violating (lower panel) cases. In each panel, from up to down the solid curves represent annihilation to final states $c\bar{c}$, $b\bar{b}$ and $\tau\bar{\tau}$, assuming 100% branching ratio.

We mainly are interested in the final state neutrinos from the annihilation which can be detected by the neutrino telescopes such as the Super-K experiment. We use the results of Ref. [18] to obtain the neutrino spectrum $(dN_{\nu_i}/dE_{\nu_i})_F$ per process, taking into account of hadronization, hadron stopping, neutrino absorption and assuming the effect of neutrino oscillation to the earth averages the three neutrino flavors [19]. Here F denotes the annihilation product of the DM. For light DM, the important final states are $\tau\bar{\tau}$, $b\bar{b}$ and $c\bar{c}$, which can further decay to neutrinos.

The Super-K experiment [20] measures the Cherenkov radiation of energetic muons generated in the charge-current interactions. The effective area of detection is around $A_{\rm eff} = 900\,{\rm m}^2$, and the $\tau = 1679.6$ live days measurement allows at most 11 events other than originating from the atmosphere neutrino background [16] at 95% confidence level. We use this as the upper bound on the number events from DM annihilations in the Sun.

We have plotted the constraints on DM-nucleon cross section in Fig. 1, including both direct detections and indirect detection via neutrinos from the Sun. We focus on the low mass DM region in light of the recent direct detection excitement. As was noticed in [13], the positive signals from DAMA and CoGeNT can be reconciled by including isospin-violating DM-nucleon interactions. Isospin violation effect can also relieve the tension with the null results of XENON experiments, but cannot remove the constraints from CDMS which uses the same material as CoGeNT.

An interesting finding is that the indirect detection with neutrinos from DM annihilation in

Journal of Physics: Conference Series **384** (2012) 012021 doi:10.1088/1742-6596/384/1/012021

the Sun imposes a stronger constraint if the annihilation final states are neutrino-rich, i.e., $\tau\bar{\tau}$ or $b\bar{b}$ (or marginally $c\bar{c}$), as shown by the solid curves in the Fig. 1. The annihilation to light quarks or muon is still allowed, since they would lose most energy before decay, due to relatively longer lifetimes. Therefore, the qualified IVDM candidates should annihilate preferably into light flavors, which brings challenge for the IVDM model building.

3. Leptobigenesis: a common origin for matter and dark matter

The asymmetric dark matter (ADM) in contrast does not suffer the constraints from indirect detection results, since the ADM doesn't self-annihilate. The ADM models build a bridge connecting the abundance of the dark matter with the baryon asymmetry in the Universe and therefore may answer the question why the densities of matter and dark matter are in the same order [21].

The observed dark matter density would represent an asymmetry between dark matter and anti-dark matter densities exactly as the case for the observed asymmetry between familiar matter and anti-matter. If these two asymmetries could arise from a common mechanism, it would be a major step towards understanding why their contributions to Ω are of the same order. An elegant way to generate the Baryon-antibaryon asymmetry in the universie (BAU) is through leptogenesis in the framework of seesaw mechanism which naturally explains the smallness of the observed neutrino masses. In these models, lepton asymmetry is generated through the out-of-equilibrium decay of the very heavy right-handed (RH) neutrinos which then get converted to baryon asymmetry through the non-perturbative $B + L$ violating electroweak sphaleron process. The appealing mechanism of baryogenesis via leptogenesis combined with the fact that the relic abundances of baryons and dark matter are of the same order of magnitude inspire us to think whether genesis of dark matter could also have its origin in a manner similar to leptogenesis.

3.1. The model [8]

We propose a model of asymmetric dark matter where the dark sector is an identical copy of both forces and matter of the standard model (SM) as in the mirror universe models. The two sectors communicate with each other by gravity and possibly some SM singlet interactions which are very weak at the current age of the Universe. There is a dark baryon and lepton number in the mirror sector which is the exact analog of the familiar baryon and lepton number. The main hypothesis is that the same leptogenesis mechanism that could be producing matter-anti-matter asymmetry, is also producing asymmetry of dark matter-anti-dark-matter. This then links the dark matter energy density of the Universe to that contributed by matter making them of the same order.

A key ingredient in our attempt to connect the matter asymmetry to dark matter asymmetry is the assumption that the visible and the mirror sectors talk to each other not only through gravity but also through a common set of three right-handed neutrinos coupled to leptons and Higgs fields in each sector through Yukawa couplings [22], as shown in Fig. 2. Since the RH neutrinos are standard model singlets, this is consistent with gauge invariance. Also mirror symmetry makes the $N\ell H$ couplings on both sides equal. The out-of-equilibrium decays of right-handed neutrinos in the early universe can then produce lepton number asymmetries in both sectors, which are then transferred to baryon and dark baryon numbers through the sphaleron processes in each sector. If one imposes exact mirror symmetry on the theory, the primordial lepton asymmetries generated in each sector are equal and after sphaleron interaction produce the same number density for baryons and dark baryons in the early universe. Since we expect the symmetry breaking pattern in both sectors to be different for the model to be consistent with cosmology (see below), the resulting energy density contributions can be different and in the ratio $\Omega_B : \Omega_{DM} \approx 1 : 5$ if we require that mass of the dark baryons is five times the mass of

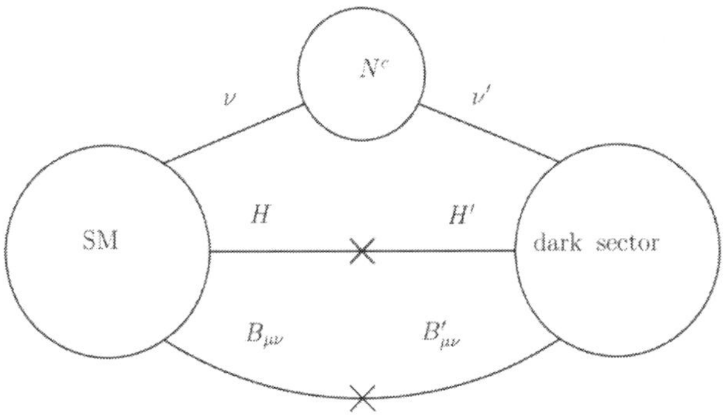

Figure 2. The connection between the standard model and the dark sector.

the familiar SM baryons. This mass difference can arise from the difference in the scales of two $SU(3)_c$ strong interactions ($\Lambda_{QCD}, \Lambda'_{QCD}$). It turns out that this difference depends on the ratio of two electroweak scales $v_{\rm wk}$ and $v'_{\rm wk}$, with $v_{\rm wk} : v'_{\rm wk} \sim 1 : 10^3$ giving the required difference between Ω_B and Ω_{DM}.

In the SM, the neutron is slightly heavier than the proton due to $m_u < m_d$ and a free neutron will decay to a proton through beta decay. With the exact mirror Yukawa couplings, the dark neutron is expected to be heavier than the dark proton as in the SM. The situation could be different for two Higgs doublets $H^{(\prime)}_{u,d}$, with $\tan\beta = v_u/v_d, \tan\beta' = v'_u/v'_d$ different. $H^{(\prime)}_u$ only couples to up-type (mirror) quarks and (mirror) neutrinos while $H^{(\prime)}_d$ to down-type (mirror) quarks and charged (mirror) leptons, respectively, by imposed Z_2 symmetries in both the two sectors. When $\tan\beta'/\tan\beta > m_d/m_u$, the dark neutron is lighter than the dark proton and then taken as the dark matter candidate.

The spectra of the SM neutrinos ν_α and the dark neutrinos ν'_α are determined by the inverse seesaw and type-I seesaw mechanisms, respectively, with the mixing between the ν_α and ν'_α given by the ratio $v_{\rm wk}/v'_{\rm wk}$. The dark neutrinos can decay into the SM particles due to this mixing.

In addition to the common set right-handed neutrinos, the SM sector and the mirror sector can also be connected through Higgs interaction and the kinetic mixing between the $U(1)$ gauge bosons consistent with gauge invariance (as shown in Fig. 2). The photon sector mixing is necessary for the model to be consistent with Big Bang Nucleosynthesis (BBN). In this work we assume the photon in the mirror sector acquires a mass around 50 MeV through the spontaneous symmetry breaking so that the mirror-electrically charged particles which are heavier, pair annihilates into it before the BBN, and the mirror photon itself decays to the electron-positron pair through the kinetic mixing. To generate this small mass two Higgs doublets are needed in the mirror sector. The kinetic mixing between the photon and mirror photon also plays an important role in the direct detection.

3.2. Energy dependence of the direct detection cross section [9]

The dark baryon (ADM) can be detected directly by observing the nucleus recoil at low background experiments. In this section, we explore how the dark nucleon can be tested in direct search experiments. In particular, we point out that if the dark matter happens to be the mirror neutron, the direct detection cross section has the unique feature that it increases at low recoil energy unlike the case of conventional WIMPs.

The interaction of nucleons with the mirror photon can then be written as

$$\mathcal{L} = \varepsilon_\gamma e \bar{p}\gamma^\mu p A'_\mu + \varepsilon_\gamma \frac{\mu_N}{2} \bar{N}\sigma^{\mu\nu} N F'_{\mu\nu} \,, \tag{2}$$

where $N = p, n$ stands for proton and neutron, respectively, and μ_N is the anomalous magnetic dipole of the nucleons.

Consider a particle from the mirror sector as the dark matter candidate, and it carries vanishing mirror electric charge. Therefore, it interacts with the mirror photon through its anomalous magnetic dipole moment or other higher dimensional operators. In analogy to the effective field theories of nucleons in QCD, we write down all possible operators up to dimension six.

$$\mathcal{L}' = c_1 \frac{e}{2m_\chi} \bar{\chi}\sigma^{\mu\nu}\chi F'_{\mu\nu} + c_2 \frac{e}{2m_\chi^2} \bar{\chi}\gamma^\mu \chi \partial^\nu F'_{\mu\nu} + c_3 \frac{e}{m_\chi^2} \bar{\chi}\gamma^\mu \partial^\nu \chi F'_{\mu\nu} + \text{h.c.} \,, \tag{3}$$

where $\mu_\chi = c_1 e/m_\chi$ is defined as the anomalous mirror magnetic dipole moment of the mirror neutron. It is easy to check that other operators such as $(e/m_\chi^2)\varepsilon^{\mu\nu\rho\sigma}\bar{\chi}\gamma_\mu\gamma_5\partial_\nu\chi F'_{\rho\sigma}$ can be decomposed into linear combinations of the above three.

The matrix element of the low-energy scattering between the nucleon and dark matter can be obtained by integrating out the mirror photon.

$$\begin{aligned} \mathcal{M}_{\text{eff}} &= \varepsilon_\gamma \frac{c_1}{m_\chi} \frac{e^2}{m_{\gamma'}^2} (\bar{p}\gamma^\mu p) q^\nu (\bar{\chi}\sigma_{\mu\nu}\chi) + i\varepsilon_\gamma \frac{c_1\mu_N}{m_\chi} \frac{e}{m_{\gamma'}^2} (\bar{N}\sigma^{\mu\nu}N) q_\mu q^\alpha (\bar{\chi}\sigma_{\alpha\nu}\chi) \\ &+ i\varepsilon_\gamma \frac{c_2}{2m_\chi^2} \frac{e^2}{m_{\gamma'}^2} (\bar{p}\gamma^\mu p) q^2 (\bar{\chi}\gamma_\mu\chi) + i\varepsilon_\gamma \frac{c_3}{m_\chi^2} \frac{e^2}{m_{\gamma'}^2} (\bar{p}\gamma^\mu p) q^\nu \left[\bar{\chi}(\gamma_\mu P_\nu - \gamma_\nu P_\mu)\chi\right] , \end{aligned} \tag{4}$$

where q is the momentum transfer and P is the sum of momenta of the initial and final nucleons.

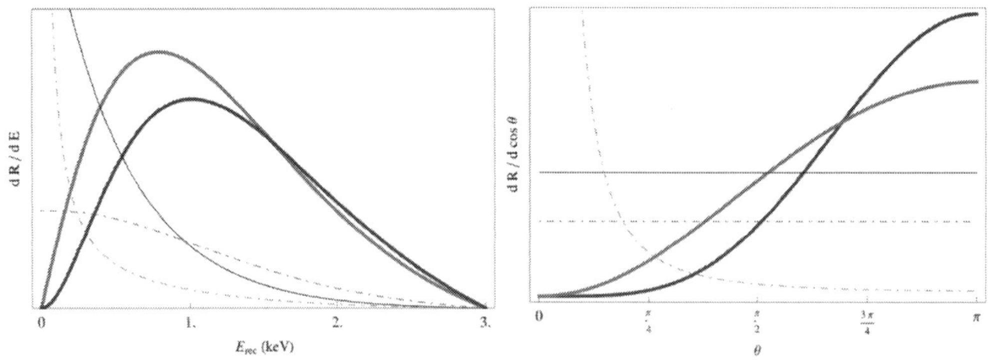

Figure 3. These two graphs display the spectral and angular distribution of SI (blue solid line) and SD (red dashed line) differential cross sections. The dot-dashed blue (SI) and red (SD) lines represent the special cases when $m_{\gamma'} = 0$, while the black thin solid line stands for the conventional SI (SD) interactions. We have chosen dark matter mass to be 5 GeV, $c_2 = 0$ and used an arbitrary scale in making the above plots.

We found that the spectral distribution of the cross sections are quite different from the conventional SI and SD interactions, as shown in Fig. 3. We also plot the SI and SD differential cross sections as a function of the scattering angle θ in the CM frame. This is a distinct feature of the new type of interactions which could be tested in low threshold direction sensitive DM detectors [23].

4. Summary

In summary, we discuss some aspects of the low mass DM. We studied the capture of low mass isospin-violating DM in the Sun and the corresponding neutrinos flux from their subsequent annihilation. We find that the indirect detection of neutrino signals through the neutrino telescope Super-K sets stronger constraints on the DM-nucleon interactions and brings further tension to such explanation, if the IVDM particles annihilate into neutrino-rich final states, e.g., tau leptons or bottom quarks. The asymmetric dark matter (ADM) doesn't suffer such constraints from the indirect detection results. We propose a model of asymmetric dark matter where the matter and dark matter share the common origin. The dark matter of the Universe be identified with the lightest baryon of a possible mirror duplicate of the standard model with the only difference between the two sectors being in the symmetry breaking patterns. The mirror photon in our model is massive but mixes with the normal photon to avoid the BBN constraints. We also explored how this model can be tested in direct search experiments. There is an energy dependence in the direct detection cross section as well as an angular dependence different from the usual symmetric WIMP case.

Acknowledgments

This contribution is based on work done in collaboration with H. An, R. Mohapatra, S. Nussinov and Y. Zhang. The work of SLC is partially supported by the NSFC under Grant No. 11145001, 11175069 and the MOE with the program NCET.

References

[1] M. W. Goodman, E. Witten, Phys. Rev. **D31**, 3059 (1985).
[2] R. Bernabei *et al.* [DAMA Collaboration], Eur. Phys. J. C **56**, 333 (2008) [arXiv:0804.2741 [astro-ph]].
[3] C. E. Aalseth *et al.* [CoGeNT collaboration], Phys. Rev. Lett. **106**, 131301 (2011) [arXiv:1002.4703 [astro-ph.CO]]; C. E. Aalseth *et al.*, arXiv:1106.0650 [astro-ph.CO].
[4] A. K. Drukier, K. Freese and D. N. Spergel, Phys. Rev. D **33**, 3495 (1986).
[5] D. S. Akerib *et al.* [CDMS Collaboration], Phys. Rev. D **82**, 122004 (2010) [arXiv:1010.4290 [astro-ph.CO]]; Z. Ahmed *et al.* [CDMS-II Collaboration], Phys. Rev. Lett. **106**, 131302 (2011) [arXiv:1011.2482 [astro-ph.CO]].
[6] E. Aprile *et al.* [XENON100 Collaboration], Phys. Rev. Lett. **105**, 131302 (2010) [arXiv:1005.0380 [astro-ph.CO]]; E. Aprile *et al.* [XENON100 Collaboration], arXiv:1104.2549 [astro-ph.CO].
[7] S. -L. Chen and Y. Zhang, Phys. Rev. D **84**, 031301 (2011) [arXiv:1106.4044 [hep-ph]].
[8] H. An, S. -L. Chen, R. N. Mohapatra and Y. Zhang, JHEP **1003**, 124 (2010) [arXiv:0911.4463 [hep-ph]].
[9] H. An, S. -L. Chen, R. N. Mohapatra, S. Nussinov and Y. Zhang, Phys. Rev. D **82**, 023533 (2010) [arXiv:1004.3296 [hep-ph]].
[10] J. Kopp, T. Schwetz, J. Zupan, JCAP **1002**, 014 (2010). [arXiv:0912.4264 [hep-ph]].
[11] S. Chang, J. Liu, A. Pierce, N. Weiner, I. Yavin, JCAP **1008**, 018 (2010). [arXiv:1004.0697 [hep-ph]];
[12] A. Kurylov, M. Kamionkowski, Phys. Rev. **D69**, 063503 (2004). [hep-ph/0307185];
[13] J. L. Feng, J. Kumar, D. Marfatia, D. Sanford, [arXiv:1102.4331 [hep-ph]].
[14] M. T. Frandsen, F. Kahlhoefer, J. March-Russell, C. McCabe, M. McCullough and K. Schmidt-Hoberg, arXiv:1105.3734 [hep-ph].
[15] E. Del Nobile, C. Kouvaris and F. Sannino, arXiv:1105.5431 [hep-ph].
[16] D. Hooper, F. Petriello, K. M. Zurek and M. Kamionkowski, Phys. Rev. D **79**, 015010 (2009) [arXiv:0808.2464 [hep-ph]].
[17] A. H. G. Peter, Phys. Rev. D **79**, 103532 (2009) [arXiv:0902.1347 [astro-ph.HE]].
[18] M. Cirelli, N. Fornengo, T. Montaruli, I. A. Sokalski, A. Strumia, F. Vissani, Nucl. Phys. **B727**, 99-138 (2005). [hep-ph/0506298].
[19] G. Jungman, M. Kamionkowski, K. Griest, Phys. Rept. **267**, 195-373 (1996). [hep-ph/9506380].
[20] S. Desai *et al.* [Super-Kamiokande Collaboration], Phys. Rev. D **70**, 083523 (2004) [Erratum-ibid. D **70**, 109901 (2004)] [arXiv:hep-ex/0404025].

[21] S. Nussinov, Phys. Lett. B **165**, 55 (1985); S. M. Barr, Phys. Rev. D **44**, 3062 (1991); D. B. Kaplan, Phys. Rev. Lett. **68**, 741 (1992); D. E. Kaplan, M. A. Luty and K. M. Zurek, Phys. Rev. D **79**, 115016 (2009) [arXiv:0901.4117 [hep-ph]].
[22] L. Bento and Z. Berezhiani, Phys. Rev. Lett. **87**, 231304 (2001)
[23] G. Sciolla, Mod. Phys. Lett. A **24**, 1793 (2009) [arXiv:0811.2764 [astro-ph]].

Neutrinos and dark matter

J. W. F. Valle

AHEP Group, Institut de Física Corpuscular – C.S.I.C./Universitat de València
Edificio Institutos de Paterna, Apt. 22085, E–46071 Valencia, Spain

E-mail: valle@ific.uv.es

Abstract.
Neutrino mass and dark matter may be closely connected. I briefly review the status of neutrino oscillations along with neutrino mass generation schemes which suggest dark matter candidates, briefly discussing their properties and relevant direct/indirect/collider detection prospects.

1. Introduction

The mounting evidence in favor of the existence of some sort of non-baryonic dark matter [1], together with the historic discovery of neutrino oscillations [2] and their profound implications for neutrino properties [3, 4, 5], both indicate the need for new physics, beyond the Standard Model. In other words, the simplest Standard Model picture of matter in terms of three generations of fermions interacting by $SU(3) \otimes SU(2) \otimes U(1)$ gauge boson exchange certainly requires amendment.

Underpinning the origin of neutrino masses and mixing and the elusive nature of dark matter constitute two of the most important challenges of elementary particle physics and modern cosmology. As already mentioned in preceding talks, in the framework of supersymmetry, for example in the Minimal Supersymmetric Standard Model (MSSM), there is a popular dark matter candidate, namely the lightest supersymmetric particle, typically a neutralino. Its stability [6] follows from assuming an *ad hoc* symmetry associated to R-parity conservation.

Here I show how, irrespective of the existence of supersymmetry in nature, the problem of neutrino mass and the explanation of dark matter may have a common origin [7]. As we illustrate, some of the the associated dark matter candidates are potentially detectable through nuclear recoil, while others require indirect methods.

2. Status of neutrino oscillations

The basic tool in terms of which to describe oscillations is the lepton mixing matrix, whose general structure has been characterized long ago [8]. This matrix has new phases with no counterpart in the quark sector. These are intrinsic to the Majorana nature of neutrinos and do not affect (standard) oscillations. However they play a crucial role in the description of lepton-number-violating processes like neutrinoless double beta decay. The latter also highlights the conceptual advantage of the original [8] symmetric parametrization [9] over its equivalent PDG

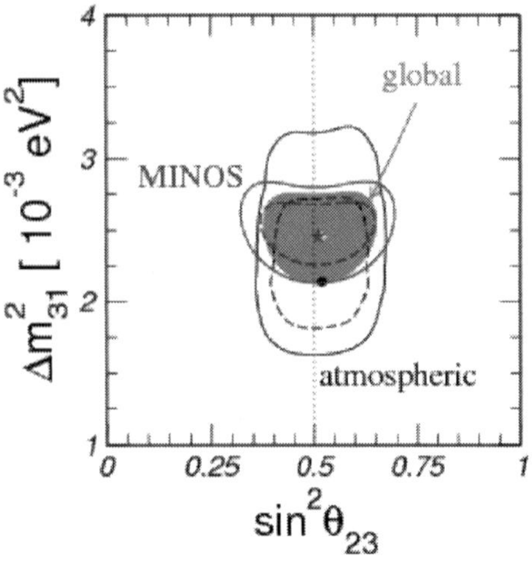

Figure 1. Complementarity of SK atmospheric data (black) and MINOS disappearance data (blue). The combined region is the red/shaded one, for normal hierarchy at 90% CL (dashed) and 3 σ (solid) [3].

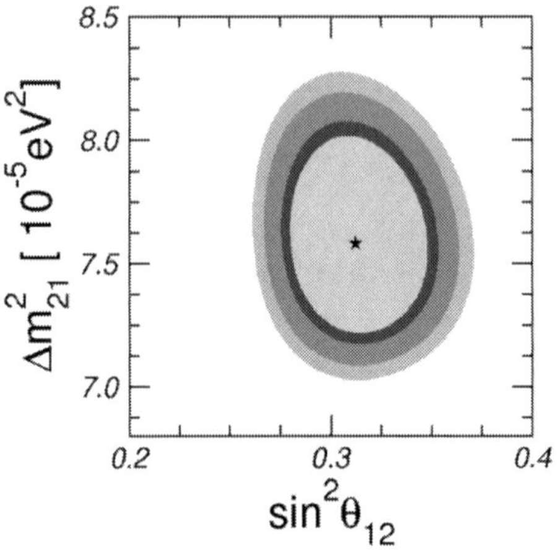

Figure 2. $\sin^2\theta_{12}$ – Δm^2_{21} regions allowed at 90%, 95%, 99%, 99.73% CL in our recommended analysis of solar + KamLAND data including short-baseline reactor results [3].

presentation [2] [1].

Neutrino oscillation data come from a variety of solar (Homestake, SAGE, GALLEX/GNO, Super-K and SNO), atmospheric (mainly Super-K), reactor (mainly Chooz and short-baseline experiments) and accelerator experiments (mainly MINOS and recently T2K) [2]. To describe them one assumes the simplest unitary form for the lepton mixing matrix and uses state-of-the-art solar and atmospheric flux calculations.

Figs. 1,2 show the regions of oscillation parameters allowed by current neutrino oscillation data. The curves in Fig. 1 delimit the regions in the $\sin^2\theta_{23}$ – Δm^2_{31} plane consistent with Super-K I, II and III atmospheric data and long-baseline neutrino oscillation appearance data, while the shaded area gives the region allowed by all global data. On the other hand the regions in $\sin^2\theta_{12}$ – Δm^2_{21} allowed by solar + KamLAND data, after minimizing over θ_{13}, are shown in Fig. 2. These depend upon reactor flux assumptions, the results displayed here correspond to our recommended analysis of solar + KamLAND data including short-baseline reactor results [3]. Other alternative choices are also given in Ref. [3], and also the comparison with the previous solar + KamLAND data analysis [11].

Fig. 3 shows the regions in $\sin^2\theta_{13} - \delta$ allowed by latest global neutrino oscillation data, after minimizing over all undisplayed oscillation parameters. The left (right) panels correspond to normal (inverted) neutrino mass hierarchy types. The triangle (star) corresponds to the best fit point of the global long-baseline neutrino oscillation appearance analysis [4]. Marginalizing over the CP phase δ and remaining oscillation parameters we find that [4]

$$\begin{array}{lll} \sin^2\theta_{13} = 0.013^{+0.007}_{-0.005}, & \Delta\chi^2 = 10.1\,(3.2\sigma) & \text{(normal)}, \\ \sin^2\theta_{13} = 0.016^{+0.008}_{-0.006}, & \Delta\chi^2 = 10.1\,(3.2\sigma) & \text{(inverted)}. \end{array} \tag{1}$$

where we display the best fit, one-sigma errors, and the significance for $\theta_{13} > 0$.

[1] In seesaw-type schemes lepton mixing has a rectangular form, which leads to lepton-flavour violation and non-standard neutrino interactions [10].

Journal of Physics: Conference Series **384** (2012) 012022 doi:10.1088/1742-6596/384/1/012022

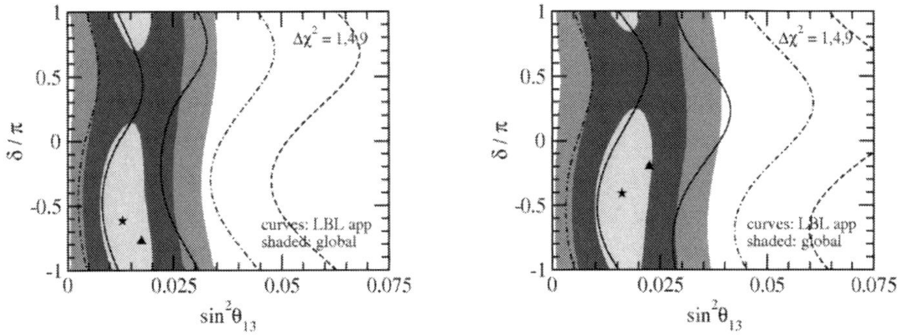

Figure 3. Shaded regions in $\sin^2\theta_{13} - \delta$ are allowed by latest global neutrino oscillation data. Curves delimit regions consistent with long-baseline appearance data. The left (right) panels are for normal (inverted) neutrino mass hierarchy. The triangle (star) corresponds to the best fit point of the LBL app (global) analysis, see Ref. [4] for more details.

As expected the upper bound on $\sin^2\theta_{13}$ is dominated by global data without long-baseline appearance data, whereas the lower bound comes mainly from T2K. We eagerly wait for more data in order to settle the situation and to strengthen the prospects for CP violation studies in the next generation of neutrino oscillation experiments [12, 13].

3. Origin of neutrino mass

The origin of neutrino mass remains as elusive as ever. The simplest way it can arise is from Weinberg's dimension-five operator in the left panel of Fig. 4.

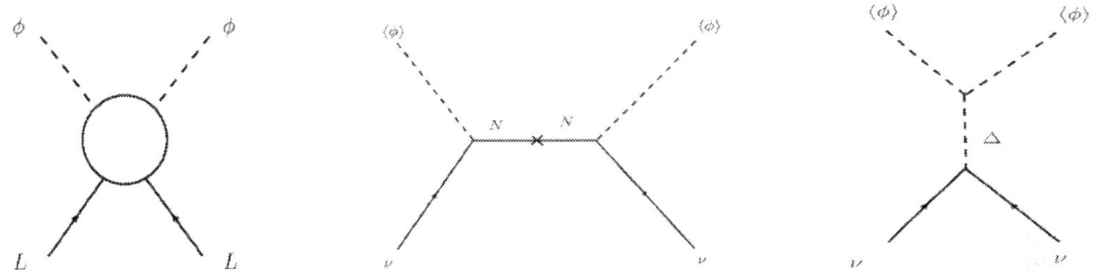

Figure 4. Weinberg operator and type-1/type-2 seesaw schemes.

The seesaw remains an attractive mechanism to ascribe the observed smallness of neutrino mass relative to other standard model fermions to the presence of a large mass scale [8, 14, 15, 16, 17, 18, 19, 20] characterizing the violation of lepton number. Effective neutrino masses arise from the exchange of a heavy $SU(3) \otimes SU(2) \otimes U(1)$ singlet "right"-handed neutrino or a heavy scalar triplet scalar boson, or both, as indicated in the two middle panels in Fig. 4.

Thanks to the arbitrary number of gauge singlets that may be added in the seesaw [8] it can be realized also at low-scale. Examples of low-scale seesaw schemes are the inverse [21, 22] and linear seesaw schemes [23, 24, 25]. In this case, in the presence of supersymmetry the role of dark matter may be played by a sneutrino-like state, instead of the standard neutralino [26, 27].

If the breaking of lepton number is spontaneous there is additional dynamics associated to neutrino mass generation. For example in gauged lepton number schemes [23, 24, 25] [28] there is an additional neutral gauge boson coupled to neutrinos. In contrast, if lepton number is not a gauge symmetry, like the $SU(2) \otimes U(1)$ seesaw scheme, there is a remnant physical Goldstone boson, generically called majoron [29, 18]. As we will see below in the presence of gravitational interactions this may act as dark matter [7, 30].

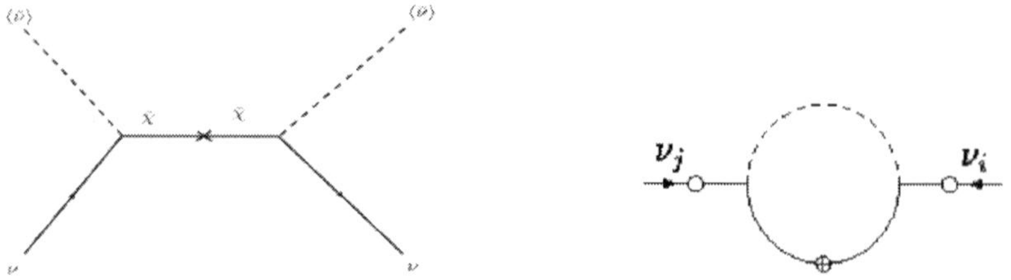

Figure 5. Neutrino mass generation diagrams from spontaneous R-parity violation.

On the other hand there are alternative schemes based on weak-scale physics, such as supersymmetry without a conserved R-parity [31], where one may naturally arrange for the breaking of R-parity to be spontaneous [32, 33]. This leads to a very predictive scheme called bilinear R-parity violation (BRpV) with the atmospheric neutrino mass scale generated by tree seesaw-like diagrams such as the one in the left plot in Fig. 5 involving supersymmetric fermion exchange, and the solar scale induced radiatively (right plot) [34]. This accounts naturally for the smallness of solar with respect to atmospheric squared mass splittings. In this case the neutralino can not be dark matter, as it decays typically inside collider experiments detectors, such as those at the Large Hadron Collider (LHC). However one can show that the gravitino can play the role of dark matter [35].

4. Neutrino mass dark matter connection

It has been suggested that these two apparently unrelated issues may be closely inter-linked [7]. I now scan over a few examples of "neutrino-motivated" dark matter candidates. While they are all cold insofar as the properties of the Cosmic microwave background (CMB) are concerned, some, like the majoron, may behave as warm dark matter regarding structure formation. When it comes to detection, some are ideal for direct detection, while others require indirect, like the majoron and the gravitino.

4.1. Sneutrino-like dark matter in inverse seesaw

As a first example we consider the case of supersymmetric dark matter in a model where neutrinos acquire mass through a low-scale seesaw mechanism. For definiteness we take the simplest SU(3) ⊗ SU(2) ⊗ U(1) inverse seesaw mechanism [21, 22] within the constrained Minimal Supersymmetric Standard Model (CMSSM) inspired by minimal supergravity [26]. The low-scale breaking of lepton number may arise dynamically as discussed in Ref. [27].

One can show that when neutrino masses are generated this way it is more likely to have a sneutrino-like state as the lightest superparticle than the conventional neutralino [26], as illustrated in the left plot in Fig. 6. In this region the neutralino behaves as LSP in both standard and modified mSUGRA within the white region (plotted for a given choice for $\tan\beta = 35$, $A_0 = 0$ and $\mu > 0$). However in the shaded (red) and light (yellow) areas one finds that, for a suitable choice of parameters associated to the inverse seesaw scheme a sneutrino-like state is the LSP [26]. Note also that the light (yellow) region includes all the region where the $\tilde{\tau}$ is the LSP in the standard mSUGRA case. On the other hand the dark (blue) region is excluded by experimental and theoretical constraints. This region is now substantially larger thanks to the recent results from the ATLAS and CMS collaborations [36], as illustrated by the extended region in dark at the left part of the left plot in Fig. 6.

One can demonstrate that such schemes also naturally reconcile the small neutrino masses with the correct relic sneutrino dark matter abundance, as shown in the mid panel in Fig. 6, which gives the sneutrino-like LSP relic abundance Ωh^2 as a function of the LSP sneutrino mass m_1, when the supersymmetric parameters are scanned within the ranges $100\,\text{GeV} < m_0 < 3\,\text{TeV}$, $100\,\text{GeV} < m_{1/2} < 3\,\text{TeV}$, $A_0 = 0$, $3 < \tan\beta < 50$, and similarly for the inverse seesaw parameters (see Ref. [26] for details). Finally, the right panel in Fig. 6 illustrates how one can obtain an accessible direct detection rates in nuclear recoil experiments.

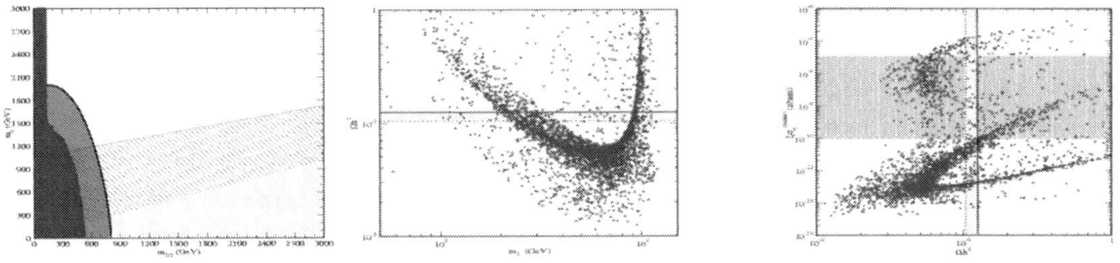

Figure 6. The blue points in the left panel give the LSP relic abundance Ωh^2 as a function of the "sneutrino" LSP mass. The right panel gives the "sneutrino"–nucleon scattering cross section vs. its relic abundance, for the same scan of parameters as in the left panel. The horizontal (light blue) band denotes the current sensitivity of direct detection experiments. The yellow bands delimit the WMAP cold dark matter 3 σ range [37].

4.2. Dark matter stabilized by flavour symmetry

A remarkable feature of current neutrino oscillation data described in Sec. 2 is that they imply a pattern of mixing angles radically different from that which characterizes the quark sector, with one maximal mixing angle, one large but non-maximal and one small [3, 4]. Attempts to understand this in terms of basic flavour symmetries have brought in the idea that perhaps the symmetry explaining flavour may also account for the required stability of dark matter. This opens an attractive link between neutrino physics and dark matter; two sectors that show a clear need for physics beyond the Standard Model.

A mechanism realizing the idea of stabilizing dark matter through a discrete unbroken subgroup of a non-abelian flavour symmetry has been recently proposed in Ref. [38].

The model is based on an A4 symmetry extending the Higgs sector of the Standard Model with three scalar doublets. After electroweak symmetry breaking two of the scalars of the model acquire vacuum expectation values (vevs) which spontaneously break A4, leaving a residual parity symmetry. The lightest neutral odd scalar is then protected to be stable and can play the role of dark matter candidate. The phenomenology of this model has been studied in [39]. The dark matter particle is thermally produced via a Higgs portal (left plot in Fig. 7) and is therefore a typical WIMP. All constraints from current laboratory experiments and astrophysical observations are satisfied and direct detection prospects are promising for the near future as many collaborations are closing in on the low mass region of the WIMP parameter space (right plot in Fig. 7).

Fig. 7 gives the spin-independent dark matter scattering cross section off-protons as a function of the dark matter mass. The orange regions delimited by the dashed (solid) line show the DAMA/LIBRA annual modulation regions including (neglecting) the channeling effect [40]. The green region corresponds to the COGENT data [41]. Dashed and dotted lines correspond to the upper bound from CDMS (respectively from [42] and [43]). XENON100 bounds [44] are shown as a solid black line.

Coming to indirect detection, one may also compute the annihilation cross section times velocity as a function of the dark matter mass [39]. One finds that the sensitivity of current Fermi-LAT observations are starting to probe the model for low-intermediate dark matter masses (see e.g. [45, 46]).

From the neutrino side one finds an inverted neutrino mass hierarchy, hence a neutrinoless double beta decay rate accessible to upcoming searches, while $\theta_{13} = 0$ giving no CP violation in neutrino oscillations. This scheme is just an example of a possibly deeper connection between neutrino and dark matter physics, the details of which are highly model dependent.

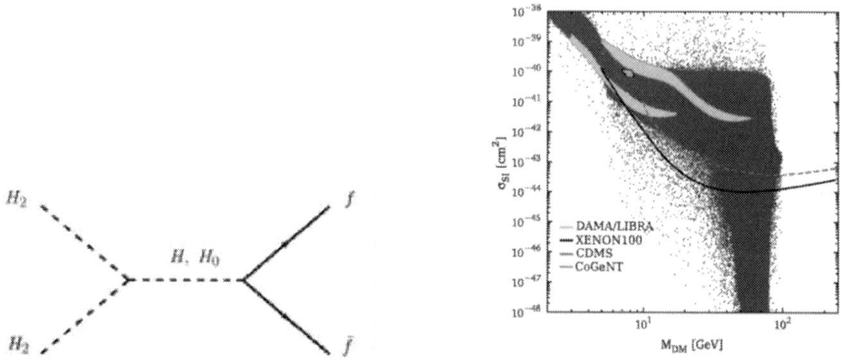

Figure 7. Higgs portal (left plot), and spin-independent dark matter scattering cross section off-protons as a function of the dark matter mass versus current experimental sensitivities (right plot).

4.3. Majoron as decaying dark matter

In a large class of models neutrino masses may arise from spontaneous breaking of ungauged lepton number [29, 18]. Due to quantum gravity effects [47] the associated Goldstone boson - the majoron - will pick up a mass.

By construction, the majoron couples to, and hence will decay to, a pair of neutrinos [48]. The lifetime and mass required by cosmic microwave background observations can be determined so that the massive majoron provides the observed dark matter of the Universe [30]. This scenario fits nicely in models where neutrino masses arise *a la seesaw*, and may lead to other possible cosmological implications associated to structure formation. Indeed, even though the majoron in this mass range behaves as cold dark matter particle as far as the CMB is concerned, its small thermal velocities might play a role in structure formation.

In a number of schemes the decaying majoron has also a sub-leading decay mode into two photons see, e.g. Refs. [7] and [49, 50]. Upper limits exist on the decay rate arising from NGC3227 (red), the Milky Way halo observed with a prototype cryogenic spectrometer (salmon), XMM observations of the Milky Way (sand), Chandra observations of the Bullet Cluster and M31 (orange), HEAO-1 observations of the diffuse X-ray background (aquamarine), INTEGRAL SPI line search in the Milky Way halo (blue). Shaded regions are excluded. These have been compiled and extended in Ref. [49] where details and original experimental references can be found. For a specific decaying dark matter seesaw model see Ref. [50]. Improved sensitivities are expected by the proposed XENIA mission [51].

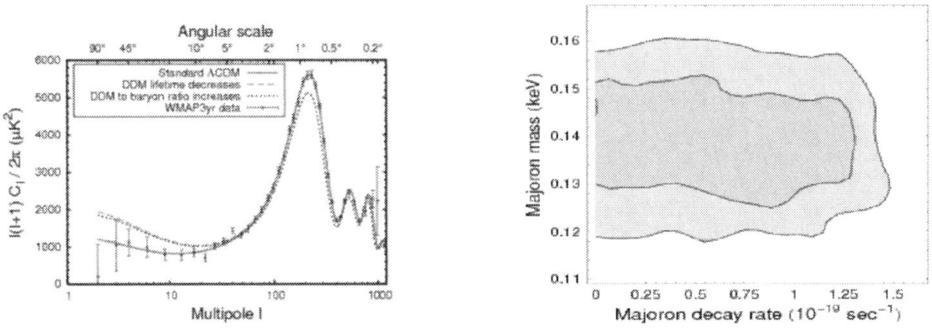

Figure 8. Left panel: (Color online) Distortion of CMB anisotropy spectrum arising from decaying dark matter majoron, see Ref. [30] for details. Right panel: (Color online) Contours of the 68% (green/dark) and 95% (yellow/light) CL allowed mass-decay-rate regions.

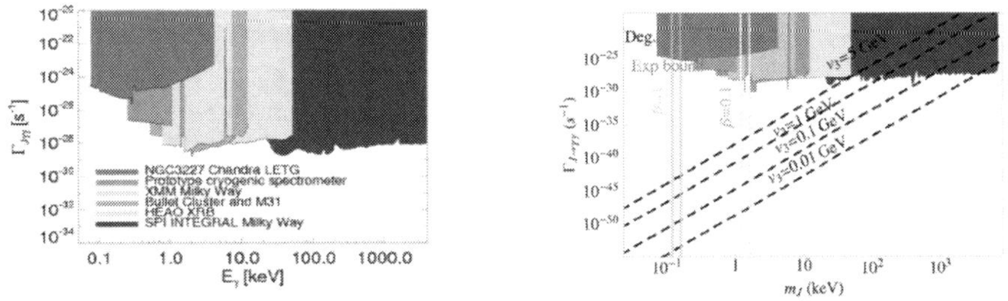

Figure 9. Majoron decay rate to photons as a function of the majoron mass, for different values of the triplet vev, v_3. We assume the invisible decay bound to be saturated and a (quasi)degenerate neutrino mass spectrum. The shaded regions are excluded by observations [49] and the vertical lines characterize the cosmological majoron production mechanism [30].

4.4. Decaying gravitino as dark matter in BRPV

An attractive way to induce the dimension-five operator responsible for neutrino mass generation and lepton mixing angles is to assume the weak-scale supersymmetry without R-parity [31], i.e. to assume that the origin of neutrino mass is intrinsically supersymmetric, as indicated in Fig. 5. In the presence of $SU(3) \otimes SU(2) \otimes U(1)$ singlet "right" neutrino superfields one can indeed induce the spontaneous arrange for the breaking of R-parity in agreement with LEP measurements of the invisible Z width [32, 33].

This leads to effective bilinear R-parity violation, or BRpV [34], the simplest extension of the minimal supersymmetric standard model (MSSM) which includes lepton number violation and which can successfully reproduce the present neutrino oscillation parameters [52, 53].

We describe how it leads to a successful phenomenological model for neutrino masses, but one in which the lightest neutralino can not be dark matter, as it decays typically inside collider detectors, as illustrated in Fig. 10.

An alternative way to relate dark matter with neutrino properties in a scenario where supersymmetry is the origin of neutrino mass is to assume that the LSP is the gravitino [35]. In this picture the same lepton number violating superpotential terms that generate neutrino masses and mixing also induce gravitino decays, into a gamma plus a neutrino, which also breaks R parity. The latter is doubly suppressed, first by the smallness of the R-parity violating couplings, and in addition by the Planck scale. The allowed gravitino mass-lifetime region consistent with neutrino oscillation data and astrophysical bounds from searches for gamma-ray

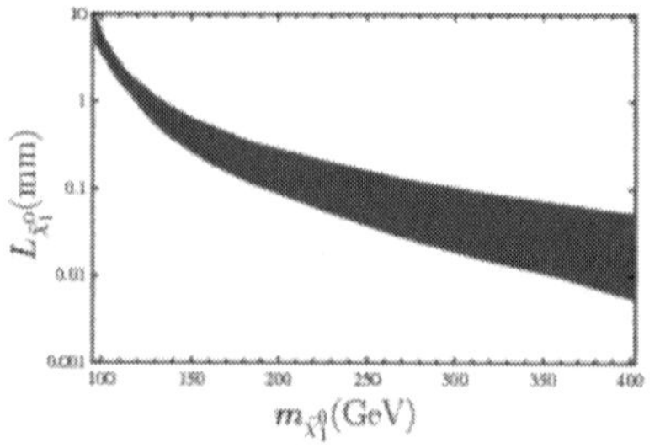

Figure 10. Neutralino NLSP decay length as a function of its mass. For illustration we show the results of a scan with A_0=100 GeV, sign(μ)=+1, $\tan\beta$=10, 200$\leq M_0 \leq$1000 GeV and 240$\leq m_{1/2} \leq$1000 GeV.

lines from dark matter decay is indicated in the left plot in Fig. 11. The yellow region is excluded by gamma-ray line searches at Fermi and EGRET. The lower and upper black lines correspond to $m_{1/2}$=240 and 3000 GeV respectively. The right plot shows the parameters in terms of reheat temperature and gravitino mass [35]. In the orange area a consistent gravitino relic abundance would require gluino masses already excluded by present collider searches. In contrast, the blue area is viable but requires too large gluino masses, M_3 >6000 GeV, while the yellow region is excluded by astrophysical gamma-ray line searches.

In contrast to seesaw models, the BRpV gravitino dark matter model can be probed at future collider experiments, like the LHC or the Next Linear Collider, since the decay pattern of the next-to-lightest supersymmetric particle provides a direct connection with the lepton mixing angles determined by neutrino experiments [54, 55, 56, 57].

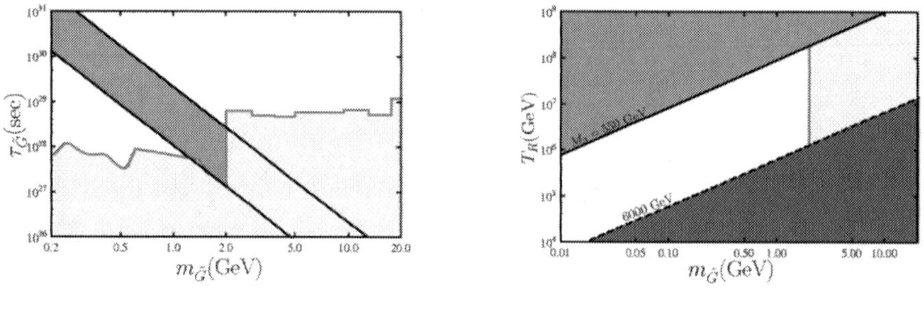

Figure 11. Allowed gravitino mass-lifetime region (grey color) consistent with neutrino oscillation data and astrophysical bounds on gamma-ray lines from dark matter decay (left plot). The white region in the right plot is consistent with neutrino oscillations, gravitino dark matter, gamma-ray line searches, and gluino searches at the LHC.

5. Acknowledgments

I thank the organizers for the hospitality at Beijing, and all my collaborators, in particular C. Arina, S. Boucenna, M. Taro's and M. Tortola, for help in updating the material with respect to the original publications. This work was supported by the Spanish MEC under grants FPA2008-00319/FPA, FPA2011-22975 and MULTIDARK CSD2009-00064 (Consolider-Ingenio 2010 Programme), by Prometeo/2009/091 (Generalitat Valenciana), by the EU ITN UNILHC PITN-GA-2009-237920.

References

[1] C. Munoz, Int.J.Mod.Phys. **A19**, 3093 (2004), and talk at these proceedings.
[2] K. Nakamura *et al.*, Journal of Physics G: Nuclear and Particle Physics **37**, 075021 (2010).
[3] T. Schwetz, M. Tortola and J. W. F. Valle, New J. Phys. **13**, 063004 (2011).
[4] T. Schwetz, M. Tortola and J. W. F Valle, New J.Phys. **13**, 109401 (2011), [1108.1376].
[5] For a review see M. Maltoni, T. Schwetz, M. A. Tortola and J. W. F. Valle, New J. Phys. **6**, 122 (2004).
[6] T. Hambye, PoS **IDM2010**, 098 (2011), [1012.4587].
[7] V. Berezinsky and J. W. F. Valle, Phys. Lett. **B318**, 360 (1993), [hep-ph/9309214].
[8] J. Schechter and J. W. F. Valle, Phys. Rev. **D22**, 2227 (1980).
[9] W. Rodejohann and J. W. F. Valle, Phys.Rev. **D84**, 073011 (2011), [1108.3484].
[10] D. Forero, S. Morisi, M. Tortola and J. Valle, JHEP **1109**, 142 (2011), [1107.6009].
[11] T. Schwetz, M. Tortola and J. W. F. Valle, New J. Phys. **10**, 113011 (2008), [0808.2016].
[12] H. Nunokawa, S. J. Parke and J. W. F. Valle, Prog. Part. Nucl. Phys. **60**, 338 (2008).
[13] ISS Physics Working Group, A. Bandyopadhyay *et al.*, Rept.Prog.Phys. **72**, 106201 (2009), [0710.4947].
[14] P. Minkowski, Phys. Lett. **B67**, 421 (1977).
[15] M. Gell-Mann, P. Ramond and R. Slansky, (1979), Print-80-0576 (CERN).
[16] T. Yanagida, (KEK lectures, 1979), ed. O. Sawada and A. Sugamoto (KEK, 1979).
[17] R. N. Mohapatra and G. Senjanovic, Phys. Rev. **D23**, 16 (1981).
[18] J. Schechter and J. W. F. Valle, Phys. Rev. **D25**, 774 (1982).
[19] G. Lazarides, Q. Shafi and C. Wetterich, Nucl. Phys. **B181**, 287 (1981).
[20] J. W. F. Valle, J. Phys. Conf. Ser. **53**, 473 (2006), [hep-ph/0608101], Review lectures at Corfu.
[21] R. N. Mohapatra and J. W. F. Valle, Phys. Rev. **D34**, 1642 (1986).
[22] M. C. Gonzalez-Garcia and J. W. F. Valle, Phys. Lett. **B216**, 360 (1989).
[23] E. Akhmedov *et al.*, Phys. Lett. **B368**, 270 (1996).
[24] E. Akhmedov *et al.*, Phys. Rev. **D53**, 2752 (1996).
[25] M. Malinsky, J. C. Romao and J. W. F. Valle, Phys. Rev. Lett. **95**, 161801 (2005).
[26] C. Arina, F. Bazzocchi, N. Fornengo, J. Romao and J. W. F. Valle, Phys. Rev. Lett. **101**, 161802 (2008).
[27] F. Bazzocchi, D. Cerdeno, C. Munoz and J. W. F. Valle, Phys.Rev. **D81**, 051701 (2010).
[28] J. W. F. Valle, Phys. Lett. **B196**, 157 (1987).
[29] Y. Chikashige, R. N. Mohapatra and R. D. Peccei, Phys. Lett. **B98**, 265 (1981).
[30] M. Lattanzi and J. W. F. Valle, Phys. Rev. Lett. **99**, 121301 (2007).
[31] G. G. Ross and J. W. F. Valle, Phys. Lett. **B151**, 375 (1985).
[32] A. Masiero and J. W. F. Valle, Phys. Lett. **B251**, 273 (1990).
[33] J. C. Romao, C. A. Santos and J. W. F. Valle, Phys. Lett. **B288**, 311 (1992).
[34] M. Hirsch and J. W. F. Valle, New J. Phys. **6**, 76 (2004), [hep-ph/0405015].
[35] D. Restrepo, M. Taoso, J. W. F. Valle and O. Zapata, 1109.0512.
[36] Atlas Collaboration, G. Aad *et al.*, Phys.Lett. **B701**, 186 (2011), [1102.5290].
[37] WMAP Collaboration, E. Komatsu *et al.*, Astrophys. J. Suppl. **180**, 330 (2009), [0803.0547].
[38] M. Hirsch, S. Morisi, E. Peinado and J. W. F. Valle, Phys. Rev. **D82**, 116003 (2010), [1007.0871].
[39] M. Boucenna *et al.*, JHEP **1105**, 037 (2011), [1101.2874].
[40] N. Fornengo, S. Scopel and A. Bottino, Phys. Rev. **D83**, 015001 (2011), [1011.4743].
[41] C. Aalseth *et al.*, Phys.Rev.Lett. **107**, 141301 (2011), [1106.0650].
[42] CDMS-II Collaboration, Z. Ahmed *et al.*, Science **327**, 1619 (2010), [0912.3592].
[43] CDMS-II Collaboration, Z. Ahmed *et al.*, Phys.Rev.Lett. **106**, 131302 (2011), [1011.2482].
[44] XENON100 Collaboration, E. Aprile *et al.*, Phys.Rev.Lett. **107**, 131302 (2011), [1104.2549].
[45] Fermi-LAT collaboration, M. Ackermann *et al.*, Phys.Rev.Lett. **107**, 241302 (2011), [1108.3546].
[46] A. Geringer-Sameth and S. M. Koushiappas, Phys.Rev.Lett. **107**, 241303 (2011), [1108.2914].
[47] S. R. Coleman, Nucl. Phys. **B310**, 643 (1988).
[48] J. Schechter and J. W. F. Valle, Phys. Rev. **D25**, 774 (1982).
[49] F. Bazzocchi *et al.*, JCAP **0808**, 013 (2008), [0805.2372].
[50] J. Esteves *et al.*, Phys.Rev. **D82**, 073008 (2010), [1007.0898].
[51] J. W. den Herder *et al.*, 0906.1788.
[52] M. Hirsch *et al.*, Phys. Rev. **D62**, 113008 (2000), [hep-ph/0004115], Err-ibid. **D65**:119901,2002.
[53] M. A. Diaz *et al.*, Phys. Rev. **D68**, 013009 (2003), [hep-ph/0302021].
[54] A. Datta, B. Mukhopadhyaya and F. Vissani, Phys. Lett. **B492**, 324 (2000), [hep-ph/9910296].
[55] W. Porod *et al.*, Phys. Rev. **D63**, 115004 (2001).
[56] F. de Campos *et al.*, JHEP **05**, 048 (2008).
[57] F. De Campos *et al.*, Phys. Rev. **D82**, 075002 (2010), [1006.5075].

Journal of Physics: Conference Series **384** (2012) 012023 doi:10.1088/1742-6596/384/1/012023

Predicted rates for direct WIMP searches

J. D. Vergados

Theoretical Physics Division, University of Ioannina, Ioannina, Gr 451 10, Greece

E-mail: vergados@uoi.gr

Abstract. The differential event rate for direct detection of dark matter, both the time averaged and the modulated one due to the motion of the Earth, are discussed. The calculations focus on relatively light cold dark matter candidates (WIMP) and low energy transfers. It is shown that for some WIMP masses the modulation amplitude may change sign. This effect can be exploited to yield information about the mass of the dark matter candidate.

1. Introduction

The combined MAXIMA-1 [1], BOOMERANG [2], DASI [3] and COBE/DMR Cosmic Microwave Background (CMB) observations [4] imply that the Universe is flat [5] and that most of the matter in the Universe is Dark [6, 7], i.e. exotic. Combining the data of these quite precise experiments one finds:

$$\Omega_b = 0.0456 \pm 0.0015, \quad \Omega_{\rm CDM} = 0.228 \pm 0.013, \quad \Omega_\Lambda = 0.726 \pm 0.015.$$

Since any "invisible" non exotic component cannot possibly exceed 40% of the above $\Omega_{\rm CDM}$ [8], exotic (non baryonic) matter is required and there is room for cold dark matter candidates or WIMPs (Weakly Interacting Massive Particles).

Even though there exists firm indirect evidence for a halo of dark matter in galaxies from the observed rotational curves, see e.g the review [9], it is essential to directly detect such matter. The possibility of such detection, however, depends on the nature of the dark matter constituents and their interactions.

Since the WIMP's are expected to be extremely non relativistic, with average kinetic energy $\langle T \rangle \approx 50$ keV$(m_{\rm WIMP}/100$ GeV$)$, they are not likely to excite the nucleus. So they can be directly detected mainly via the recoiling of a nucleus (A,Z) in elastic scattering. The event rate for such a process can be computed from the following ingredients[10] : i) The elementary nucleon cross section. ii) knowledge of the relevant nuclear matrix elements [11, 12], obtained with as reliable as possible many body nuclear wave functions, iii) knowledge of the WIMP density in our vicinity and its velocity distribution.

The nucleon cross sections can also be extracted from the data of event rates, if and when such data become available. From limits on the event rates, one can obtain exclusion plots on the nucleon cross sections as functions of the WIMP mass. In the standard nuclear recoil experiments, first proposed more than 30 years ago [13], one has to face the problem that the reaction of interest does not have a characteristic feature to distinguish it from the background. So for the expected low counting rates the background is a formidable problem. Some special features of the WIMP-nuclear interaction can be exploited to reduce the background problems. Such are:

i) the modulation effect: this yields a periodic signal due to the motion of the earth around the sun. Unfortunately this effect, also proposed a long time ago [14] and subsequently studied by many authors [15, 16, 17, 18, 19, 20, 21, 22, 23], is small and becomes even smaller than 2% due to cancelations arising from nuclear physics effects,

ii) backward-forward asymmetry expected in directional experiments, i.e. experiments in which the direction of the recoiling nucleus is also observed. Such an asymmetry has also been predicted a long time ago [24], but it has not been exploited, since such experiments have been considered very difficult to perform, but they now appear to be feasible[10]. iii) transitions to excited states: in this case one need not measure nuclear recoils, but the de-excitation γ rays. This can happen only in very special cases since the average WIMP energy is too low to excite the nucleus. It has, however, been found that in the special case of the target ^{127}I such a process is feasible [25] with branching ratios around 5%, (iv) detection of electrons produced during the WIMP-nucleus collision [26, 27] and v) detection of hard X-rays produced when the inner shell holes are filled[28].

In the present paper we will limit our attention to the standaerd recoil experiments and study the differential event rates, both time averaged and modulated, in the region of low energy transfers, as in the DAMA experiment [29, 30], focusing our attention on relatively light WIMPS [31, 32, 33]. Such light WIMPs can be accommodated in some SUSY models [34]. We will also present some results on the total rates as well. We will employ here the standard Maxwell-Boltzmann (M-B) distribution for the WIMPs of our galaxy and we will not be concerned with other distributions [35, 36, 37, 38], even though some of them may affect the modulation. The latter will be studied elsewhere. We will explicitly show that the modulation amplitude, entering both the differential and the total rates, changes sign for certain reduced WIMP-nuclear masses. As a result such an effect, if and when the needed data become available, may be exploited to infer the WIMP mass.

2. The formalism for the WIMP-nucleus differential event rate

This formalism adopted in this work is well known (see e.g. the recent reviews [39, 40]). So we will briefly discuss its essential elements here. The differential event rate can be cast in the form:

$$\frac{dR}{dQ}|_A = \frac{dR_0}{dQ}|_A + \frac{d\tilde{H}}{dQ}|_A \cos\alpha \qquad (1)$$

where the first term represents the time averaged (non modulated) differential event rate, while the second gives the time dependent (modulated) one due to the motion of the Earth (see below). Furthermore

$$\begin{aligned} \frac{dR_0}{dQ}|_A &= \frac{\rho_\chi}{m_\chi}\frac{m_t}{Am_p}\sigma_n\left(\frac{\mu_r}{\mu_p}\right)^2\sqrt{<v^2>}A^2\frac{1}{Q_0(A)}\frac{dt}{du} \\ \frac{d\tilde{H}}{dQ}|_A &= \frac{\rho_\chi}{m_\chi}\frac{m_t}{Am_p}\sigma_n\left(\frac{\mu_r}{\mu_p}\right)^2\sqrt{<v^2>}A^2\frac{1}{Q_0(A)}\frac{dh}{du} \end{aligned} \qquad (2)$$

with with μ_r (μ_p) the WIMP-nucleus (nucleon) reduced mass, A is the nuclear mass number and σ_n is the elementary WIMP-nucleon cross section. m_χ is the WIMP mass and m_t the mass of the target. Furthermore one can show that

$$\frac{dt}{du} = \sqrt{\frac{2}{3}}a^2F^2(u)\Psi_0(a\sqrt{u}), \quad \frac{dh}{du} = \sqrt{\frac{2}{3}}a^2F^2(u)\Psi_1(a\sqrt{u}) \qquad (3)$$

with $a = (\sqrt{2}\mu_r b \upsilon_0)^{-1}$, υ_0 the velocity of the sun around the center of the galaxy and b the nuclear harmonic oscillator size parameter characterizing the nuclear wave function. u is the energy transfer Q in dimensionless units given by

$$u = \frac{Q}{Q_0(A)} \quad , \quad Q_0(A) = [m_p A b^2]^{-1} = 40A^{-4/3} \text{ MeV} \tag{4}$$

and $F(u)$ is the nuclear form factor. Note that the parameter a depends both on the WIMP , the target and the velocity distribution. Note also that for a given energy transfer Q the quantity u depends on A.
The functions $\Psi_0(a\sqrt{u})$ and $\Psi_1(a\sqrt{u})$ for a M-B distribution take the following form:

$$\Psi_0(x) = \frac{1}{2}(\text{erf}(1-x) + \text{erf}(x+1) + \text{erfc}(1-y_{esc}) + \text{erfc}(y_{esc}+1) - 2) \tag{5}$$

$$\begin{aligned}\Psi_1(x) &= \frac{1}{4}\delta\left(-\text{erf}(1-x) - \text{erf}(x+1) - \text{erfc}(1-y_{esc}) - \text{erfc}(y_{esc}+1)\right. \\ &\left. + \frac{2e^{-(x-1)^2}}{\sqrt{\pi}} + \frac{2e^{-(x+1)^2}}{\sqrt{\pi}} - \frac{2e^{-(y_{esc}-1)^2}}{\sqrt{\pi}} - \frac{2e^{-(y_{esc}+1)^2}}{\sqrt{\pi}} + 2\right)\end{aligned} \tag{6}$$

where $\text{erf}(x)$ and $\text{erfc}(x)$ are the error function and its complement respectively, $\delta \approx 0.135$ is the ratio of the velocity of the Earth around the sun to that of the sun around the center of the galaxy and α is the phase of the Earth ($\alpha = 0$, around June 3nd).

Sometimes we will write the differential rate as:

$$\frac{dR}{dQ}|_A = \frac{\rho_\chi}{m_\chi}\frac{m_t}{Am_p}\sigma_n\left(\frac{\mu_r}{\mu_p}\right)^2\sqrt{<v^2>}A^2\frac{1}{Q_0(A)}\left(\frac{dt}{du}(1 + H(a\sqrt{u})\cos\alpha\right) \tag{7}$$

In this formulation $H(a\sqrt{u})$, the ratio of the modulated to the non modulated differential rate, gives the relative differential modulation amplitude.

The function $H(a\sqrt{u})$ is shown in Fig. 1. It is independent of the nuclear physics and depends only on the reduced mass and the velocity distribution. They are thus the same for both the coherent and the spin mode. Note that $H(a\sqrt{u})$ can take both positive and negative values, which affects the location of the maximum.

3. Some results on differential rates

We will apply the above formalism in the case of NaI, a target used in the DAMA experiment [29, 30]. The results for the Xe target are similar [31]. The differential rates $\frac{dR}{dQ}|_A$ and $\frac{d\tilde{H}}{dQ}|_A$, for each component ($A = 127$ and $A = 23$) are exhibited in Fig. 2-3. Following the practice of the DAMA experiment we express the energy transfer is in keVee using the phenomenological quenching factor [41, 42]. The nuclear form factor has been included (for a heavy target, like ^{127}I or ^{131}Xe, its effect is sizable even for an energy transfer[10] of 10 keV). The differential rate for the spin mode for low energy transfers is similar to those exhibited in Figs 2-3, since the spin form factors are similar. They are, of course, simply scaled down by A^2, if one takes the spin cross section, a combination of the nuclear spin ME and the nucleon spin amplitudes, to be the same with the coherent nucleon cross section, i.e. $\sigma^{spin}_{nuclear} = 10^{-7}$pb. For the actual spin nucleon cross sections extracted from experiment see [43] and [44, 45, 46].

The functions $H(a\sqrt{u})\cos\alpha$ for each target component are shown in Figs 4- 5 as a function of α for various low energy transfers. The corresponding quantities for the spin

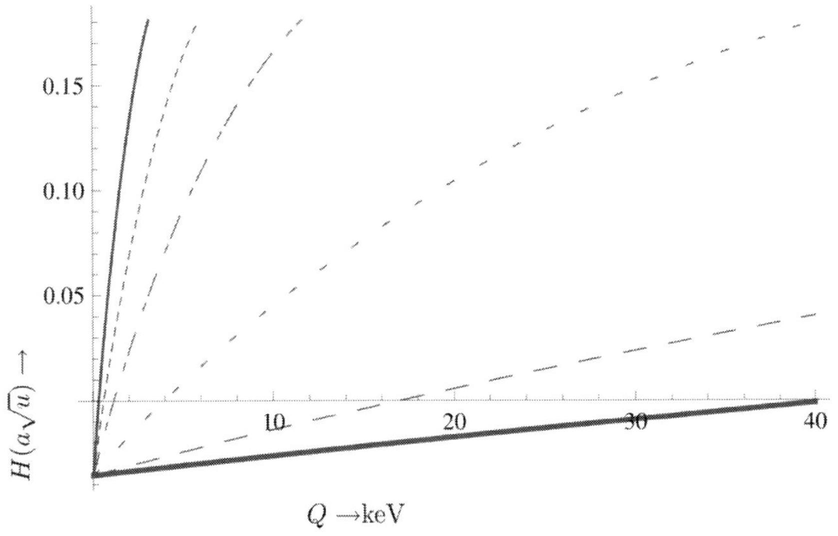

Figure 1: The function $H(a\sqrt{u})$ entering the modulated differential rate as a function of the recoil energy for a heavy target, e.g. ^{127}I. Note that this is independent of the form factor. The solid, dotted, dot-dashed, dashed, long dashed and thick solid lines correspond to 5, 7, 10, 20, 50 and 100 GeV WIMP masses.

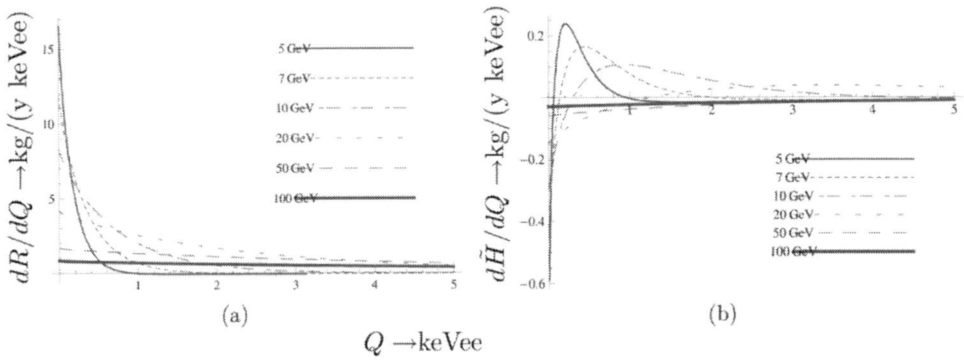

Figure 2: The differential rate $\frac{dR}{dQ}$, as a function of the recoil energy for a heavy target, e.g. ^{127}I (a) and the amplitude for the modulated differential rate $\frac{d\tilde{H}}{dQ}$ (b), assuming a nucleon cross section of 10^{-7}pb. The solid, dotted, dot-dashed, dashed, long dashed and thick solid lines correspond to 5, 7, 10, 20, 50 and 100 GeV WIMP masses. Note that $\frac{d\tilde{H}}{dQ}$ is given in absolute units.

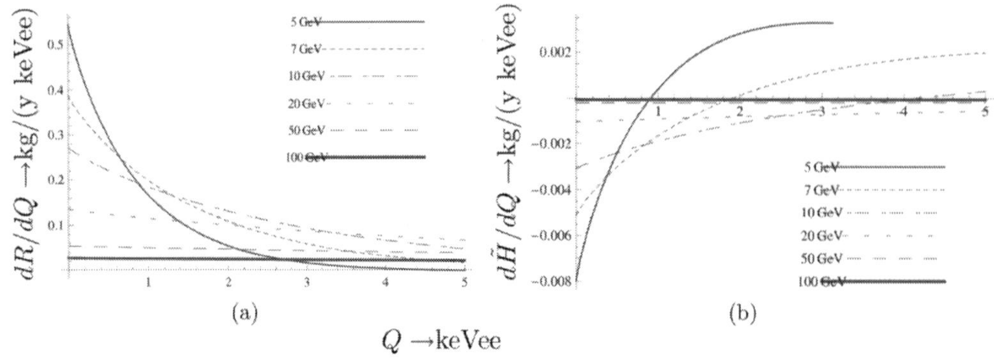

Figure 3: The same as in Fig. 2 for the target ^{23}Na.

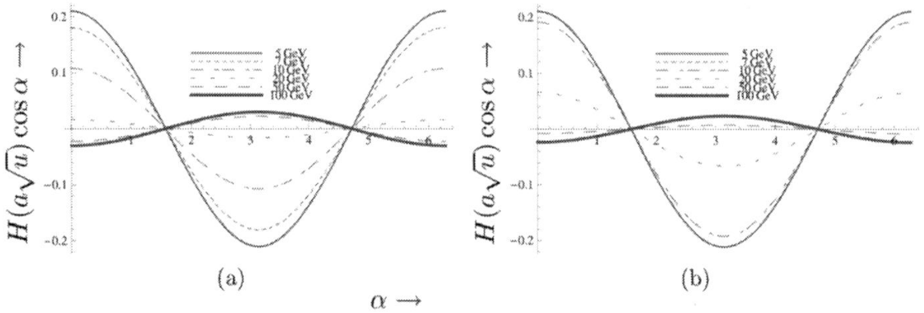

Figure 4: The modulation $H(a\sqrt{u})\cos\alpha$ with an energy transfer of 2 keVee (a) and 5 keVee (b) for a heavy target (I or Xe). The solid, dotted, dot-dashed, dashed, long dashed and thick solid lines correspond to 5, 7, 10, 20, 50 and 100 GeV WIMP masses. Note that for some wimp masses on June 2nd the amplitude becomes negative (location of minimum rate). Note that the modulation is given relative to the time averaged rate.

mode are almost identical. We see that for certain values of the WIMP mass the modulation amplitude changes sign. This may perhaps by exploited to extract information on the WIMP mass from the data. A similar behavior has been found by considering various halo models and different minimum WIMP velocities [22, 23]. Sometimes, as is the case for the DAMA experiment, the target has many components. In such cases the above formalism can be applied as follows:

$$\frac{dR}{dQ}|_A \rightarrow \sum_i X_i \frac{dR}{dQ}|_{A_i}, \quad u \rightarrow u_i, \quad X_i = \text{the fraction of the component } A_i \text{ in the target} \tag{8}$$

The function $H(a\sqrt{u})\cos\alpha$ for NaI as a function of α is exhibited in Fig. 6.

4. Some results on total rates

For completeness and comparison we will briefly present our results on the total rates. Integrating the differential rates discussed in the previous section we obtain the total time averaged rate R_0, the total modulated rate $\tilde{H}$ and the relative modulation amplitude$h$ given

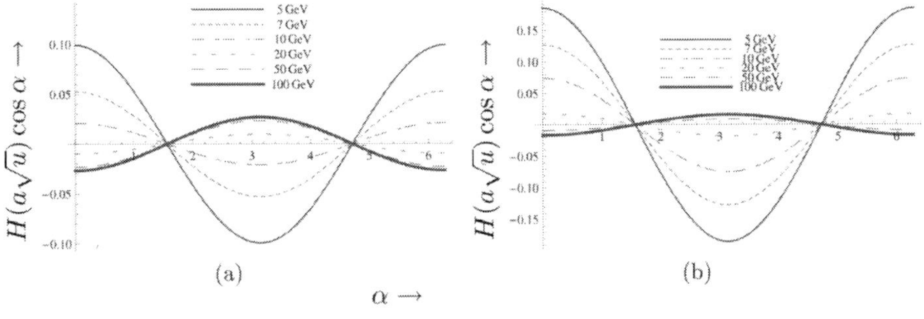

Figure 5: The same as in Fig. 4 for a light target (Na or F).

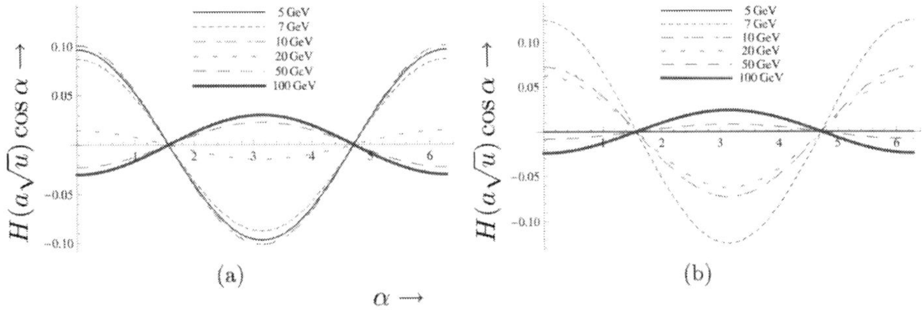

Figure 6: The same as in Fig. 4 for a NaI target.

Table 1: Some total event rates for some special WIMP masses and energy thresholds. The coherent nucleon cross section of $\sigma_n = 10^{-7}$pb was employed.

E_{th} (keVee)	m_{WIMP} GeV	R_0(I) kg-y	$\tilde{H}$(I) kg-y	h(I)	R_0(Na) kg-y	$\tilde{H}$(Na) kg-y	h(Na)	R_0(NaI) kg-y	$\tilde{H}$(NaI) kg-y	h(NaI)
0	80	16.3	-0.311	-0.019	1.518	0.028	0.019	14.0	-0.259	-0.018
0	20	25.8	0.285	0.011	2.35	0.050	0.021	22.2	0.249	0.019
0	10	18.4	0.356	0.019	2.045	0.046	0.022	15.9	0.309	0.019
5	80	7.00	-0.042	-0.006	1.133	0.038	0.034	6.11	-0.030	-0.005
5	20	2.72	0.247	0.091	1.07	0.065	0.060	2.47	0.219	0.089
5	10	0.008	0.001	0.187	0.303	0.031	0.103	0.053	0.006	0.114

by:

$$R = R_0 + \tilde{H}\cos\alpha \text{ or } R = R_0\,(1 + h\cos\alpha) \tag{9}$$

Some special results in the case of low WIMP mass are exhibited in Tables 1-2. From table 2 it becomes clear that, for low mass WIMPs, large nucleon cross sections can accommodate the data. A similar interpretation holds for the [32] data. In the case of non zero threshold one notices the strong dependence of the rime averaged rate on the WIMP mass. Also in this case the relative modulation h substantially increases, the difference between the maximum and the minimum can reach 20%. This however occurs at the expense of the number of counts, since both the time averaged and the time dependent part decrease, but the time averaged part decreases faster. So their ratio increases. This can be understood by noticing that the cancellation of the negative and positive parts in the differential modulated amplitide, becomes less effective in this case.

Table 2: The same as in table 1 for $\sigma_n = 2 \times 10^{-4}$pb relevant for the DAMA region. One sees that, for very low mass WIMPs, large nucleon cross sections are required to obtain the rates claimed by the DAMA experiment [30].

E_{th} (keVee)	m_{WIMP} GeV	R_0(I) kg-y	$\tilde{H}$(I) kg-y	h(I)	R_0(Na) kg-y	$\tilde{H}$(Na) kg-y	h(Na)	R_0(NaI) kg-y	$\tilde{H}$(NaI) kg-y	h(NaI)
0	80	4.07×10^4	-776	-0.019	3.80×10^3	70.2	0.019	3.50×10^4	-647	-0.018
0	20	6.43×10^4	712	0.011	5.87×10^3	126	0.021	5.54×10^4	622	0.011
0	10	4.61×10^4	891	0.019	5.11×10^3	115	0.022	3.98×10^4	772	0.019
5	80	1.75×10^4	-105	-0.006	4.83×10^3	95.0	0.034	1.53×10^4	-74.6	-0.005
5	20	6.80×10^3	617	0.091	2.69×10^3	162	0.060	6.17×10^3	547	0.089
5	10	19.4	3.62	0.187	757	78.1	0.103	132	15.0	0.114

5. Discussion

In the present paper we obtained results on the differential event rates, both modulated and time averaged, focusing our attention on small energy transfers and relatively light WIMPS. We found that:

- The relative modulation amplitude crucially depends on the WIMP mass. For small masses it exhibits normal behavior, but for large masses it changes sign (minimum in June). This effect is more pronounced in the case of heavy targets. We thus suggest an analysis of the experiments, in particular the DAMA experiment, along these lines to establish the location of the maximum on the α-axis.
- The relative modulation amplitude depends somewhat on the energy transfer, especially at low transfers.
- For WIMP masses less than 10 GeV, the difference between the maximum and the minimum could reach between 20% and 40% for a heavy target, but it is a bit less for a light target, depending on the enegy transfer.
- The relative modulation amplitude for NaI is the weighted average of its two components, and in the low energy regime, between 1 and 6 keVee, it does not change much with the energy transfer.
- Once it is established that one actually observes the modulation effect, the sign of the modulation may be exploited to infer the WIMP mass.

For low WIMP mass the total rates depend strongly on the threshold energy, especially for a heavy target. The relative modulation in the presence a threshold gets quite large ($h \approx 0.2$), but, unfortunately, this occurs at the expense of the number of counts. It is important to compare the relative total modulation in a least one light and one heavy target. For very low energy thresholds, if the signs are opposite, one may infer that the WIMP is heavy, $m_{\text{WIMP}} \geq 100$ GeV.

Acknowledgments

The author is indebted to the CERN Theory Division and KITPC/ITP-CAS for their hospitality and support. This work was partially supported by UNILHC PITN-GA-2009-237920.

References

[1] S. Hanary *et al*: *Astrophys. J.* **545**, L5 (2000);
J.H.P Wu *et al*: *Phys. Rev. Lett.* **87**, 251303 (2001);
M.G. Santos *et al*: *Phys. Rev. Lett.* **88**, 241302 (2002).

Journal of Physics: Conference Series **384** (2012) 012023 doi:10.1088/1742-6596/384/1/012023

[2] P. D. Mauskopf *et al*: *Astrophys. J.* **536**, L59 (2002);
S. Mosi *et al*: *Prog. Nuc.Part. Phys.* **48**, 243 (2002);
S. B. Ruhl *al*, astro-ph/0212229 and references therein.

[3] N. W. Halverson *et al*: Astrophys. J. **568**, 38 (2002)
L. S. Sievers *et al*: astro-ph/0205287 and references therein.

[4] G. F. Smoot and *et al* (COBE Collaboration). *Astrophys. J.*, 396:L1, 1992.

[5] A. H. Jaffe and *et al*. *Phys. Rev. Lett.*, 86:3475, 2001.

[6] D. N. Spergel and *et al*. *Astrophys. J. Suppl.*, 148:175, 2003.

[7] D.N. Spergel et al. *Astrophys. J. Suppl.*, 170:377, 2007. [arXiv:astro-ph/0603449v2].

[8] D. P. Bennett and *et al*. *Phys. Rev. Lett.*, 74:2867, 1995.

[9] P. Ullio and M. Kamioknowski. *JHEP*, 0103:049, 2001.

[10] For a recent summary see, e.g., J.D. Vergados, Predicted modulated differential rates for direct WIMP searches at low energy transfers, arXiv:1108.4768 (hep-ph).

[11] M. T. Ressell *et al*., *Phys. Rev. D* **48**, 5519 (1993); M.T. Ressell and D. J. Dean, Phys. Rev. C **56**, 535 (1997).

[12] P. C. Divari, T. S. Kosmas, J. D. Vergados, and L. D. Skouras. *Phys. Rev. C*, 61:054612–1, 2000.

[13] M. W. Goodman and E. Witten. *Phys. Rev. D*, 31:3059, 1985.

[14] A. Drukier, K. Freeze, and D. Spergel. *Phys. Rev. D*, 33:3495, 1986.

[15] J. R. Primack, D. Seckel, and B. Sadoulet. *Ann. Rev. Nucl. Part. Sci.*, 38:751, 1988.

[16] A. Gabutti and K. Schmiemann. *Phys. Lett. B*, 308:411, 1993.

[17] R. Bernabei. *Riv. Nouvo Cimento*, 18 (5):1, 1995.

[18] J. D. Lewin and P. F. Smith. *Astropart. Phys.*, 6:87, 1996.

[19] D. Abriola et al. *Astropart. Phys.*, 10:133, 1999. arXiv:astro-ph/9809018.

[20] F. Hasenbalg. *Astropart. Phys.*, 9:339, 1998. arXiv:astro-ph/9806198.

[21] J. D. Vergados. *Phys. Rev. D*, 67:103003, 2003. hep-ph/0303231.

[22] A.M. Green. *Phys. Rev. D*, 68:023004, 2003. ibid: D **69** (2004) 109902; arXiv:astro-ph/0304446.

[23] C. Savage, K. Freese, and Paolo Gondolo. *Phys. Rev. D*, 74:043531, 2006. arXiv:astro-ph/0607121.

[24] D. Spergel. *Phys. Rev. D*, 37:1353, 1988.

[25] J. D. Vergados, P. Quentin, and D. Strottman. *IJMPE*, 14:751, 2005. hep-ph/0310365.

[26] J. D. Vergados and H. Ejiri. *Phys. Lett. B*, 606:305, 2005. hep-ph/0401151.

[27] Ch. C. Moustakidis, J. D. Vergados, and H. Ejiri. *Nucl. Phys. B*, 727:406, 2005. hep-ph/0507123.

[28] Ch. C. Moustakidis, J.D. Vergados, and H. Ejiri. *Nucl. Phys.*, B 727:406, 2005.

[29] R. Bernabei and Others. *Eur. Phys. J. C*, 56:333, 2008. [DAMA Collaboration]: [arXiv:0804.2741 [astro-ph]].

[30] P. Belli *et al*, arXiv:1106.4667 [astro-ph.GA].

[31] J. Angle *et al*, arXiv:1104.3088 [hep-ph].

[32] C.E. Aalseth et al. *Phys. Rev. Lett.*, 106:131301, 2011. CoGeNT collaboration arXiv:10002.4703 [astro-ph.CO].

[33] M. Farina, D. Pappadopulo, A. Strumia, T. Volansky, Can CoGeNT and DAMA Modulations Be Due to Dark Matter? arXiv:1107.0715 [hep-ph].

[34] Junjie Cao, Ken-ichi Hikasa, Wenyu Wang, Jin Min Yang, arXiv:1104.1754 [hep-ph].

[35] J.D. Vergados and D. Owen. *Phys. Rev.*, D 75:043503, 2007.

[36] J.D. Vergados. *Astronomical Journal*, 137:10, 2009. [arXiv:0811.0382 (astro-ph)].

[37] N. Tetradis, J.D. Vergados, and Amand Faessler. *Phys. Rev.*, D 75:023504, 2007.

[38] J. D. Vergados, S. H. Hansen, and O. Host. *Phys. Rev. D*, D 77:023509, 2008.

[39] J. D. Vergados. *Lect. Notes Phys.*, 720:69, 2007. hep-ph/0601064.

[40] J. D. Vergados and Ch. C. Moustakidis. *Eur. J. Phys.*, 9(3):628, 2011. arXiv:0912.3121 [astro-ph.CO].

[41] J. Lidhart *et al*, Mat. Phys. Medd. Dan. Vid. Selsk, 33 (1963) 1.

[42] E. Simon. *et al, Nucl. Instr. Meth.*, A 507:643, 2003.

[43] J. D. Vergados. *Nuc. Phys. B*, 829:383, 2010. arXiv:0907.3587 [hep-ph].

[44] S. Archambault et al. *Phys. Lett. B*, 682:185, 2009. collaboration PICASSO, arXiv:0907.0307 [astro-ex].

[45] E. Behnke et al. *Phys. Rev, Lett.*, 106:021303, 2011. collaboration COUPP, arXiv:1008.3518 [astro-phCO].

[46] M. Felizardo et al. *Phys. Rev, Lett.*, 105:211301, 2010. collaboration SIMPLE, arXiv:1003.2987 [astro-phCO].

Dark matter conversion as a source of boost factor

Ze-Peng Liu, Yue-Liang Wu, and Yu-Feng Zhou
State Key Laboratory of Theoretical Physics,
Kavli Institute for Theoretical Physics China,
Institute of Theoretical Physics,
Chinese Academy of Sciences, Beijing, 100190, P.R. China
E-mail: yfzhou@itp.ac.cn

Abstract. In interacting multi-component dark matter (DM) models, the interactions between the DM components can covert relatively heavy DM components into lighter ones at late times after the thermal decoupling. As a consequence, the relic density of the lightest DM component can be greatly enhanced at late times, which can lead to an alternative source of boost factor required to explain the positron and electron excesses reported by the recent DM indirect search experiments.

In this talk, we discuss an alternative origin of the boost factor from the late time dark matter conversion processes [1]. In the recent years, a number of experiments such as PAMELA [2], ATIC [3], Fermi-LAT [4] and HESS [5] etc. have reported excesses in the high energy spectrum of cosmic-ray positrons and electrons over the backgrounds estimated from the standard astrophysics. The dark matter (DM) annihilation or decay provides possible exciting explanations from particle physics. If the DM particle is a thermal relic such as the weakly interacting massive particle (WIMP), the thermally averaged product of its annihilation cross section with the relative velocity at the time of thermal freeze out is typically $\langle\sigma v\rangle_F \simeq 3\times 10^{-26}\text{cm}^3\text{s}^{-1}$. The positron or electron flux produced by the DM annihilation can be parametrized by

$$\Phi_e = B\overline{N}_e\frac{\rho_0^2\langle\sigma v\rangle_F}{m_D^2}, \tag{1}$$

where ρ_0 is the smooth local halo DM energy density estimated from astrophysics, $\overline{N}_e$ is the averaged electron number produced per DM annihilation which depends on DM models and propagation parameters, and m_D is the mass of the DM particle. The boost factor B is defined as $B \equiv (\rho/\rho_0)^2\langle\sigma v\rangle/\langle\sigma v\rangle_F$ with ρ the true local DM density and $\langle\sigma v\rangle$ the DM annihilation cross section multiplied by the relative velocity and averaged over the DM velocity distribution today. Both the PAMELA and Fermi-LAT results indicate that a large boost factor is needed [6, 7]. For a typical DM mass of $\sim$1(1.6) TeV the required boost factor B is $\sim$ 500(1000) for DM annihilating directly into $\mu^+\mu^-$ and ρ fixed at $\rho_0 = 0.3$ GeV $\cdot$ cm^{-3} [7].

A large boost factor may arise from the non-uniformity of the DM distribution in the DM halo. The N-body simulations show however that the local cumps of dark matter density are unlikely to contribute to a large enough ρ/ρ_0 [8,9]. An other possibility of enhancing the boost factor is that the DM annihilation cross section may be velocity-dependent which grows at low velocity. The DM annihilation cross section today may be much larger than that at the time of thermal freeze out, and thus is not constrained by the DM relic density. Some enhancement mechanisms have been proposed along this line, such as the Sommerfeld enhancement [10–18] and the resonance enhancement [19–21] etc. In some non-thermal DM scenarios, the number density of the DM particle can be enhanced by the out of equilibrium decay of some heavier unstable particles if the DM particle is among the decay products of the decaying particle [22,23]. The decay of the unstable particle must take place at very late time. Otherwise the DM particles with the enhanced number density will annihilate into the Standard Model (SM) particles again, which washes out the effect of the enhancement.

In this talk, we discuss an alternative possibility for generating a boost factor, which does not require the velocity-dependent annihilation cross section or the decay of unstable particles. We show that in the scenarios of interacting Milt-component DM, the interactions among the DM components may convert the heavier DM components into the lighter ones, which is not sensitive to the details of the conversion interaction. If the interactions are strong enough and the DM components are nearly degenerate in mass, the conversion can enhance the number density of the lighter DM components at late time after the thermal decoupling. Eventually, the whole DM today in the Universe can be dominated by the lightest DM component with enhanced number density, which corresponds to a large boost factor. The scenarios of multi-component DM have been discuss previously in Refs. [24–33]. Note however that the models with simply mixed non-interacting multi-component DM cannot generate large boost factors.

Let us consider a generic model in which the whole cold DM contains N components χ_i $(i = 1, \ldots, N)$, with masses m_i and internal degrees of freedom g_i respectively. The DM components are labeled such that $m_i < m_j$ for $i < j$, thus χ_1 is the lightest DM particle. We are interested in the case that χ_i are nearly degenerate in mass, namely the relative mass differences between χ_i and χ_1 satisfy $\varepsilon_i \equiv (m_i - m_1)/m_1 \ll 1$. In this case, we shall show that the interactions between the DM components lead to the DM conversion. The thermal evolution of the DM number density normalized to the entropy density $Y_i \equiv n_i/s$ with respect to the rescaled temperature $x \equiv m_1/T$ is govern by the following Boltzmann equation

$$\frac{dY_i(x)}{dx} = -\frac{\lambda}{x^2}\left[\langle\sigma_i v\rangle(Y_i^2 - Y_{ieq}^2) - \sum_j \langle\sigma_{ij} v\rangle(Y_i^2 - r_{ij}^2 Y_j^2)\right], \tag{2}$$

where $\lambda \equiv xs/H(T)$ is a combination of x, the entropy density s and the Hubble parameter $H(T)$ as a function of temperature T. $Y_{ieq} \simeq (g_i/s)[m_i T/(2\pi)]^{3/2}\exp(-\varepsilon_i x)$ is the equilibrium number density normalized to entropy density for non-relativistic particles. $\langle\sigma_i v\rangle$ are the thermally averaged cross sections multiplied by the DM relative velocity for the process $\chi_i\chi_i \to XX'$ with XX' standing for the light SM particles which are in thermal equilibrium, and $\langle\sigma_{ij} v\rangle$ are the ones for the DM conversion process $\chi_i\chi_i \to \chi_j\chi_j$. The quantity

$$r_{ij}(x) \equiv \frac{Y_{ieq}(x)}{Y_{jeq}(x)} = \left(\frac{g_i}{g_j}\right)\left(\frac{m_i}{m_j}\right)^{3/2} \exp[-(\epsilon_i - \epsilon_j)x] \tag{3}$$

is the ratio between the two equilibrium number density functions, In Eq. (2) we have assumed kinetic equilibrium. The first term in the r.h.s. of Eq.(2) describes the change of number density of χ_i due to the annihilation into the SM particles, and the second term describes the change due to the conversion to other DM particles.

In the case that the cross section of the conversion process $\langle\sigma_{ij} v\rangle$ is large enough, the DM particle χ_i can be kept in thermal equilibrium with χ_j for a long time after both χ_i and χ_j have decoupled from the thermal equilibrium with the SM particles. In this case, the number densities of $\chi_{i,j}$ satisfy a simple relation

$$\frac{Y_i(x)}{Y_j(x)} \approx \frac{Y_{ieq}(x)}{Y_{jeq}(x)} = r_{ij}(x). \tag{4}$$

Even when χ_i is in equilibrium with χ_j the ratio of the number density $Y_i(x)/Y_j(x)$ can be quite different from unity and can vary with temperature. For instance, if $g_i \gg g_j$ and $0 < (\epsilon_i - \epsilon_j) \ll 1$, from Eq. (3) and (4) one obtains $Y_i(x) \gg Y_j(x)$ at the early time when $(\epsilon_i - \epsilon_j)x \ll 1$. However, at the late time when $(\epsilon_i - \epsilon_j)x \gg 1$, one gets $Y_i(x) \ll Y_j(x)$, which is due to the Boltzmann suppression factor $\exp[-(\epsilon_i - \epsilon_j)x]$ in the expression of r_{ij}. Thus the heavier particles can be gradually converted into lighter ones through this temperature-dependent equilibrium between χ_i and χ_j.

An interesting limit to consider is that the rates of DM conversion are large compared with that of the individual DM annihilation into the SM particles, i.e. $\langle\sigma_{ij}v\rangle \gtrsim \langle\sigma_i v\rangle$. In this limit, after both the DM particles have decoupled from the thermal equilibrium with the SM particles, which take place at a typical temperature $x = x_{dec} \approx 25$, the strong interactions of conversion will maintain an equilibrium between χ_i and χ_j for a long time until the rate of the conversion cannot compete with the expansion rate of the Universe. Making use of Eq. (4), the evolution of the total density $Y(x) \equiv \sum_{i=1}^{N} Y_i(x)$ can be written as

$$\frac{dY}{dx} = -\frac{\lambda}{x^2}\langle\sigma_{eff}v\rangle\left(Y^2 - Y_{eq}^2\right), \tag{5}$$

where $\langle\sigma_{eff}v\rangle$ is the effective thermally averaged product of DM annihilation cross section and the relative velocity which can be written as

$$\langle\sigma_{eff}v\rangle = \frac{\sum_{i=1}^{N} w_i g_i^2(1+\varepsilon_i)^3\exp(-2\varepsilon_i x)}{g_{eff}^2}\langle\sigma_1 v\rangle, \tag{6}$$

where $w_i \equiv \langle\sigma_i v\rangle/\langle\sigma_1 v\rangle$ is the annihilation cross section relative to that of the lightest one. The total equilibrium number density can be written as

$$Y_{eq} \equiv \sum_{i=1}^{N} Y_{ieq}(x) \approx g_{eff}\left(\frac{m_1 T}{2\pi}\right)^{3/2}\exp(-x), \tag{7}$$

with effective degrees of freedom $g_{eff} = \sum_i g_i(1+\varepsilon_i)^{3/2}\exp(-\varepsilon_i x)$. Note that the conversion terms do not show up explicitly in Eq. (5). Through the conversion processes $\chi_i\chi_i \to \chi_j\chi_j$ the slightly heavier components will be converted into the lighter ones, because the factor $r_{ij}(x)$ is proportional to $\exp[-(m_i - m_j)/T]$ which suppresses the density of the heavier components at lower temperature. If the conversion cross section is large enough, most of the DM components will be converted into the lightest χ_1 before the interaction of conversion decouples, which may result in a large enhancement of the relic density of χ_1 and leads to a large boost factor.

As an example, let us consider a generic DM model with only two components. For relatively large conversion cross section $u \equiv \langle\sigma_{21}v\rangle/\langle\sigma_1 v\rangle \gtrsim 1$, The effective total cross section is given by $\langle\sigma_{eff}v\rangle = \langle\sigma_1 v\rangle[1 + wg^2\exp(-2\varepsilon x)]/[1 + g\exp(-\varepsilon x)]^2]$, where $w \equiv w_2$, $g \equiv g_2/g_1$ and $\varepsilon \equiv \varepsilon_2$. Because of the x-dependence in $\langle\sigma_{eff}v\rangle$, the thermal evolution of $Y(x)$ differs significantly from that of the standard WIMP. In the case that χ_2 has large degrees of freedom but a small annihilation cross section, namely $g \gg 1$, $w \ll 1$ and $wg^2 \ll 1$, the thermal evolution of the total density Y can be simplified. The thermal evolution of the total number density can be roughly divided into four stages: i) At high temperature region where $3 \lesssim x \ll x_{dec}$, both the DM components are in thermal equilibrium with the SM particles. $Y_i(x)$ must closely track $Y_{ieq}(x)$ which decrease exponentially as x increases. However, since $g \gg 1$ and $\epsilon \ll 1$, the number density of χ_2 is much higher than that of χ_1, i.e. $Y_2(x) \gg Y_1(x)$. ii) When the temperature goes down and x is close to the decoupling point x_{dec}, both the DM components start to decouple from the thermal equilibrium. In the region $x_{dec} \lesssim x \ll 1/\varepsilon$, $\langle\sigma_{eff}v\rangle$ is nearly a constant and

$\langle\sigma_{eff}v\rangle \approx \langle\sigma_1 v\rangle/(1+g)^2 \ll \langle\sigma_1 v\rangle$, the total density $Y(x)$ behaves just like that of an ordinary WIMP which converges quickly to $Y(x) \approx x_{dec}/(\lambda\langle\sigma_1 v\rangle)$. iii) As x continues growing, the suppression factor $\exp(-\varepsilon x)$ in $\langle\sigma_{eff}v\rangle$ becomes relevant. The value of $\langle\sigma_{eff}v\rangle$ grows rapidly especially after x reaches the point $\varepsilon x \approx \mathcal{O}(1)$, which leads to the further reduction of $Y(x)$. In this stage, although both $\chi_{1,2}$ have decoupled from the thermal equilibrium with the SM particles. The strong conversion interaction $\chi_2\chi_2 \leftrightarrow \chi_1\chi_1$ maintains an equilibrium between the two DM components. According to Eq. (4), the relative number density $Y_2(x)/Y_1(x)$ decreases with x increasing, which corresponds to the conversion from the heavier DM component into the lighter one. At the point $x_c = (1/\varepsilon)\ln g$ one has $Y_2(x) \approx Y_1(x)$. For the region $x > x_{dec}$ and x is not close to x_c, because of $Y_{eq}(x) \ll Y(x)$ and $g\exp(-\varepsilon x) \gg 1$, the total number density can be analytically integrated out, and $Y(x)$ in this region can be approximated by

$$Y(x) \approx \frac{g^2 x_{dec}}{\lambda\langle\sigma_1 v\rangle}\left[1+\left(\frac{x_{dec}}{x}\right)\frac{\exp(2\varepsilon x)}{2\varepsilon x}\right]^{-1}. \tag{8}$$

iv) When x becomes very large $\varepsilon x \gg \mathcal{O}(1)$, $\langle\sigma_{eff}v\rangle$ quickly approaches $\langle\sigma_1 v\rangle$, and becomes independent of x again. The evolution of $Y(x)$ in this region can be obtained by a simple integration as it was done in the stage ii). The solution of $Y(x)$ shows a second decoupling. Finally when the conversion rate cannot compete with the expansion rate of the Universe at some point x_F corresponding to $sY_2\langle\sigma_{21}v\rangle/H \approx 1$, both $Y_1(x)$ and $Y_2(x)$ remain unchanged as relics. The whole DM can be dominated by χ_1 if the conversion is efficient enough.

By matching the analytic solutions of $Y(x)$ in different regions near the points x_{dec} and x_c, and requiring that the final total relic density is equivalent to the observed $\Omega_{CDM}h^2 \approx 0.11$, we obtain the following approximate expression of the boost factor

$$B \approx g^2\left[1+\left(\frac{x_{dec}}{x_c}\right)\left(\frac{\exp(2\varepsilon x_c)}{2\varepsilon x_c}+g^2\right)\right]^{-1}. \tag{9}$$

As expected, the enhancement essentially comes from the conversion of the degrees of freedom. Thus the maximum enhancement is g^2. The two terms in the r.h.s of the above equation correspond to the reduction of $Y(x)$ during the late time conversion stages. For large enough g, the boost factor can be approximated by $B \approx g^2/(1+\varepsilon g^2 x_{dec}/\ln g)$. In order to have a large boost factor, a small $\varepsilon \ll \ln g/(g^2 x_{dec})$ is also required. As shown in Eq. (9) the boost factor is not sensitive to the exact values of the cross sections as long as the conditions $w \ll 1$ and $u \gg 1$ are satisfied.

We numerically calculate the thermal evolution of $Y_i(x)$ and the boost factor without using approximations for a generic two-component DM model. The results for $w = 10^{-4}$, $u = 10$ and $\varepsilon = 2\times10^{-4}$ is shown in Fig. 1. The value of $\langle\sigma_2 v\rangle$ is adjusted such that the final total DM relic abundance is always equal to the observed value $\Omega_{CDM}h^2$. The mass of the light DM particle is set to $m_1 = 1$ TeV. For an illustration the ratio between the internal degrees of freedom is set to be large $g = 60$. From the figure, the four stages of the thermal evolution of $Y(x)$ as well as the crossing point can be clearly seen. The crossing point at $x = x_c \approx 2\times10^{-4}$ indicates the time when the number density of χ_1 start to surpass that of χ_2 and eventually dominant the whole DM relic density. In this parameter set a large boost factor $B \approx \langle\sigma_1 v\rangle/\langle\sigma v\rangle_F \approx 585$ is obtained which is in a remarkable agreement with Eq. (9) with error less than $\sim 5\%$. For a comparison, in Fig. 1 we also show the cases without conversions.

The whole DM in the universe necessarily contains multiple components, as the lightest active neutrino already contributes to a small fraction of the DM relic density. It is easy to construct models with more stable neutrinos or neutrinos with lifetime longer than that of the universe. For instance, in fourth generation models with right-handed neutrinos, extra stable neutrinos may be the keV scale sterlile neutrinos and the heavy Majorana neutrinos which are stable due

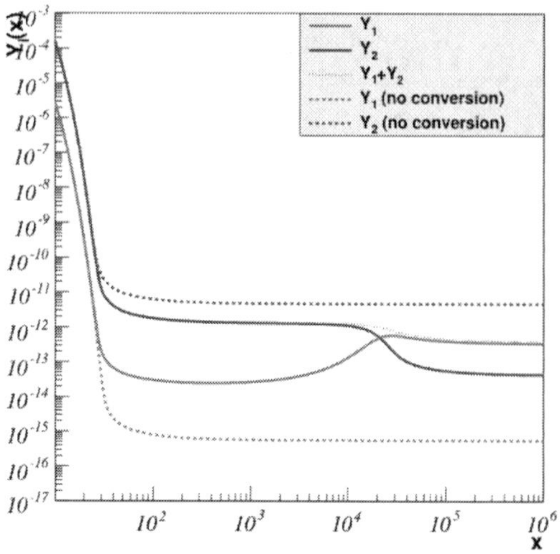

Figure 1. Thermal evolution of the number densities $Y_1(x)$ (red solid) and $Y_2(x)$ (blue solid) with respect to x. The solid (dashed) curves correspond to the case with (without) DM conversions. The green dotted curve corresponds to the sum of Y_1 and Y_2, for parameters $g = 60$, $m_1 = 1\text{TeV}$, $\varepsilon = 2 \times 10^{-4}$, $w = 10^{-4}$ and $u = 10$ respectively.

to additional symmetries [34]. For models with multiple DM components, it is possible that there exists interactions among the DM components which may lead to the conversions among them. In this talk we consider a simple interacting two-component DM model by adding to the standard model (SM) with two SM gauge singlet fermionic DM particles $\chi_{1,2}$. The particles $\chi_{1,2}$ are charged under a local $U(1)$ symmetry which is broken spontaneously by the vacuum expectation value (VEV) of a scalar field ϕ. The corresponding massive gauge boson is denoted by A which may cause the reaction $\bar{\chi_2}\chi_2 \leftrightarrow \bar{\chi_1}\chi_1$. The stability of $\chi_{1,2}$ is protected by two different global $U(1)$ number symmetries. An SM gauge singlet pseudo-scalar η is introduced as a messenger field which couples to both the dark sector and the SM sector. In order to have the leptophilic nature of DM annihilation, we also introduce an SM $SU(2)_L$ triplet field Δ with the SM quantum number $(1,3,1)$ and flavor contents $\Delta = (\delta^{++}, \delta^+, \delta^0)$. The triplet carries the quantum number $B - L$=2 such that it can couple to the SM left-handed leptons ℓ_L through Yukawa interactions $\bar{\ell}^c_L \Delta \ell_L$, but cannot couple to quarks directly. The VEV of the triplet has to be very small around eV scale, which is required by the smallness of the neutrino masses. As a consequence, the couplings between one triplet and two SM gauge bosons such as $\delta^{\pm\pm}W^{\mp}W^{\mp}$, $\delta^{\pm}W^{\mp}Z^0$ and $\delta^0 Z^0 Z^0$ are strongly suppressed as they are all proportional to the VEV of the triplet, which makes it difficult for the triplet to decay even indirectly into quarks through SM gauge bosons [35–40]. If η has a stronger coupling to Δ than that to the SM Higgs boson H and ϕ then the annihilation products of the dark matter particles $\chi_{1,2}$ will be mostly leptons.

The Lagrangian of the model can be written as $\mathcal{L} = \mathcal{L}_{SM} + \mathcal{L}_1$ The new interactions in $\mathcal{L}_1$ which are relevant to the DM annihilation and conversion are given by

$$\begin{aligned}
\mathcal{L}_1 \supset\ & \bar{\chi}_i(i\not{D} - m_i)\chi_i + (D_\mu\phi)^\dagger(D^\mu\phi) - m_\phi^2\phi^\dagger\phi \\
& + \frac{1}{2}\partial_\mu\eta\partial^\mu\eta - \frac{1}{2}m_\eta^2\eta^2 - y_i\bar{\chi}_i i\gamma_5\eta\chi_i - y_\ell\bar{\ell}^c_L\Delta\ell_L + \text{h.c} \\
& -(\mu\eta + \xi\eta^2)\left[\text{Tr}(\Delta^\dagger\Delta) + \kappa(H^\dagger H) + \zeta(\phi^\dagger\phi)\right], \quad (i = 1,2) \qquad (10)
\end{aligned}$$

Note that ϕ and η do not directly couple to the SM fermions. After the spontaneous symmetry breaking in $V(\phi)$, the scalar ϕ obtains a nonzero VEV $\langle\phi\rangle = v_\phi/\sqrt{2}$ which generates the mass of the gauge boson $m_A = g_A v_\phi$. At the tree level, the three components of the triplet δ^{++}, δ^+ and δ^0 are degenerate in mass, i.e. $m_{\delta^{++}} = m_{\delta^+} = m_{\delta^+} \equiv m_\Delta$.

We assume that χ_2 has large internal degrees of freedom relative to that of χ_1, i.e., $g_2 \gg g_1$, which can be realized if χ_2 belongs to a multiplet of the product of some global nonabelian groups. For instance $g_2 = 4\tilde{g}_2$ with $\tilde{g}_2$ =16, 8, and 4 if it belongs to the spinor representation of a single group of $SO(8)$, $SO(6)$ and $SO(4)$ respectively. When χ_2 belongs to a representation of the product of these groups, its internal degrees of freedom can be very large.

At the early time when the temperature of the Universe is high enough, the triplet Δ can be kept in thermal equilibrium with SM particles through the SM gauge interactions. The DM particles χ_i are in thermal equilibrium by annihilating into the triplet through the intermediate particle η. The annihilation $\bar{\chi}_2\chi_2 \to \eta^* \to \delta^{\pm\pm}\delta^{\mp\mp}, \delta^\pm\delta^\mp, \delta^0\delta^{0*}$ is an s-wave process which is the dominant contribution . The ratio of the two annihilation cross sections is $w = (y_2/y_1)^2(g_1/g_2)$. It is easy to get a very small w provided that $y_2 \ll y_1$ and $g_1 \ll g_2$. In order to have a large enough $\langle\sigma_1 v\rangle \gg \langle\sigma v\rangle_F$ the product of the coupling constants $y_1\mu$ must be large enough, or the squared mass of η is close to s. The cross section of the conversion process $\bar{\chi}_2\chi_2 \to A^* \to \bar{\chi}_1\chi_1$ is suppress by g_1/g_2 and also the phase space factor $\sqrt{1-4m_1^2/s}$ when s is close to $4m_2^2$ at the vary late time of the thermal evolution. However, the cross section can be greatly enhanced if m_A is close to a resonance when the relation $s \simeq m_A^2$ is satisfied. In the numerical calculations, we find that for the following selected parameters: m_1 = 1TeV, $\epsilon = 1 \times 10^{-4}$, $g_1 = 1$, $g_2 = 60$, m_Δ = 500 GeV, m_η = 1.5 TeV, m_A = 2.02 TeV, $y_1 = 3$, $y_2 = 0.07$, $\mu/m_1 = 3$, and $g_A = 2.5$, the following ratio of the cross section can be obtained

$$w \simeq 1 \times 10^{-5}, \;\; u \simeq 0.5, \text{ and } \langle\sigma_1 v\rangle/\langle\sigma v\rangle_F \simeq 500.$$

In this parameter set the relative mass difference between m_A and $2m_2$ is around 1%. The corresponding boost factor is $B \sim 500$, which is large enough to account for the PAMELA data for the dark matter mass around TeV.

In summary, We have considered an alternative mechanism for obtaining boost factors from DM conversions which does not require the velocity-dependent annihilation cross section or the decay of unstable particles. We have shown that if the whole DM is composed of multiple components, the relic density of each DM component may not necessarily be inversely proportional to its own annihilation cross section. We demonstrate the possibility that the number density of the lightest DM component can get enhanced in late time through DM conversation processes, and finally dominates the whole relic abundance, which corresponds to a boost factor needed to explain the excesses in cosmic-ray positron and electrons reported by the recent experiments.

Acknowledgments
This work is supported in part by the National Basic Research Program of China (973 Program) under Grants No. 2010CB833000; the National Nature Science Foundation of China (NSFC) under Grants No. 10975170, No. 10821504 and No. 10905084; and the Project of Knowledge Innovation Program (PKIP) of the Chinese Academy of Science.

References

[1] Liu Z P, Wu Y L and Zhou Y F 2011 *Eur. Phys. J.* **C71** 1749 (*Preprint* 1101.4148)
[2] Adriani O *et al.* (PAMELA) 2009 *Nature* **458** 607–609 (*Preprint* 0810.4995)

[3] Chang J *et al.* 2008 *Nature* **456** 362–365
[4] Abdo A A *et al.* (The Fermi LAT) 2009 *Phys. Rev. Lett.* **102** 181101 (*Preprint* 0905.0025)
[5] Aharonian F *et al.* (H.E.S.S.) 2009 *Astron. Astrophys.* **508** 561 (*Preprint* 0905.0105)
[6] Cholis I, Goodenough L, Hooper D, Simet M and Weiner N 2009 *Phys. Rev.* **D80** 123511 (*Preprint* 0809.1683)
[7] Bergstrom L, Edsjo J and Zaharijas G 2009 *Phys. Rev. Lett.* **103** 031103 (*Preprint* 0905.0333)
[8] Springel V *et al.* 2008 (*Preprint* 0809.0894)
[9] Diemand J *et al.* 2008 *Nature* **454** 735–738 (*Preprint* 0805.1244)
[10] Sommerfeld A 1931 *Annalen der Physik* **403** 257
[11] Hisano J, Matsumoto S and Nojiri M M 2003 *Phys. Rev.* **D67** 075014 (*Preprint* hep-ph/0212022)
[12] Hisano J, Matsumoto S and Nojiri M M 2004 *Phys. Rev. Lett.* **92** 031303 (*Preprint* hep-ph/0307216)
[13] Cirelli M, Strumia A and Tamburini M 2007 *Nucl. Phys.* **B787** 152–175 (*Preprint* 0706.4071)
[14] Arkani-Hamed N, Finkbeiner D P, Slatyer T R and Weiner N 2009 *Phys. Rev.* **D79** 015014 (*Preprint* 0810.0713)
[15] Pospelov M and Ritz A 2009 *Phys. Lett.* **B671** 391–397 (*Preprint* 0810.1502)
[16] March-Russell J D and West S M 2009 *Phys. Lett.* **B676** 133–139 (*Preprint* 0812.0559)
[17] Iengo R 2009 *JHEP* **05** 024 (*Preprint* 0902.0688)
[18] Cassel S 2010 *J. Phys.* **G37** 105009 (*Preprint* 0903.5307)
[19] Feldman D, Liu Z and Nath P 2009 *Phys. Rev.* **D79** 063509 (*Preprint* 0810.5762)
[20] Ibe M, Murayama H and Yanagida T T 2009 *Phys. Rev.* **D79** 095009 (*Preprint* 0812.0072)
[21] Guo W L and Wu Y L 2009 *Phys. Rev.* **D79** 055012 (*Preprint* 0901.1450)
[22] Fairbairn M and Zupan J 2009 *JCAP* **0907** 001 (*Preprint* 0810.4147)
[23] Feldman D, Liu Z, Nath P and Nelson B D 2009 *Phys. Rev.* **D80** 075001 (*Preprint* 0907.5392)
[24] Boehm C, Fayet P and Silk J 2004 *Phys. Rev.* **D69** 101302 (*Preprint* hep-ph/0311143)
[25] Hur T, Lee H S and Nasri S 2008 *Phys. Rev.* **D77** 015008 (*Preprint* 0710.2653)
[26] Adibzadeh M and Hung P Q 2008 *Nucl. Phys.* **B804** 223–249 (*Preprint* 0801.4895)
[27] Feng J L and Kumar J 2008 *Phys. Rev. Lett.* **101** 231301 (*Preprint* 0803.4196)
[28] Zurek K M 2009 *Phys. Rev.* **D79** 115002 (*Preprint* 0811.4429)
[29] Batell B, Pospelov M and Ritz A 2009 *Phys. Rev.* **D79** 115019 (*Preprint* 0903.3396)
[30] Profumo S, Sigurdson K and Ubaldi L 2009 *JCAP* **0912** 016 (*Preprint* 0907.4374)
[31] Zhang H, Li C S, Cao Q H and Li Z 2010 *Phys. Rev.* **D82** 075003 (*Preprint* 0910.2831)
[32] Gao X, Kang Z and Li T 2010 *Eur. Phys. J.* **C69** 467–480 (*Preprint* 1001.3278)
[33] Feldman D, Liu Z, Nath P and Peim G 2010 *Phys. Rev.* **D81** 095017 (*Preprint* 1004.0649)
[34] Zhou Y F 2011 (*Preprint* 1110.2930)
[35] Gogoladze I, Okada N and Shafi Q 2009 *Phys. Lett.* **B679** 237–241 (*Preprint* 0904.2201)
[36] Guo W L, Wu Y L and Zhou Y F 2011 *Int. J. Mod. Phys.* **D20** 1389–1397
[37] Guo W L, Wu Y L and Zhou Y F 2010 *Phys. Rev.* **D82** 095004 (*Preprint* 1008.4479)
[38] Guo W L, Wu Y L and Zhou Y F 2010 *Phys. Rev.* **D81** 075014 (*Preprint* 1001.0307)
[39] Guo W L, Wang L M, Wu Y L, Zhou Y F and Zhuang C 2009 *Phys. Rev.* **D79** 055015 (*Preprint* 0811.2556)
[40] Wu Y L and Zhou Y F 2008 *Sci. China* **G51** 1808–1825 (*Preprint* 0709.0042)

Journal of Physics: Conference Series **384** (2012) 012025 doi:10.1088/1742-6596/384/1/012025

Higgs Boson Mass in Yukawa Unified SUSY SO(10)

Qaisar Shafi

Bartol Research Institute, Department of Physics and Astronomy, University of Delaware, Newark, DE 19716, USA

E-mail: shafi@bartol.udel.edu

Abstract. We employ third family Yukawa unification, predicted by simple supersymmetric SO(10) models, to estimate the lightest MSSM Higgs boson mass. For $\mu > 0$ (or $\mu < 0$) and $m_t = 173.1$ GeV, the Higgs mass is estimated to lie close to 123-124 GeV. The theoretical uncertainty in this estimate is ± 3 GeV. We highlight some LHC testable benchmark points which also display the presence of neutralino-stau coannihilation channel.

1. Introduction

Supersymmetric (SUSY) $SO(10)$ grand unified theory (GUT), in contrast to its non-SUSY version, yields third family (t-b-τ) Yukawa unification via the unique renormalizable Yukawa coupling $16 \cdot 16 \cdot 10$, if the Higgs 10-plet is assumed to contain the two Higgs doublets H_u and H_d of the minimal supersymmetric standard model (MSSM) [1]. The matter 16-plet contains the 15 chiral superfields of MSSM as well as the right handed neutrino superfield. The implications of this Yukawa unification condition at $M_{\rm GUT} \sim 2 \times 10^{16}$ GeV have been extensively explored over the years [1, 2]. In $SO(10)$ Yukawa unification with $\mu > 0$ and universal gaugino masses, the gluino is the lightest colored sparticle [3, 4], which is now being tested [5] at the Large Hadron Collider (LHC). The squarks and sleptons, especially those from the first two families, turn out to have masses in the multi-TeV range. Moreover, it is argued in [3, 4] that the lightest neutralino is not a viable cold dark matter candidate in $SO(10)$ Yukawa unification with $\mu > 0$ and universal gaugino masses at $M_{\rm GUT}$.

Spurred by these developments we have investigated t-b-τ Yukawa unification [4, 6, 7] in the framework of SUSY $SU(4)_c \times SU(2)_L \times SU(2)_R$ [8] (4-2-2, for short). The 4-2-2 structure allows us to consider non-universal gaugino masses while preserving Yukawa unification. An important conclusion reached in [4, 6] is that with same sign non-universal gaugino soft terms, Yukawa unification in 4-2-2 is compatible with neutralino dark matter and gluino coannihilation [4, 5, 6, 9] being a unique dark matter scenario for $\mu > 0$.

Encouraged by the abundance of solutions and coannihilation channels available in the case of Yukawa unified 4-2-2 with $M_2 < 0$ and $\mu < 0$, it seems natural to explore Yukawa unification in SO(10) GUT (with $M_2 < 0$ and $\mu < 0$). It has been pointed out [10] that non-universal MSSM gaugino masses at $M_{\rm GUT}$ can arise from non-singlet F-terms, compatible with the underlying GUT symmetry such as SU(5) and SO(10). The SSB gaugino masses in supergravity [11] can arise, say, from the following dimension five operator:

$$\frac{F^{ab}}{2M_{\rm P}} \lambda^a \lambda^b + \rm cc \qquad (1)$$

Here λ^a is the two-component gaugino field, F^{ab} denotes the F-component of the field which breaks SUSY, the indices a, b run over the adjoint representation of the gauge group, and $M_P = 2.4 \times 10^{18}$ GeV is the reduced Planck mass. The resulting gaugino mass matrix is $\langle F^{ab} \rangle / M_{\rm P}$ where the supersymmetry breaking parameter $\langle F^{ab} \rangle$ transforms as a singlet under the MSSM gauge group $SU(3)_c \times SU(2)_L \times U(1)_Y$. The F^{ab} fields belong to an irreducible representation in the symmetric part of the direct product of the adjoint representation of the unified group.

In SO(10), for example,

$$(45 \times 45)_S = 1 + 54 + 210 + 770 \tag{2}$$

If F transforms as a 54 or 210 dimensional representation of SO(10) [10], one obtains the following relation among the MSSM gaugino masses at $M_{\rm GUT}$:

$$M_3 : M_2 : M_1 = 2 : -3 : -1, \tag{3}$$

where M_1, M_2, M_3 denote the gaugino masses of $U(1)$, $SU(2)_L$ and $SU(3)_c$ respectively. The low energy implications of this relation have recently been investigated in [12] without imposing Yukawa unification. To our surprise, we find that this class of t-b-τ Yukawa unification models make a rather sharp prediction for the mass of the lightest SM-like Higgs boson. In addition, lower mass bounds on the masses of the squarks and gluino are obtained.

2. Phenomenological Constraints and Scanning Procedure

We employ the ISAJET 7.80 package [13] to perform random scans over the fundamental parameter space. In this package, the weak scale values of gauge and third generation Yukawa couplings are evolved to $M_{\rm GUT}$ via the MSSM renormalization group equations (RGEs) in the $\overline{DR}$ regularization scheme. We do not strictly enforce the unification condition $g_3 = g_1 = g_2$ at $M_{\rm GUT}$, since a few percent deviation from unification can be assigned to unknown GUT-scale threshold corrections [14]. The deviation between $g_1 = g_2$ and g_3 at $M_{\rm GUT}$ is no worse than $3-4\%$. For simplicity we do not include the Dirac neutrino Yukawa coupling in the RGEs, whose contribution is expected to be small.

The various boundary conditions are imposed at $M_{\rm GUT}$ and all the SSB parameters, along with the gauge and Yukawa couplings, are evolved back to the weak scale M_Z. In the evaluation of Yukawa couplings the SUSY threshold corrections [15] are taken into account at the common scale $M_{\rm SUSY} = \sqrt{m_{\tilde{t}_L} m_{\tilde{t}_R}}$ where $m_{\tilde{t}_L}$ and $m_{\tilde{t}_R}$ are the third generation left and right handed stop quarks. For further details about iteration of the entire parameter set between M_Z and $M_{\rm GUT}$, see [13].

An important constraint comes from limits on the cosmological abundance of stable charged particles [16]. This excludes regions in the parameter space where charged SUSY particles become the lightest supersymmetric particle (LSP). We accept only those solutions for which one of the neutralinos is the LSP and saturates the WMAP bound on relic dark matter abundance.

The MSSM Higgs doublets reside in the 10 dimensional representation of $SO(10)$ and fermions of the third family belong to the 16 dimensional representation of SO(10), which implies Yukawa coupling unification at $M_{\rm GUT}$. At low scale, Higgs mass is estimated with the approximate error of $\pm$3GeV, which could stem from experimental uncertainties in the values of m_t and α_s.

We have performed random scans for the following parameter range:

$$\begin{array}{lcr} 0 \leq & \mathrm{m}_{16} & \leq 5\,\mathrm{TeV} \\ 0 \leq & \mathrm{m}_{10} & \leq 5\,\mathrm{TeV} \\ 0 \leq & \mathrm{M}_{1/2} & \leq 2\,\mathrm{TeV} \\ 35 \leq & \tan\beta & \leq 55 \\ -3 \leq & \mathrm{A}_0/m_{16} & \leq 3 \end{array} \tag{4}$$

Here m_{16} is the universal SSB mass for MSSM sfermions, m_{10} is the universal SSB mass term for up and down MSSM Higgs masses, $M_{1/2}$ is the gaugino mass parameter, $\tan\beta$ is the ratio of the vacuum expectation values (VEVs) of the two MSSM Higgs doublets, A_0 is the universal SSB trilinear scalar interaction (with corresponding Yukawa coupling factored out). We use $m_t = 173.1\,\text{GeV}$ [17]. Note that our results are not too sensitive to one or two sigma variation in the value of m_t [18]. We use $m_b(m_Z) = 2.83$ GeV which is hard-coded into ISAJET.

Employing the boundary condition from Eq.(3) one can define the MSSM gaugino masses at M_{GUT} in terms of the mass parameter $M_{1/2}$:

$$M_1 = -M_{1/2}\ , \quad M_2 = -3M_{1/2}\ , \quad M_3 = 2M_{1/2} \tag{5}$$

Note that M_2 and M_3 have opposite signs which, as we show, is important implementing Yukawa coupling unification to a high accuracy.

In order to obtain the correct sign for the desired contribution to $(g-2)_\mu$, we consider two possible combinations of $SU(2)_L$ gaugino and the sign of μ term: (a) $\mu < 0$ and $M_2 < 0$, (b) $\mu > 0$ and $M_2 > 0$.

In scanning the parameter space, we employ the Metropolis-Hastings algorithm as described in [19]. The data points collected all satisfy the requirement of REWSB, with the neutralino in each case being the LSP. After collecting the data, we impose the mass bounds on all the particles [16] and use the IsaTools package [20] to implement the various phenomenological constraints. We successively apply the following experimental constraints on the data that we acquire from ISAJET:

$$\begin{array}{rll}
m_h \text{ (lightest Higgs mass)} & \geq 114.4 \text{ GeV} & [21] \\
BR(B_s \to \mu^+\mu^-) & < 1.2 \times 10^{-8} & [22] \\
2.85 \times 10^{-4} \leq BR(b \to s\gamma) & \leq 4.24 \times 10^{-4}\ (2\sigma) & [23] \\
0.15 \leq \frac{BR(B_u \to \tau\nu_\tau)_{\text{MSSM}}}{BR(B_u \to \tau\nu_\tau)_{\text{SM}}} & \leq 2.41\ (3\sigma) & [23] \\
\Omega_{\text{CDM}} h^2 & = 0.1123 \pm 0.0035\ (5\sigma) & [24] \\
0 \leq \Delta(g-2)_\mu/2 & \leq 55.6 \times 10^{-10} & [25]
\end{array}$$

3. Yukawa Unification, Higgs Boson Mass and Sparticle Spectrum

In order to quantify Yukawa coupling unification, we define the quantity R as,

$$R = \frac{\max(y_t, y_b, y_\tau)}{\min(y_t, y_b, y_\tau)} \tag{6}$$

In Figure 1 we show the results in the $R - m_{16}$, $R - M_{1/2}$, $R - m_{10}$ and $R - \tan\beta$ planes. The left panels correspond to the following choice of parameters: $\mu < 0$, $M_2 < 0$ and $M_3 : M_2 : M_1 = 2 : -3 : -1$. The right panels are for $\mu > 0$, $M_2 > 0$ and $M_3 : M_2 : M_1 = -2 : 3 : 1$. The gray points are consistent with REWSB and neutralino LSP. The green points satisfy the particle mass bounds and constraints from $BR(B_s \to \mu^+\mu^-)$, $BR(b \to s\gamma)$ and $BR(B_u \to \tau\nu_\tau)$. In addition, the green points do no worse than the SM in terms of $(g-2)_\mu$. The brown points belong to a subset of the green points and satisfy the WMAP bounds on neutralino dark matter abundance.

In the R - m_{16} plane of Figure 1 we see that with $\mu > 0$, $M_2 > 0$ and $M_3 < 0$, we can realize perfect Yukawa unification consistent with all constraints mentioned in Section 2. This is possible because we can implement Yukawa unification for relatively small values of m_{16} ($\sim 1\,\text{TeV}$). This is more than an order of magnitude reduction in the m_{16} values required for Yukawa unification, compared with the case $\mu > 0$ and universal gaugino masses. We also see

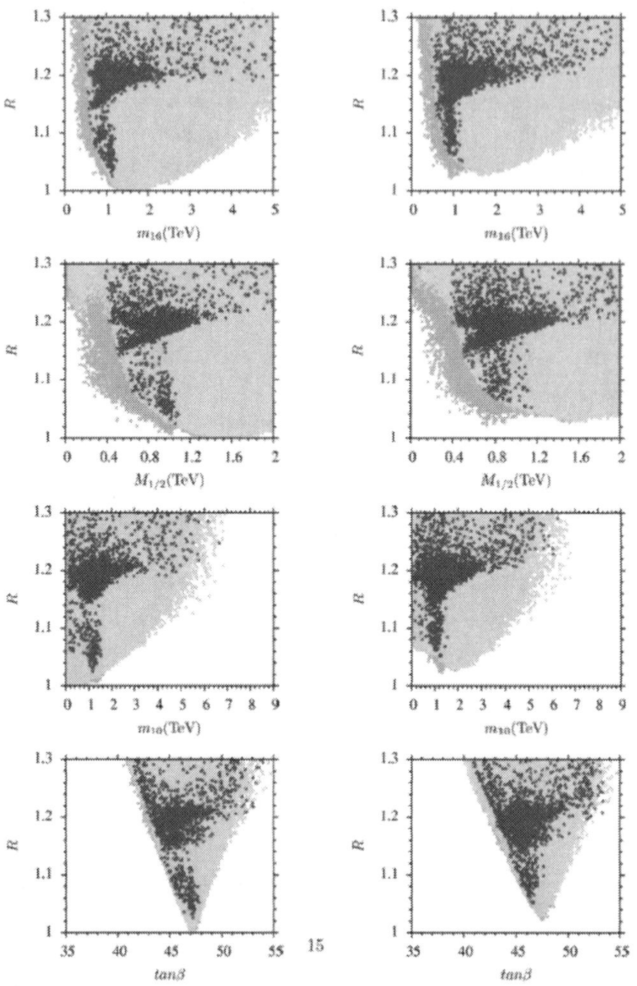

Figure 1. Plots in $R - m_{16}$, $R - M_{1/2}$ and $R - \tan\beta$ planes. Left side panels correspond to the following choice of parameters: $\mu < 0$, $M_2 < 0$ and $M_3 : M_2 : M_1 = 2 : -3 : -1$. Right side panels are for $\mu > 0$, $M_2 > 0$ and $M_3 : M_2 : M_1 = -2 : 3 : 1$. Gray points are consistent with REWSB and neutralino LSP. Green points satisfy particle mass bounds and constraints from $BR(B_s \to \mu^+\mu^-)$, $BR(b \to s\gamma)$ and $BR(B_u \to \tau\nu_\tau)$. In addition, we require that green points do no worse than the SM in terms of $(g-2)_\mu$. Brown points belong to a subset of green points and satisfy the WMAP bounds on neutralino dark matter abundance.

that the minimum R values, close to 1.02, can be reached for $m_{16} \simeq 2$ TeV for the case $\mu < 0$, $M_2 < 0$ and $M_3 > 0$. In the $R - M_{1/2}$ plane of Figure 1, we see that employing the boundary conditions for gauginos presented in Eq. (5), perfect t-b-τ Yukawa unification prefers heavier (> 1.2 TeV) values for $M_{1/2}$. We can also predict that $\tan\beta \approx 47$. In this case for Yukawa unification up to 5% level, $45 \le \tan\beta \le 48$.

In Figure 2 we show the results in the $R - m_h$, $m_{\tilde{q}} - m_{\tilde{g}}$, $m_{\tilde{\tau}} - m_{\tilde{\chi}^0}$ planes. The left and right panels, as in Figure 1, correspond to $\mu < 0$ and $\mu > 0$ respectively, with the sign of M_2 suitably chosen. It is most interesting to note that demanding precise t-b-τ Yukawa unification for $\mu > 0$ allows us to predict the light CP even Higgs mass in a very narrow interval, 122 GeV

$< m_h <$ 124 GeV. For Yukawa unification of up to 5%, it becomes 120 GeV $< m_h <$ 125 GeV. We expect a theoretical uncertainty of about 3 GeV in the calculation of the light Higgs mass. Similarly, for $\mu < 0$, the best unification solutions predict a Higgs mass close to 123 GeV (± 3 GeV).

The $m_{\tilde{q}} - m_{\tilde{g}}$ panel shows that t-b-τ Yukawa unification in our scenario predicts masses for the gluino and the first two family squarks which lie somewhat beyond the current ATLAS [26] and CMS [27] bounds.

In the present framework, the WMAP constraint on the relic dark matter abundance is only satisfied by the neutralino-stau coannihilation scenario. From the plot in $m_{\tilde{\tau}} - m_{\tilde{\chi}^0}$ plane, we see that the LSP neutralino mass is greater than or of order 300 GeV.

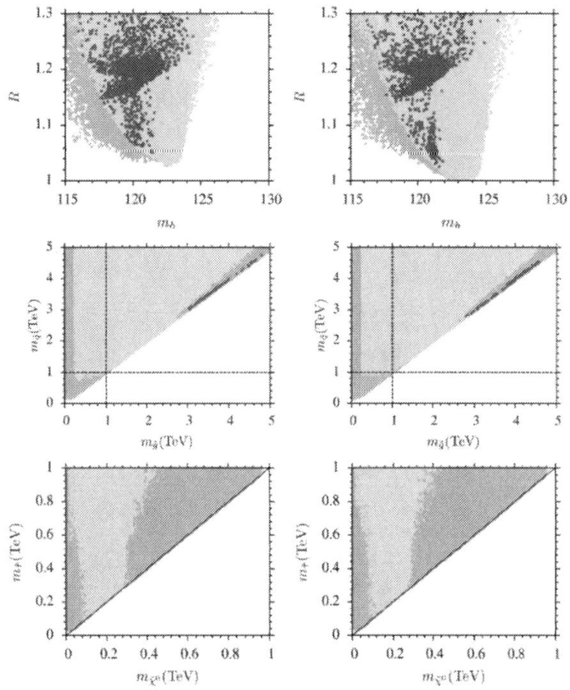

Figure 2. Plots in $R - m_h$, $m_{\tilde{q}} - m_{\tilde{g}}$, $m_{\tilde{\tau}} - m_{\tilde{\chi}^0}$ planes. Left and right panels, as in Figure 1, correspond to two different choices of sign of μ with suitable choice of sign for M_2.

Finally, in Tables 1 we present some benchmark points with $\mu > 0$ (on left side) and $\mu < 0$ (on right side) respectively. All of these benchmark points satisfy the various constraints, except possibly for a small discrepancy with the WMAP bound on relic dark density. Points 1's on both sides depict solutions corresponding to minimum R and best $\Omega_{CDM}h^2$ values. Each Point 1 predicts the CP-even light Higgs boson mass $\sim$ 124 GeV and the relic density abundance is satisfied by the neutralino-stau coannihilation scenario. However; these points predict stau mass about 850 GeV. Point 2's on both sides display solutions for neutralino-stau coannihilation scenario with relatively lighter stau mass ($<$ 500 GeV). The CP-even light Higgs boson mass is about 122 GeV, if $m_{\tilde{\tau}} <$ 500 GeV.

4. Conclusion

We have reconsidered t-b-τ Yukawa unification within a slightly revised framework in this paper. The main difference from most previous investigations stems from the assumptions

Journal of Physics: Conference Series **384** (2012) 012025 doi:10.1088/1742-6596/384/1/012025

$\mu > 0$	Point 1	Point 2
m_{16}	1920	1153
M_1	1768	1060
M_2	5304	3180
M_3	-3536	-2120
m_{10}	240.2	1149.8
$\tan\beta$	46.9	47.2
A_0/m_{16}	2.73	1.22
m_t	173.1	173.1
μ	3478.7	1693.3
B_μ	26.9	17.6
m_h	124	121.6
m_H	559.5	377.7
m_A	555.8	375.2
$m_{H^\pm}$	568.3	390.2
$m_{\tilde{\chi}^0_{1,2}}$	833.4, 3487.6	492.4, 1703
$m_{\tilde{\chi}^0_{3,4}}$	3490.7, 4503.2	1706.6, 2695.7
$m_{\tilde{\chi}^\pm_{1,2}}$	3533.3, 4458.5	1729.5, 2668.3
$m_{\tilde{g}}$	7217.9	4472.3
$m_{\tilde{u}_{L,R}}$	7176.2, 6386	4443.8, 3974
$m_{\tilde{t}_{1,2}}$	4967.6, 6120.1	3165.2, 3846.3
$m_{\tilde{d}_{L,R}}$	7176.7, 6378.7	4444.6, 3969
$m_{\tilde{b}_{1,2}}$	5168.1, 6088.1	3286.9, 3824.4
$m_{\tilde{\nu}_1}$	3844.3	2330.4
$m_{\tilde{\nu}_3}$	3605.8	2187.4
$m_{\tilde{e}_{L,R}}$	3849.1, 2016.8	2334.7, 1211.3
$m_{\tilde{\tau}_{1,2}}$	842.4, 3621.2	492.7, 2196.3
$\Delta(g-2)_\mu$	0.35×10^{-10}	0.10×10^{-9}
σ_{SI}(pb)	0.46×10^{-10}	0.89×10^{-9}
σ_{SD}(pb)	0.63×10^{-9}	0.15×10^{-7}
$\Omega_{CDM}h^2$	0.55	0.13
R	1.01	1.02

$\mu < 0$	Point 1	Point 2
m_{16}	2189	1162
M_1	-1972	-980
M_2	-5916	-2940
M_3	3944	1960
m_{10}	2558.9	1449
$\tan\beta$	46.9	46.7
A_0/m_{16}	-1.67	-1.43
m_t	173.1	173.1
μ	-2729.5	1464.7
B_μ	-40.7	-18.9
m_h	123.6	121.1
m_H	774.9	445.6
m_A	769.9	442.7
$m_{H^\pm}$	781.2	456.2
$m_{\tilde{\chi}^0_{1,2}}$	935, 2754.4	453.8, 1474.4
$m_{\tilde{\chi}^0_{3,4}}$	2756, 5039.3	1478, 2493.2
$m_{\tilde{\chi}^\pm_{1,2}}$	2800.6, 4989.5	1498.2, 2467.1
$m_{\tilde{g}}$	7989.4	4161.9
$m_{\tilde{u}_{L,R}}$	7960.1, 7078.2	4159.3, 3728.5
$m_{\tilde{t}_{1,2}}$	5542.8, 6830.2	2916.4, 3558.7
$m_{\tilde{d}_{L,R}}$	7960.6, 7068.2	4160.1, 3724.1
$m_{\tilde{b}_{1,2}}$	5791.2, 6794.3	3047.6, 3537.1
$m_{\tilde{\nu}_1}$	4305.6	2207.2
$m_{\tilde{\nu}_3}$	2756	2056.2
$m_{\tilde{e}_{L,R}}$	4310.7, 2294.3	2211.2, 1211.9
$m_{\tilde{\tau}_{1,2}}$	947.4, 4041.7	488.5, 2065.3
$\Delta(g-2)_\mu$	0.32×10^{-10}	0.12×10^{-9}
σ_{SI}(pb)	0.42×10^{-10}	0.72×10^{-9}
σ_{SD}(pb)	0.26×10^{-8}	0.30×10^{-7}
$\Omega_{CDM}h^2$	0.72	0.84
R	1.04	1.04

Table 1. Sparticle and Higgs masses (in GeV). The columns on left are for $\mu > 0$ while those on right for $\mu < 0$. On both sides, Point 1's depict solutions for the heaviest mass for the CP-even light Higgs boson corresponding to minimum R and best $\Omega_{CDM}h^2$ values. Each Point 1 predicts the CP-even light Higgs boson mass $\sim$ 124 GeV and relic dark matter abundance is satisfied by the neutralino-stau coannihilation scenario. For these points stau mass is about 850 GeV. Point2's on both sides seek for a solution for neutralino-stau coannihilation scenario with relatively lighter stau mass ($\sim$ 500 GeV). The maximum mass of the CP-even Higgs boson mass is about 122 GeV, if $m_{\tilde{\tau}} < 500$ GeV.

we make related to the soft supersymmetry breaking parameters. First, we assume that the MSSM gauginos have non-universal masses, which are related to one another at $M_{\rm GUT}$ by some appropriate SO(10) group theory factors. Second, we set the masses of the two MSSM Higgs doublets to be equal at $M_{\rm GUT}$. Overall, this means that we effectively have one parameter less than in the standard approach to SO(10) Yukawa unification.

The ramifications of these slightly different assumptions for TeV scale physics turn out to be quite startling, with the low energy predictions being very different from previous studies. We find, for instance, that t-b-τ Yukawa unification solutions at 5% level or better exist in our model for both $\mu > 0$ and $\mu < 0$.

These solutions, obtained using the ISAJET software, are compatible with all experimental observations, as well as the WMAP dark matter constraint (through stau coannihilation). The masses of the gluino and first two family squarks are found to lie in the 2.7 - 5 TeV range, while the lightest stop (top squark) weighs at least 2 TeV or so. Finally, with 5% or better Yukawa unification, the lightest Higgs boson is predicted to have a mass of around 120 - 125 GeV mass range, (with an uncertainty of ± 3 GeV). The MSSM parameter $\tan\beta$ is around 45 - 47.

Acknowledgments
I would like to thank Professors Yue- Liang Wu and Yu- Feng Zhou and members of the local organizing committee for their warm hospitality.

References

[1] B. Ananthanarayan, G. Lazarides and Q. Shafi, Phys. Rev. D **44**, 1613 (1991) and Phys. Lett. B **300**, 24 (1993)5; Q. Shafi and B. Ananthanarayan, Trieste HEP Cosmol.1991:233-244.

[2] L. J. Hall, R. Rattazzi and U. Sarid, Phys. Rev. D **50**, 7048 (1994); M. Olechowski and S. Pokorski, Phys. Lett. B **214**, 393 (1988); T. Banks, Nucl. Phys. B **303**, 172 (1988); V. Barger, M. Berger and P. Ohmann, Phys. Rev. D **49**, (1994) 4908; M. Carena, M. Olechowski, S. Pokorski and C. Wagner, Nucl. Phys. B **426**, 269 (1994); B. Ananthanarayan, Q. Shafi and X. Wang, Phys. Rev. D **50**, 5980 (1994); G. Anderson et al. Phys. Rev. D **47**, (1993) 3702 and Phys. Rev. D **49**, 3660 (1994); R. Rattazzi and U. Sarid, Phys. Rev. D **53**, 1553 (1996); T. Blazek, M. Carena, S. Raby and C. Wagner, Phys. Rev. D **56**, 6919 (1997); T. Blazek, S. Raby and K. Tobe, Phys. Rev. D **62**, 055001 (2000); H. Baer, M. Diaz, J. Ferrandis and X. Tata, Phys. Rev. D **61**, 111701 (2000); H. Baer, M. Brhlik, M. Diaz, J. Ferrandis, P. Mercadante, P. Quintana and X. Tata, Phys. Rev. D **63**, 015007(2001); C. Balazs and R. Dermisek, JHEP **0306**, 024 (2003); C. Pallis, Nucl. Phys. B **678**, 398 (2004); U. Chattopadhyay, A. Corsetti and P. Nath, Phys. Rev. D **66** 035003, (2002); T. Blazek, R. Dermisek and S. Raby, Phys. Rev. Lett. **88**, 111804 (2002) and Phys. Rev. D **65**, 115004 (2002); M. Gomez, T. Ibrahim, P. Nath and S. Skadhauge, Phys. Rev. D **72**, 095008 (2005); K. Tobe and J. D. Wells, Nucl. Phys. B **663**, 123 (2003); I. Gogoladze, Y. Mimura, S. Nandi, Phys. Lett. **B562**, 307 (2003); W. Altmannshofer, D. Guadagnoli, S. Raby and D. M. Straub, Phys. Lett. B **668**, 385 (2008); S. Antusch and M. Spinrath, Phys. Rev. D **78**, 075020 (2008); H. Baer, S. Kraml and S. Sekmen, JHEP **0909**, 005 (2009); S. Antusch and M. Spinrath, Phys. Rev. D **79**, 095004 (2009); D. Guadagnoli, S. Raby and D. M. Straub, JHEP **0910**, 059 (2009); K. Choi, D. Guadagnoli, S. H. Im and C. B. Park, JHEP **1010**, 025 (2010); S. Dar, I. Gogoladze, Q. Shafi and C. S. Un, Phys. Rev. D **84**, 085015 (2011); N. Karagiannakis, G. Lazarides and C. Pallis, Phys. Lett. B **704**, 43 (2011); I. Gogoladze, Q. Shafi and C. S. Un, Phys. Lett. B **704**, 201 (2011); M. Badziak, M. Olechowski and S. Pokorski, JHEP **1108**, 147 (2011); S. Antusch, L. Calibbi, V. Maurer, M. Monaco and M. Spinrath, arXiv:1111.6547 [hep-ph]; J. S. Gainer, R. Huo and C. E. M. Wagner, arXiv:1111.3639 [hep-ph].

[3] H. Baer, S. Kraml, S. Sekmen and H. Summy, JHEP **0803**, 056 (2008); H. Baer, M. Haider, S. Kraml, S. Sekmen and H. Summy, JCAP **0902**, 002 (2009).

[4] I. Gogoladze, R. Khalid and Q. Shafi, Phys. Rev. D **79**, 115004 (2009).

[5] H. Baer, S. Kraml, A. Lessa and S. Sekmen, JHEP **1002**, 055 (2010); D. Feldman, Z. Liu and P. Nath, Phys. Rev. D **80**, 015007 (2009); M. A. Ajaib, T. Li, Q. Shafi and K. Wang, JHEP **1101**, 028 (2011).

[6] I. Gogoladze, R. Khalid and Q. Shafi, Phys. Rev. D **80**, 095016 (2009);

[7] I. Gogoladze, R. Khalid, S. Raza and Q. Shafi, JHEP **1012**, 055 (2010); arXiv:1008.2765 [hep-ph].

[8] J. C. Pati and A. Salam, Phys. Rev. D **10**, 275 (1974).

[9] S. Profumo and C. E. Yaguna, Phys. Rev. D **69**, 115009 (2004); D. Feldman, Z. Liu and P. Nath, Phys. Rev. D **80**, 015007 (2009).

[10] See, for instance, S. P. Martin, Phys. Rev. **D79**, 095019 (2009); U. Chattopadhyay, D. Das and D. P. Roy, Phys. Rev. D **79**, 095013 (2009); B. Ananthanarayan, P. N. Pandita, Int. J. Mod. Phys. **A22**, 3229-3259 (2007); S. Bhattacharya, A. Datta and B. Mukhopadhyaya, JHEP **0710**, 080 (2007); A. Corsetti and P. Nath, Phys. Rev. D **64**, 125010 (2001) and references therein.

[11] A. Chamseddine, R. Arnowitt and P. Nath, Phys. Rev. Lett. **49** (1982) 970; R. Barbieri, S. Ferrara and C. Savoy, Phys. Lett. **B119** (1982) 343; N. Ohta, Prog. Theor. Phys. **70** (1983) 542; L. J. Hall, J. D. Lykken

and S. Weinberg, Phys. Rev. **D27** (1983) 2359; for a review see S. Weinberg, *The Quantum Theory of Fields: Volume 3, Supersymmetry, Cambridge University Press (2000) 442p.*

[12] N. Okada, S. Raza and Q. Shafi, Phys. Rev. D **84**, 095018 (2011) [arXiv:1107.0941 [hep-ph]].

[13] F. E. Paige, S. D. Protopopescu, H. Baer and X. Tata, hep-ph/0312045.

[14] J. Hisano, H. Murayama , and T. Yanagida, Nucl. Phys. **B402** (1993) 46. Y. Yamada, Z. Phys. **C60** (1993) 83; J. L. Chkareuli and I. G. Gogoladze, Phys. Rev. D **58**, 055011 (1998).

[15] D. M. Pierce, J. A. Bagger, K. T. Matchev, and R.-j. Zhang, Nucl. Phys. **B491** (1997) 3.

[16] K. Nakamura *et al.* [Particle Data Group Collaboration], J. Phys. G **G37**, 075021 (2010).

[17] [Tevatron Electroweak Working Group and CDF Collaboration and D0 Collab], arXiv:0903.2503 [hep-ex].

[18] I. Gogoladze, R. Khalid, S. Raza and Q. Shafi, JHEP **1106** (2011) 117.

[19] G. Belanger, F. Boudjema, A. Pukhov and R. K. Singh, JHEP **0911**, 026 (2009); H. Baer, S. Kraml, S. Sekmen and H. Summy, JHEP **0803**, 056 (2008).

[20] H. Baer, C. Balazs, and A. Belyaev, JHEP **03** (2002) 042; H. Baer, C. Balazs, J. Ferrandis, and X. Tata Phys. Rev. **D64** (2001) 035004.

[21] S. Schael *et al.* Eur. Phys. J. C **47**, 547 (2006).

[22] T. Aaltonen *et al.* [CDF Collaboration], Phys. Rev. Lett. **100**, 101802 (2008).

[23] E. Barberio *et al.* [Heavy Flavor Averaging Group], arXiv:0808.1297 [hep-ex].

[24] E. Komatsu *et al.* [WMAP Collaboration], Astrophys. J. Suppl. **180**, 330 (2009).

[25] G. W. Bennett *et al.* [Muon G-2 Collaboration], Phys. Rev. D **73**, 072003 (2006).

[26] G. Aad *et al.* [ATLAS Collaboration], arXiv:1109.6572 [hep-ex].

[27] S. Chatrchyan *et al.* [CMS Collaboration], arXiv:1109.2352 [hep-ex].

Journal of Physics: Conference Series **384** (2012) 012026 doi:10.1088/1742-6596/384/1/012026

Dark matter and radiative neutrino masses

Yi Cai[1], Xiao-Gang He[1,2], Michael Ramsey-Musolf[3,4] and Lu-Hsing Tsai[2]

[1] INPAC, Department of Physics, Shanghai Jiao Tong University, Shanghai, China
[2] Department of Physics and Center for Theoretical Sciences, National Taiwan University, Taipei, Taiwan
[3] Department of Physics, University of Wisconsin-Madison, Madison, WI USA
[4] California Institute of Technology, Pasadena, CA USA

E-mail: yicai@sjtu.edu.cn

Abstract. The RνMDM model relates radiative seesaw and minimal dark matter mass scales without imposing any beyond the Standard Model (SM) gauge symmetry. The model contains, besides the SM particles, a Majorana fermion multiplet N_R and a scalar multiplet χ which transform under the SM gauge group as $(1, 5, 0)$ and $(1, 6, -1/2)$. We study the dark matter physics of this model and the relevant lepton flavor violating processes.

1. Introduction

The existence of dark matter has been firmly established for decades by various astrophysical and cosmological experiments [1]. The nature of dark matter, however, has remained elusive since then, except for the fact that dark matter make up 23.3% of the universe [1]. Physicists have proposed many theoretical models accommodating this ingredient, a large majority of which introduce certain discrete symmetry to stabilize the dark matter particle. It is also possible to stabilize the dark matter particle without imposing any new symmetry through the minimalistic approach, by adding specific SM gauge multiplets such that any dark matter decay channel is automatically forbidden by the SM gauge symmetry itself [2]. The extra SM gauge multiplet is chosen according to three conditions: (a) there is an electrically neutral component in the multiplet; (b) the neutral component has the lightest mass; (c) this component is automatically stable on cosmological time-scales. The relic abundance of the dark matter candidate is then completely determined by the SM gauge interactions, which in turn fixes the mass of the dark matter to be several TeVs, the only new parameter in these Minimal Dark Matter (MDM) models.

Besides the dark matter problem, it is also necessary to extend the SM to decipher the neutrino oscillation myths [3]. Such extensions include seesaw models [4, 5, 6] and Zee models [7], all of which share a common feature, large neutrino masses above the electroweak scale, which is not determined *a priori*. These neutrino mass scales can also be linked to the dark matter mass scales by generating neutrino masses radiatively [8]. Most of these radiative seesaw models (RSSM), however, impose discrete symmetries to stabilize the dark matter particle. Incorporating the MDM and the RSSM in the same model, the so-called RνMDM provides the possibility to solve the dark matter problem and the neutrino oscillation problem in the same frame work with no beyond the SM gauge and discrete symmetries, which also has testable phenomenological predictions.

2. The model

The extra matter content of RνMDM includes a Majorana right-handed fermion N_R and a scalar χ which transform under the SM gauge group as

$$N_R : (1, 5, 0) \qquad \chi : (1, 6, -1/2), \tag{1}$$

which are selected according to the following considerations. In order to radiatively generate neutrino masses, a Majorana mass term of N_R is necessary, which means N_R should have zero hypercharge. In the mean time, $\bar{L}HN_R$ type of Yukawa coupling should not be allowed to forbid N_R from mixing with the SM leptons and thus decaying to the SM particles. These two requirements exclude the possibility of N_R as $SU(2)_L$ singlet and triplet. The next possibilities are $(2, 0)$ and $(4, 0)$ with $SU(3)_C$ quantum number omitted. Together with the existence of $\bar{L}N_R\chi$ necessary for the RSSM ingredient, these two possibilities demand χ to be $(1, -1/2)$, $(3, -1/2)$ or $(5, -1/2)$, which makes all the component fields in both N_R and χ fractionally charged and leaves us with no dark matter candidate. Thus the minimal choice for N_R should start with $(5, 0)$, which corresponds to a choice of χ to be $(4, -1/2)$ or $(6, -1/2)$. A $H^\dagger HH\chi$ term is allowed if χ is $(4, -1/2)$, which will induce vacuum expectation value (VEV) for χ and break the accidental Z_2 symmetry that stabilize the dark matter candidate.

With this choice of matter content, the most general Lagrangian containing N_R and χ is

$$\begin{aligned} L_{new} &= \bar{N}_R i\gamma^\mu D_\mu N_R + (D^\mu\chi)^\dagger D_\mu\chi - (\bar{L}_L Y \chi N_R + \frac{1}{2}\bar{N}_R^c M N_R + h.c.) - V_{new} \\ V_{new} &= \mu_\chi^2 \chi^\dagger\chi + \lambda_\chi^\alpha(\chi\chi^\dagger\chi\chi^\dagger)_\alpha + \lambda_{H\chi}^\alpha(H^\dagger H\chi^\dagger\chi)_\alpha + \left[\tilde{\lambda}_{H\chi}(H\chi)^2 + h.c.\right], \end{aligned} \tag{2}$$

which is invariant under an accidental Z_2 symmetry even after electroweak symmetry breaking, under which only N_R and χ are odd. This discrete symmetry makes the lightest component field of N_R and χ stable.

3. Dark matter and neutrino masses

3.1. Dark matter

The dark matter candidate should be the lightest neutral component of either N_R or χ. In the minimal model, χ^0, however, is ruled out since its direct detection cross section is 2 or 3 orders of magnitude bigger than the experimental limits due to its nonzero hypercharge. N_R^0, the neutral component of N_R, has the lightest mass at one-loop level according to the formula of mass splitting

$$\begin{aligned} m_N^Q - m_N^{Q'} &= \frac{g^2 m_N}{16\pi^2}(Q^2 - Q'^2)\left[\sin^2\theta_W f(m_Z/m_N) + f(m_W/m_N) - f(m_Z/m_N)\right], \\ f(x) &= \frac{x}{2}\left\{2x^3 \ln x - 2x + (x^2-4)^{1/2}(x^2+2)\ln[(x^2 - 2 - x\sqrt{x^2-4})/2]\right\}, \end{aligned} \tag{3}$$

where Q and Q' are the electric charges of different component. So N_R^0 serves as a good dark matter candidate, whose mass is determined by the relic abundance constraint to be 9.6 TeV [2]. The direct DM detection cross section is predicted to be around 10^{-44} cm^2 which is safely below the current upper limits from CDMSII and Xenon100, but can be tested in the future by superCDMS and xenon-1ton experiments [9].

3.2. Neutrino masses and mixing

The neutrino masses are generated radiatively. The mass matrix of the neutrinos at one-loop is

$$(M_\nu)_{ll'} = \frac{1}{16\pi^2}\frac{14}{5}\frac{Y_{li}Y_{l'i}}{m_{N_i}}\tilde{\lambda}_{H\chi}v^2 I(m_k, m_m, m_{N_i}) ,$$

$$I(m_{\chi_k}, m_{\chi_m}, m_{N_i}) = \frac{m^2_{N_i}}{m^2_{\chi_k} - m^2_{\chi_m}}\left(\frac{m^2_{\chi_k}\ln(m^2_{N_i}/m^2_{\chi_k})}{m^2_{N_i} - m^2_{\chi_k}} - \frac{m^2_{\chi_m}\ln(m^2_{N_i}/m^2_{\chi_m})}{m^2_{N_i} - m^2_{\chi_m}}\right) , \quad (4)$$

where $m_{\chi_{k,m}}$ and m_{N_i} are the masses of the scalars, and fermions in the loop. The mass matrix will be diagonalized by the Pontecorvo-Maki-Nakagawa-Sakata (PMNS) mixing matrix V_{PMNS}. Thus the Yukawa coupling matrix can be solved

$$Y = \left(\frac{v^2}{16\pi^2}\frac{14}{5}\tilde{\lambda}_{H\chi}\right)^{-1/2} V_{PMNS}\hat{M}_\nu^{1/2} O \hat{M}_N^{1/2}\hat{\tilde{I}}^{-1/2} , \quad (5)$$

where $\hat{M}_n$ and $\hat{\tilde{I}}$ are diagonal matrices with $\hat{M}_N = diag(m_{N_1}, m_{N_2}, ...)$ and $\hat{\tilde{I}} = diag(\tilde{I}(z_1), \tilde{I}(z_2), ...)$. O satisfies $OO^T = I$ and can be complex in general. The Yukawa coupling matrix varies with $\tilde{\lambda}_{H\chi}$, the choice of O and different neutrino mass patterns. It can be seen later that the entries of the Yukawa coupling matrix take sizable values when constraints from flavor physics are considered.

4. Lepton flavor violation

Since the mass of the dark matter candidate is set to be around 10 TeV, it is impossible to test the theory at the LHC. Flavor physics, however, can be an effective probe. For $\ell_i \to \ell_j\gamma$, the decay amplitudes for $\ell_i \to \ell_j\gamma$ can in general be parameterized as

$$\mathcal{M} = \epsilon^{*\mu}\bar{\ell}_j i\sigma_{\mu\nu}q^\nu(\tilde{A}_R P_R + \tilde{A}_L P_L)\ell_i . \quad (6)$$

Neglecting small mass splitting of component fields within a given N_R multiplet, we obtain

$$\tilde{A}_R = \frac{m_i e}{32\pi^2 m^2_\chi}\sum_k Y^*_{ik}Y_{jk}\,F(z_k) ,$$

$$F(z) = P_\chi\frac{2z^3 + 3z^2 - 6z + 1 - 6z^2\ln z}{6(1-z)^4} - P_N\frac{z^3 - 6z^2 + 3z + 2 + 6z\ln z}{6(1-z)^4} , \quad (7)$$

where $z_k = m^2_{N_k}/m^2_\chi$. Summarizing all one loop contributions, we have $P_\chi = -5$ and $P_N = 2$. $\tilde{A}_L$ can be obtained by replacing m_i by m_j in the above. In order to have a large branching ratio for $\mu \to e\gamma$ close to the upper bound in both normal and inverted hierarchies with m_{ν_1} larger than 0.08 eV, the parameter $\tilde{\lambda}_{H\chi}$ is smaller than 10^{-11}, which is plotted in Fig. 1. The smallness of $\tilde{\lambda}_{H\chi}$ is technically natural since the model has a global $U(1)$ lepton number symmetry if this coupling is set to zero. In this regime, the elements of the Yukawa matrix Y are of order a few, which may be close to the non-perturbative region wherein the results may become less accurate compared with smaller m_{ν_1}. But for m_{ν_1} less than 0.08 eV, the perturbative predictions should be reasonably reliable.

For $\mu - e$ conversion, the effective Lagrangian from Ref. [10] is

$$\begin{aligned} L_{\text{eff}} &= -\frac{4G_F}{\sqrt{2}}\left[m_\mu\bar{e}\sigma^{\mu\nu}(A_R P_R + A_L P_L)\mu F_{\mu\nu} + h.c.\right] \\ &- \frac{G_F}{\sqrt{2}}\left[\bar{e}(g_{LS(q)}P_R + g_{RS(q)}P_L)\mu\,\bar{q}q + \bar{e}(g_{LP(q)}P_R + g_{RP(q)}P_L)\mu\,\bar{q}\gamma_5 q + h.c.\right] \end{aligned}$$

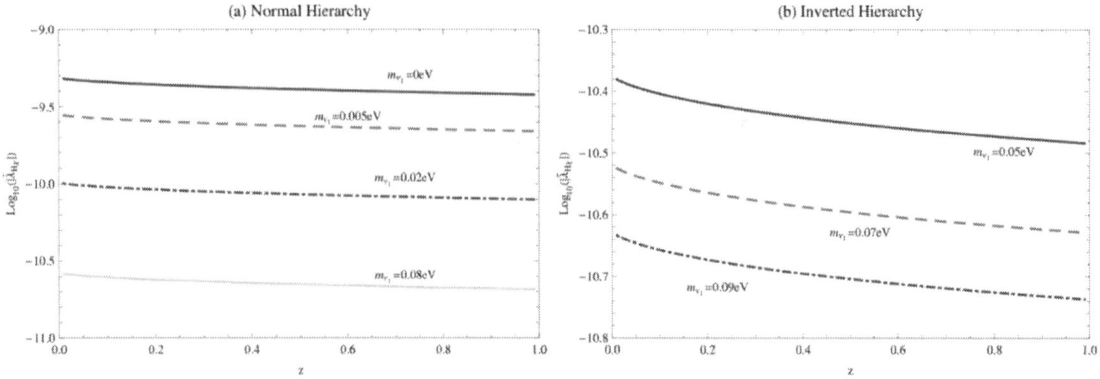

Figure 1. $\log_{10}|\tilde{\lambda}_{H\chi}|$ vs. $z = m_N^2/m_\chi^2$ for (a) normal hierarchy, and (b) inverted hierarchy for different choices of m_{ν_1}. The curves are obtained using the recent MEG upper limit $\text{Br}(\mu \to e\gamma) = 2.4 \times 10^{-12}$.

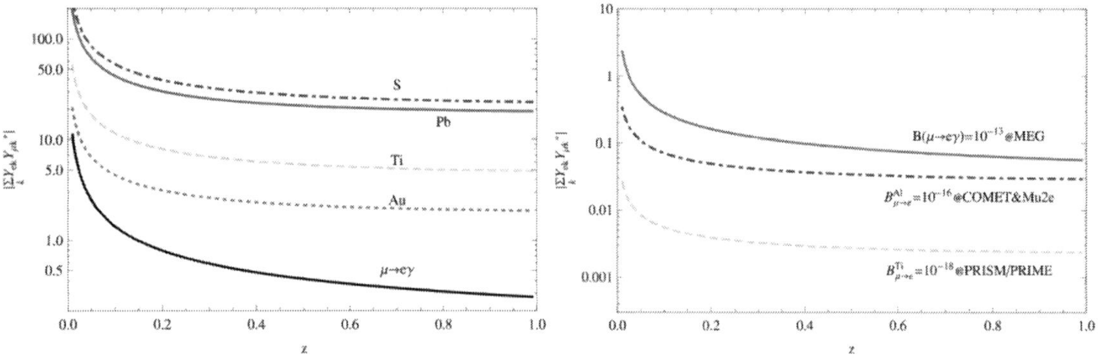

Figure 2. Comparison of present (left panel) and prospective (right panel) constraints from $\mu - e$ conversion on various nuclei and $\mu \to e\gamma$.

$$
\begin{aligned}
- & \frac{G_F}{\sqrt{2}}\left[\bar{e}(g_{LV(q)}\gamma^\mu P_L + g_{RV(q)}\gamma^\mu P_R)\mu\,\bar{q}\gamma_\mu q + \bar{e}(g_{LA(q)}\gamma^\mu P_L + g_{RA(q)}\gamma^\mu P_R)\mu\,\bar{q}\gamma_\mu\gamma_5 q + h.c.\right] \\
- & \frac{G_F}{\sqrt{2}}\left[\frac{1}{2}\bar{e}(g_{LT(q)}\sigma^{\mu\nu}P_R + g_{RT(q)}\sigma^{\mu\nu}P_L)\mu\,\bar{q}\sigma_{\mu\nu}q + h.c.\right] \,. \qquad (8)
\end{aligned}
$$

where at the one-loop level, $g_{(L,R)(S,P,A,T)(q)} = 0$, $g_{RV(q)} = 0$, and

$$
A_R = \frac{\sqrt{2}}{8}\frac{\tilde{A}_R}{G_F m_\mu}\,, \quad A_L = \frac{m_e}{m_\mu}A_R\,, \qquad (9)
$$

$$
g_{LV(q)} = -\frac{s_W^2}{2\pi^2}\sum_k \frac{m_W^2}{m_{N_k}^2} Q_q Y_{ek} Y_{\mu k}^* z_k \left[P_\chi G_\chi(z_k) + P_N G_N(z_k)\right] \,. \qquad (10)
$$

The contribution from A_L can be neglected compared with that from A_R. The constraints on $\sum_k Y_{ek}Y_{\mu k}^*$ from $\mu - e$ conversion on various nuclei and $\mu \to e\gamma$ are shown in Fig.2 on the left panel assuming that N_k are degenerate. We see that among the current available experimental limits,

$\mu \to e\gamma$ gives the strongest constraints. The current bounds on $\mu - e$ conversion do not provide significant constrains on the parameter since the Yukawa couplings can still be much larger than of order 1. On the other hand, the sensitivities of prospective future $\mu - e$ conversion searches Mu2E[11]/COMET[12] and PRISM[13] exceed that of the MEG experiment. Constraints from $\tau \to \mu(e)\gamma$ are also studied and are less stringent than those from $\mu \to e\gamma$ but may be probed by future experiment at the super B factory[14] as shown Fig. 2. Magnetic dipole moments of μ and neutrinos are also studied and the new contributions are well under current experimental limits.

5. Conclusion

In the work we have constructed a minimal model, RνMDM, incorporating the MDM and the RSSM without beyond the SM gauge symmetry. This model provides a fermionic dark matter candidate with a mass around 10 TeV that is not testable at the LHC. It predicts a a direct DM detection cross section of order 10^{-44} cm^2, which can be tested in future experiments, such as superCDMS, and Xenon-1ton. Present constraints obtained from $\mu - e$ conversion searches are weaker than those from $\mu \to e\gamma$, which will be the other way around in the future experiments. This model also can imply a sizable $\tau \to \mu\gamma$ decay rate that may be observable at super B factories.

6. Acknowledgments

Y.C, X.G.H and L.H.T were supported in part by NSC, NCTS, NNSF, SJTU Innovation Fund for Postgraduates and Postdocs. M.J.R-M was supported in part by Department of Energy contract DE-FG02-08ER41531 and by the Wisconsin Alumni Research Foundation.

References

[1] e.g., E. Komatsu *et al.*[WMAP Collaboration], *Astrophys. J. Suppl.* **192** (2011) 18[arXiv: 1001.4538[astro-ph.CO]].

[2] M. Cerilli, N. Fornengo, A. Strumia, *Nucl. Phys.* **B753**, 178 (2006); M. Cerrilli, A. Strumia, *New. J. Phys.*, 11, 105005(2009).

[3] K. Nakamura (Particle Data Group), *J. Phys.* **G37**, 075021(2010).

[4] P. Minkowski, *Phys. Lett.* **B 67**, 421 (1977); T. Yanagida, in Workshop on Unified Theories, KEK report 79-18 p.95 (1979); M. Gell-Mann, P. Ramond, R. Slansky, in Supergravity (North Holland, Amsterdam, 1979) eds. P. van Nieuwenhuizen, D. Freedman, p.315; S. L. Glashow, in 1979 Cargese Summer Institute on Quarks and Leptons (Plenum Press, New York, 1980) eds. M. Levy, J.-L. Basdevant, D. Speiser, J. Weyers, R. Gastmans and M. Jacobs, p.687; R. Barbieri, D. V. Nanopoulos, G. Morchio and F. Strocchi, *Phys. Lett.* **B 90**, 91 (1980); R. N. Mohapatra and G. Senjanovic, *Phys. Rev. Lett.* **44**, 912 (1980); G. Lazarides, Q. Shafi and C. Wetterich, *Nucl. Phys.* **B 181**, 287 (1981).

[5] W. Konetschny and W. Kummer, *Phys. Lett.* **B 70**, 433 (1977); T. P. Cheng and L. F. Li, *Phys. Rev.* **D 22**, 2860 (1980); G. Lazarides, Q. Shafi and C. Wetterich, *Nucl. Phys.* **B 181**, 287 (1981); J. Schechter and J. W. F. Valle, *Phys. Rev.* **D 22**, 2227 (1980); R. N. Mohapatra and G. Senjanovic, *Phys. Rev.* **D 23**, 165 (1981);

[6] R. Foot, H. Lew, X. G. He and G. C. Joshi, *Z. Phys.* **C 44**, 441 (1989).

[7] A. Zee, *Phys. Lett.* **B93**, 389(1980).

[8] E. Ma, *Mod. Phys. Lett.* **A23**, 721-725 (2008); E. Ma, D. Suematsu, *Mod. Phys. Lett.* **A24**, 583-589 (2009); P. -H. Gu, U. Sarkar, *Phys. Rev.* **D78**, 073012 (2008); Q. -H. Cao, E. Ma, G. Shaughnessy, *Phys. Lett.* **B673**, 152-155 (2009); D. Suematsu, T. Toma, T. Yoshida, *Phys. Rev.* **D79**, 093004 (2009); P.-F. Perez and M. Wise, *Phys. Rev.* **D80**, 053006(2009); A. Adulpravitchai, P. -H. Gu, M. Lindner, *Phys. Rev.* **D82**, 073013 (2010); T. Li, W. Chao, *Nucl. Phys.* **B843**, 396-412 (2011); S. Kanemura, O. Seto, T. Shimomura, [arXiv:1101.5713 [hep-ph]]; M. K. Parida, [arXiv:1106.4137 [hep-ph]].

[9] E. Aprile and L. Baudis (XENON Collaboration) 2Proc. Identification of Dark Matter 2008 Conf. (Stockholm, Swesen, 18-22 aug. 2008)(Trieste, Italy: Proceedings of Science) pp 1-10 [arXiv:0902.4253[astro-ph.IM]]; CDM-II Collaboration2005 eConf C041213(2004) 2529[arXiv:0503583[astro-ph]].

[10] R. Kitano, M. Koike and Y. Okada, *Phys. Rev.* **D 66**, 096002 (2002) [Erratum-ibid. D **76**, 059902 (2007)],

[11] J. P. Miller [Mu2E collabaration], *Proposal to Search for* $\mu^- N \to e^- N$ *with a Single Event Sensitivity Below* 10^{-16}.
[12] Y. Kuno et.al. [COMET collaboration], *An Experimental Search for lepton Flavor Violating* $\mu - e$ *Conversion at Sensitivity of* 10^{-16} *with a Slow-Extracted Bunched Beam.*
[13] Y. Kuno et.al. [PRISM/PRIME Group], Letter of Intent, *An Experimental Search for a* $\mu - e$ *Conversion at Sensitivity of the Order of* 10^{-18} *with a Highly Intense Muon Source: PRISM.*
[14] T. Aushev et al., Physics at uper B Factory, arXiv:1002.5012.

Journal of Physics: Conference Series **384** (2012) 012027 doi:10.1088/1742-6596/384/1/012027

Delta-gravity, Dark Energy and the accelerated expansion of the Universe

Jorge Alfaro
Pontificia Universidad Católica de Chile,
Av. Vicuña Mackenna 4860, Santiago, Chile
E-mail: jalfaro@puc.cl

Abstract. We present a model of the gravitational field based on two symmetric tensors. The equations of motion of test particles are derived: Massive particles do not follow a geodesic but massless particles trajectories are null geodesics of an effective metric. Outside matter, the predictions of the model coincide exactly with General Relativity, so all classical tests are satisfied. In Cosmology, we get accelerated expansion without a cosmological constant.

1. Introduction

General Relativity(GR) works very well at the macroscopic scales[1]. Its quantization has proved to be difficult, though. It is non renormalizable, which prevents its unification with the other forces of Nature. Trying to make sense of Quantum GR is the main physical motivation of String Theories [2]. Moreover, recent discoveries in Cosmology [3, 4] has revealed that most part of matter is in the form of unknown matter(dark matter,DM) and that the dynamics of the expansion of the Universe is governed by a mysterious component that accelerates the later stages of the expansion(dark energy,DE). Although GR is able to accommodate both DM and DE, the interpretation of the dark sector in terms of fundamental theories of elementary particles is problematic[5]. Although some candidates exists that could play the role of DM, none have been detected yet. Also, an alternative explanation based on the modification of the dynamics for small accelerations cannot be ruled out[6].

In GR, DE can be explained if a small cosmological constant(Λ) is present. At the later stages of the evolution of the Universe Λ will dominate the expansion, explaining the acceleration. Such small Λ is very difficult to generate in Quantum Field Theory (QFT) models, because in this models Λ is the vacuum energy, which is usually very large.

In recent years there has been various proposals to explain the observed acceleration of the universe. They involve the inclusion of some additional field like in quintessence, chameleon, vector dark energy or massive gravity; Addition of higher order terms in the Einsten-Hilbert action, like f(R) theories and Gauss-Bonnet terms; Modification of gravity on large scales by introduction of extra dimensions. For a review, see [7].

Less widely explored, but interesting possibilities, are the search for non-trivial ultraviolet fixed points in gravity (asymptotic safety[9]) and the notion of induced gravity[10]. The first possibility uses exact renormalization-group techniques[11] and lattice and numerical techniques such as Lorentzian triangulation analysis[12]. Induced gravity proposed that gravitation is a residual force produced by other interactions.

In a recent paper, [13] a two-dimensional field theory model explore the emergence of geometry by the spontaneous symmetry breaking of a larger symmetry where the metric is absent. Previous work in this direction can be found in [14], [15] and [16].
In this paper, we will review the results of [17]. The main observation is that GR is finite on shell at one loop[18]. In [20, 19] we presented a type of gauge theories, δ gauge theories(DGT): The main properties of DGT are: 1) The classical equations of motion are satisfied in the full Quantum theory 2) They live at one loop. 3) They are obtained through the extension of the former symmetry of the model introducing an extra symmetry that we call δ symmetry, since it is formally obtained as the variation of the original symmetry. When we apply this prescription to GR we obtain δ gravity. Quantization of δ gravity is discussed in [21].

The impact of dark energy on cosmological observations can be expressed in terms of a fluid equation of state $p = w(R)\rho$, which is to be determined studying its influence on the large-scale structure and dynamics of the Universe.

In this paper we follow the same approach. So we will not include the matter dynamics, except by demanding that the energy-momentum tensor of the matter fluid is covariantly conserved. This is required in order to respect the symmetries of the model.

The main properties of this model at the classical level are: a)It agrees with GR, outside the sources and with adequate boundary conditions. In particular, the causal structure of delta gravity in vacuum is the same as in General Relativity. So all standard test are satisfied automatically. b) When we study the evolution of the Universe, it predicts acceleration without a cosmological constant or additional scalar fields. The Universe ends in a Big Rip, similar to the scenario considered in [23]. c) The scale factor agrees with the standard cosmology at early times and show acceleration only at late times. Therefore we expect that density perturbations should not have large corrections.

It should be remarked that δ- gravity is not a metric model of gravity because massive particles do not move on geodesics. Only massless particles move on null geodesics of a linear combination of both tensor fields.

It was noticed in [20] that the Hamiltonian of delta models is not bounded from below. Phantoms cosmological models [22], [23] also have this property. Although it is not clear whether this problem will subsist in a diffeomorphism invariant model as delta gravity or not, we mention some ways out of the difficulty at the end.

In chapter 2, we write the action defining the model and the corresponding symmetries. Section 3 discusses the motion of particles in the model. In section 4 we define proper time and distances. In section 5, we obtain the newtonian limit. In section 6 we solve the equations of the model for Friedman-Robertson- Walker metric. In section 7, we find the red shift. In section 8, we define luminosity distance. In section 9, we fit the equations of section 6 to the Supernova Ia data. Section 10 contains the conclusions and brief discussions of open problems. In Appendix A, we review δ-symmetries. In Appendix B we discuss a simpler model: A delta harmonic oscillator, to illustrate the boundedness of the Hamiltonian.

2. Definition of Delta gravity
In this section we define the action as well as the symmetries of the model and derive the equations of motion.

We use the metric convention of [8]. The action of δ gravity is:

$$S(g,\tilde{g},\lambda) = \int d^dx\sqrt{-g}(-\frac{1}{2\kappa}R + \mathcal{L}_M) +$$
$$\kappa_2\int\left[\left(R_{\mu\nu} - \frac{1}{2}g_{\mu\nu}R\right) + \kappa T_{\mu\nu}\right]\sqrt{-g}\tilde{g}^{\mu\nu}d^dx \quad + \tag{1}$$
$$\kappa_2\kappa\int\sqrt{-g}\,(\lambda^{\mu;\nu} + \lambda^{\nu;\mu})\,T_{\mu\nu}d^dx$$

Here $\kappa = \frac{8\pi G}{c^4}$, κ_2 is an arbitrary constant and $T_{\mu\nu} := -\frac{2}{\sqrt{-g}}\frac{\delta(\sqrt{-g}\mathcal{L}_M)}{\delta g^{\mu\nu}}$ is the energy-momentum tensor of matter.$R_{\mu\nu}$ is the Ricci's tensor and R is the curvature scalar of $g_{\mu\nu}$. $\tilde{g}^{\mu\nu}$ is a two-contravariant tensor under general coordinate transformations.

The action (1) is obtained by applying the prescription contained in [20, 19]. That is, we add to the action of general relativity, the variation of it and consider the variation $\delta g_{\mu\nu} = \tilde{g}_{\mu\nu}$ as a new field. Similarly, the symmetries we write below are obtained as variation of the infinitesimal general coordinate transformations where the variation of the infinitesimal parameter $\delta\xi_0^\rho = \xi_1^\rho$ is the infinitesimal parameter of the new transformation δ. The last term in (1) is needed to implement the condition $T^{\mu\nu}_{;\nu} = 0$ as an equation of motion in order to implement the δ symmetry (2) off shell. This term is not needed in vacuum.

Action (1) is invariant under the following transformations(δ):

$$\delta g_{\mu\nu} = g_{\mu\rho}\xi^\rho_{0,\nu} + g_{\nu\rho}\xi^\rho_{0,\mu} + g_{\mu\nu,\rho}\xi^\rho_0 = \xi_{0\mu;\nu} + \xi_{0\nu;\mu}$$
$$\delta\tilde{g}_{\mu\nu}(x) = \xi_{1\mu;\nu} + \xi_{1\nu;\mu} + \tilde{g}_{\mu\rho}\xi^\rho_{0,\nu} + \tilde{g}_{\nu\rho}\xi^\rho_{0,\mu} + \tilde{g}_{\mu\nu,\rho}\xi^\rho_0 \tag{2}$$
$$\delta\lambda_\mu = -\xi_{1\mu} + \lambda_\rho\xi^\rho_{0,\mu} + \lambda_{\mu,\rho}\xi^\rho_0$$

From now on we will fix the gauge $\lambda_\mu = 0$. This gauge preserves general coordinate transformations but fixes completely the extra symmetry with parameter $\xi_{1\mu}$.

Equations of motion Varying $g_{\mu\nu}$ we get:

$$S^{\gamma\sigma} + \frac{1}{2}(R\tilde{g}^{\gamma\sigma} - g_{\mu\nu}\tilde{g}^{\mu\nu}R^{\gamma\sigma}) - \frac{1}{2}g^{\gamma\sigma}\frac{1}{\sqrt{-g}}\left(\sqrt{-g}\nabla_\nu\tilde{g}^{\mu\nu}\right)_{,\mu} +$$
$$\frac{1}{4}g^{\gamma\sigma}\frac{1}{\sqrt{-g}}\left(\sqrt{-g}g^{\alpha\beta}\nabla_\beta(g_{\mu\nu}\tilde{g}^{\mu\nu})\right)_{,\alpha} = \kappa\frac{\delta T_{\mu\nu}}{\delta g_{\gamma\sigma}}\tilde{g}^{\mu\nu} \tag{3}$$

where $S^{\gamma\sigma} = (U^{\sigma\beta\gamma\rho} + U^{\gamma\beta\sigma\rho} - U^{\sigma\gamma\beta\rho})_{;\rho\beta}$, $U^{\alpha\beta\gamma\rho} = \frac{1}{2}\left[g^{\gamma\rho}(\tilde{g}^{\beta\alpha} - \frac{1}{2}g^{\alpha\beta}g_{\mu\nu}\tilde{g}^{\mu\nu})\right]$

Varying $\tilde{g}^{\mu\nu}$ we get Einstein equation:

$$\left(R_{\mu\nu} - \frac{1}{2}g_{\mu\nu}R\right) + \kappa T_{\mu\nu} = 0 \tag{4}$$

Varying λ_μ we get:$T^{\mu\nu}_{;\nu} = 0$

Covariant derivatives as well as raising and lowering of indices are defined using $g_{\mu\nu}$. Notice that outside the sources($T_{\mu\nu} = 0$), a solution of (3) is $\tilde{g}^{\mu\nu} = \lambda g^{\mu\nu}$, for a constant λ, since $g^{\mu\nu}_{;\rho} = 0$ and $R_{\mu\nu} = 0$. We will have $\tilde{g}^{\mu\nu} = g^{\mu\nu}$, assuming that both fields satisfy the same boundary conditions far from the sources.

The equation for $\tilde{g}^{\mu\nu}$ is linear and of second order in the derivatives.

3. Particle motion in the gravitational field

We are aware of the presence of the gravitational field through its effects on test particles. For this reason, here we discuss the coupling of a test particle to a background gravitational field, such that the action of the particle is invariant under (2).

In δ gravity we postulate the following action for a test particle:

$$S_p = -m\int dt\sqrt{-g_{\mu\nu}\dot{x}^\mu\dot{x}^\nu} + \kappa_2'\int d^n y\sqrt{-g}\mathcal{T}_{\mu\nu}\left(\tilde{g}^{\mu\nu} + \lambda^{\mu;\nu} + \lambda^{\nu;\mu}\right)$$

where $\mathcal{T}_{\mu\nu}$ is the energy momentum tensor of the test particle:

$$\mathcal{T}_{\mu\nu}(y) = \frac{m}{2\sqrt{-g}}\int dt\frac{\dot{x}_\mu\dot{x}_\nu}{\sqrt{-g_{\alpha\beta}\dot{x}^\alpha\dot{x}^\beta}}\delta(y-x)$$

$\kappa_2' = \kappa_2\kappa$ is a dimensionless constant.

That is:

$$S_p = m\int\frac{dt}{\sqrt{-g_{\alpha\beta}\dot{x}^\alpha\dot{x}^\beta}}\left(g_{\mu\nu} + \frac{\kappa_2{}'}{2}\bar{g}_{\mu\nu}\right)\dot{x}^\mu\dot{x}^\nu \tag{5}$$

were $\bar{g}_{\mu\nu} = \tilde{g}_{\mu\nu} + \lambda^{\mu;\nu} + \lambda^{\nu;\mu}$. Notice that S_p is invariant under (2) and t-parametrizations.

From now on we work in the gauge $\lambda_\mu = 0$.

Since far from the sources, we must have free particles in Minkowski space,i.e $g_{\mu\nu} \sim \eta_{\mu\nu}, \tilde{g}_{\mu\nu} \sim \eta_{\mu\nu}$, it follows that we are describing the motion of a particle of mass $m' = m(1 + \frac{\kappa_2}{2}{}')$

Since in vacuum $\tilde{g}^{\mu\nu} = g^{\mu\nu}$, the equation of motion for test particles is the same as Einstein's. Moreover, the equation of motion is independent of the mass of the particle.

In order to include massless particles, we prefer to use the action [24]:

$$L = \frac{1}{2}\int dt\left(vm^2 - v^{-1}\left(g_{\mu\nu} + \kappa_2'\bar{g}_{\mu\nu}\right)\dot{x}^\mu\dot{x}^\nu + \frac{m^2 + v^{-2}\left(g_{\mu\nu} + \kappa_2'\bar{g}_{\mu\nu}\right)\dot{x}^\mu\dot{x}^\nu}{2v^{-3}g_{\alpha\beta}\dot{x}^\alpha\dot{x}^\beta}\left(m^2 + v^{-2}g_{\lambda\rho}\dot{x}^\lambda\dot{x}^\rho\right)\right) \tag{6}$$

This action is invariant under reparametrizations:

$$x'(t') = x(t); dt'v'(t') = dtv(t); t' = t - \varepsilon(t) \tag{7}$$

The equation of motion for v is:

$$v = -\frac{\sqrt{-g_{\mu\nu}\dot{x}^\mu\dot{x}^\nu}}{m} \tag{8}$$

Replacing (8) into (6), we get back (5).

Let us consider first the massive case. Using (7) we can fix the gauge $v = 1$. Introducing $mdt = d\tau$, we get the action:

$$L_1 = \frac{1}{2}m\int d\tau\left(1 - \left(g_{\mu\nu} + \kappa_2\bar{g}_{\mu\nu}\right)\dot{x}^\mu\dot{x}^\nu + \frac{1 + \left(g_{\mu\nu} + \kappa_2\bar{g}_{\mu\nu}\right)\dot{x}^\mu\dot{x}^\nu}{2g_{\alpha\beta}\dot{x}^\alpha\dot{x}^\beta}\left(1 + g_{\lambda\rho}\dot{x}^\lambda\dot{x}^\rho\right)\right) \tag{9}$$

plus the constraint obtained from the equation of motion for v:

$$g_{\mu\nu}\dot{x}^\mu\dot{x}^\nu = -1 \tag{10}$$

From L_1 the equation of motion for massive particles is derived. We define:$\mathfrak{g}_{\mu\nu} = g_{\mu\nu} + \frac{\kappa_2}{2}{}'\bar{g}_{\mu\nu}$.

$$\frac{d(\dot{x}^\mu\dot{x}^\nu\mathfrak{g}_{\mu\nu}\dot{x}^\beta g_{\alpha\beta} + 2\dot{x}^\beta\mathfrak{g}_{\alpha\beta})}{d\tau} - \frac{1}{2}\dot{x}^\mu\dot{x}^\nu\mathfrak{g}_{\mu\nu}\dot{x}^\beta\dot{x}^\gamma g_{\beta\gamma,\alpha} - \dot{x}^\mu\dot{x}^\nu\mathfrak{g}_{\mu\nu,\alpha} = 0 \tag{11}$$

We will discuss the motion of massive particles elsewhere.

The action for massless particles is:

$$L_0 = \frac{1}{4}\int dt\left(-v^{-1}\left(g_{\mu\nu} + \kappa_2\bar{g}_{\mu\nu}\right)\dot{x}^\mu\dot{x}^\nu\right) \tag{12}$$

In the gauge $v = 1$, we get:

$$L_0 = -\frac{1}{4}\int dt\left(g_{\mu\nu} + \kappa_2\bar{g}_{\mu\nu}\right)\dot{x}^\mu\dot{x}^\nu \tag{13}$$

plus the equation of motion for v evaluated at $v = 1$: $\left(g_{\mu\nu} + \kappa_2'\bar{g}_{\mu\nu}\right)\dot{x}^\mu\dot{x}^\nu = 0$

So, the massless particle moves in a null geodesic of $\mathfrak{g}_{\mu\nu} = g_{\mu\nu} + \kappa_2'\bar{g}_{\mu\nu}$.

4. Distances and time intervals

In this section, we define the measurement of time and distances in the model.

In GR the geodesic equation preserves the proper time of the particle along the trajectory. Equation(11) satisfies the same property: Along the trajectory $\dot{x}^\mu\dot{x}^\nu g_{\mu\nu}$ is constant.Therefore we define proper time using the original metric $g_{\mu\nu}$,

$$d\tau = \sqrt{-g_{\mu\nu}dx^\mu dx^\nu} = \sqrt{-g_{00}}dx^0(dx^i = 0) \tag{14}$$

Following [25], we consider the motion of light rays along infinitesimally near trajectories and (14) to get the three dimensional metric:

$$dl^2 = \gamma_{ij}dx^i dx^j,$$
$$\gamma_{ij} = \frac{g_{00}}{\mathfrak{g}_{00}}\left(\mathfrak{g}_{ij} - \frac{\mathfrak{g}_{0i}\mathfrak{g}_{0j}}{\mathfrak{g}_{00}}\right) \tag{15}$$

That is, we measure proper time using the metric $g_{\mu\nu}$ but the space geometry is determined by both metrics. In this model massive particles do not move on geodesics of a four dimensional metric. Only massless particles move on a null geodesic of $\mathfrak{g}_{\mu\nu}$. So, delta gravity is not a metric theory.

5. The Newtonian limit

The motion of a non relativistic particle in a weak static gravitational field is obtained using $g_{\mu\nu} = diag(-1 - 2U\epsilon, 1 - 2U\epsilon, 1 - 2U\epsilon, 1 - 2U\epsilon)$, which solves Einstein equations to first order in ϵ if $\nabla^2 U = \frac{1}{2}\kappa\rho$.

The solution for $\tilde{g}_{\mu\nu}$ is $\tilde{g}_{\mu\nu} = diag\left(\epsilon\,\tilde{U}, 1 + \epsilon\left(\tilde{U} - 2U\right), 1 + \epsilon\left(\tilde{U} - 2U\right), 1 + \epsilon\left(\tilde{U} - 2U\right)\right)$. Solving (3),to first order in ϵ we get $\nabla^2\tilde{U} = \frac{1}{2}\kappa\rho$.

To recover the Minkowsky metric far from the sources, $\rho \to 0$, we must require there:$U \to 0, \tilde{U} \to -\epsilon^{-1}$.

(11) implies $\frac{d^2x^i}{dt^2} = -\phi_{,i}$ with $\phi = U - \kappa_2'(2U + \tilde{U})$.

The Newtonian potential satisfies $\nabla^2\phi = \frac{\kappa}{2}(1 - 3\kappa_2')\rho, |\kappa_2'| \ll 1$. The whole effect is a small redefinition of Newton constant.

Gravitational red shift experiments can be used to put bounds on κ_2'. According to (14), the shift in frequency of a source located at x_1, compared to the same source located at x_2 due to the change in gravitational potential is: $\frac{\nu_2 - \nu_1}{\nu_1} = \phi_N(x_2) - \phi_N(x_1)$ where ϕ_N is the usual Newtonian potential, computed with κ as Newton constant. From [26] we get $\frac{\Delta\nu}{\nu} = (1{+}2.5{\pm}70{\times}10^{-6})(\varphi_S{-}\varphi_E{+}\ldots)$, where φ_S is the gravitational potential at the spacecraft

Journal of Physics: Conference Series **384** (2012) 012027 doi:10.1088/1742-6596/384/1/012027

position and φ_E is the gravitational potential on Earth. ... accounts for additional effects not related to the gravitational potential. We can ascribe the uncertainty of the experiment to κ_2', to get the bound:

$$|\kappa_2'| < 24 \times 10^{-6}$$

This bound is conservative because the Newton constant itself has a larger error [27]: $G = 6.67428 \pm 0.00067 \times 10^{-11} \frac{m^3}{\mathrm{kgs}^2}$

In our description of the evolution of the Universe, the value of κ_2' is not important, so we will keep it arbitrary for the time being.

6. Friedman-Robertson-Walker(FRW) metric

This is the main section of the paper. We discuss the equations of motion for the Universe described by the FRW metric. We use spatial curvature equal to zero to agree with cosmological observations.

In this paper we will deal only with a perfect fluid, since rotational and translational invariance implies that the energy-momentum tensor of the Universe has this form.The energy momentum tensor for a perfect fluid is [8]:

$$T_{\mu\nu} = pg_{\mu\nu} + (p+\rho)U_\mu U_\nu, g^{\mu\nu}U_\mu U_\nu = -1 \tag{16}$$

Then:

$$\frac{\delta T_{\mu\nu}}{\delta g_{\gamma\sigma}}\tilde{g}^{\mu\nu} = p\tilde{g}^{\gamma\sigma} + \frac{1}{2}(p+\rho)(U^\gamma U_\nu \tilde{g}^{\sigma\nu} + U^\sigma U_\nu \tilde{g}^{\gamma\nu}) \tag{17}$$

In this case, assuming flat three dimensional metric:

$$-ds^2 = dt^2 - R(t)^2\left\{dr^2 + r^2 d\theta^2 + r^2\sin^2\theta d\phi^2\right\}$$
$$-d\tilde{s}^2 = \tilde{A}(t)dt^2 - \tilde{B}(t)\left\{dr^2 + r^2 d\theta^2 + r^2\sin^2\theta d\phi^2\right\}$$

Using (11, 14), we can check that these are co-mobile coordinates and the proper time interval $d\tau$ for a co-moving clock is just dt, so t is the time measured in the rest frame of a co-moving clock. Equations (3, 17) give:

$$\begin{aligned} -\dot{R}\dot{\tilde{B}} - \frac{1}{2}pR\tilde{B} + \frac{1}{2}R^{-1}\dot{R}^2\tilde{B} - \frac{1}{6}\rho R^3\tilde{A} + \frac{3}{2}R\dot{R}^2\tilde{A} = 0 \\ -p\tilde{B} - 2\ddot{\tilde{B}} - R^{-2}\dot{R}^2\tilde{B} + 2R^{-1}\ddot{R}\tilde{B} + 2R^{-1}\dot{R}\dot{\tilde{B}} + \\ \rho R^2\tilde{A} + \dot{R}^2\tilde{A} + 2R\dot{R}\dot{\tilde{A}} + 2R\tilde{A}\ddot{R} = 0 \end{aligned} \tag{18}$$

Einstein's equations are:

$$\frac{3\left(\frac{d}{dt}R\right)^2}{R^2} = \kappa\rho \quad ,2R\left(\frac{d^2}{dt^2}R\right) + \left(\frac{d}{dt}R\right)^2 = -\kappa R^2 p$$

We use the equation of state $p = w\rho$, to get, for $w \neq -1$:

$$\begin{aligned} R = R_0 t^{\frac{2}{3(1+w)}}, \tilde{A} = 3wl_2 t^{\left(\frac{w-1}{w+1}\right)}, \\ \tilde{B} = R_0^2 l_2 t^b, b = \frac{4}{3w+3} + \frac{w-1}{w+1} \end{aligned} \tag{19}$$

l_2 is a free parameter.

7. Red Shift

To make the usual connection between redshift and the scale factor, we consider light waves traveling from $r = r_1$ to $r = 0$,along the r direction with fixed θ, ϕ. Photons moves on a null geodesic of $\mathfrak{g}$:

$$0 = -(1 + \kappa_2' \tilde{A})dt^2 + (R^2 + \kappa_2' \tilde{B})(dr^2 + r^2 d\theta^2 + r^2 \sin^2 \theta d\phi^2) \tag{20}$$

So,

$$\int_{t_1}^{t_0} dt \sqrt{\frac{1 + \kappa_2' tA}{R^2 + \kappa_2' tB}} = r_1 \tag{21}$$

A typical galaxy will have fixed r_1, θ_1, ϕ_1. If a second wave crest is emitted at $t = t_1 + \delta t_1$ from $r = r_1$, it will reach $r = 0$ at $t_0 + \delta t_0$, where

$$\int_{t_1+\delta t_1}^{t_0+\delta t_0} dt \sqrt{\frac{1 + \kappa_2' tA}{R^2 + \kappa_2' tB}} = r_1$$

Therefore, for $\delta t_1, \delta t_0$ small, which is true for light waves, we have:

$$\delta t_0 \sqrt{\frac{1 + \kappa_2' tA}{R^2 + \kappa_2' tB}}(t_0) = \delta t_1 \sqrt{\frac{1 + \kappa_2' tA}{R^2 + \kappa_2' tB}}(t_1) \tag{22}$$

Introduce:

$$\tilde{R}(t) = \sqrt{\frac{R^2 + \kappa_2' tB}{1 + \kappa_2' tA}}(t)$$

We get: $\frac{\delta t_0}{\delta t_1} = \frac{\tilde{R}(t_0)}{\tilde{R}(t_1)}$. A crucial point is that, according to equation (14), δt measure the change in proper time. That is: $\frac{\nu_1}{\nu_0} = \frac{\tilde{R}(t_0)}{\tilde{R}(t_1)}$, where ν_0 is the light frequency detected at $r = 0$ corresponding to a source emission at frequency ν_1. Or in terms of the redshift parameter z, defined as the fractional increase of the wavelength λ:

$$z = \frac{\tilde{R}(t_0)}{\tilde{R}(t_1)} - 1 = \frac{\lambda_0 - \lambda_1}{\lambda_1} \tag{23}$$

We see that $\tilde{R}$ replaces the usual scale factor R in the computation of z.

8. Luminosity distance

Let us consider a mirror of radius b that is receiving light from a distant source. The photons that reach the mirror are inside a cone of half-angle ε with origin at the source.

Let us compute ε.The light path of rays coming from a far away source at $\vec{x}_1$ is given by $\vec{x}(\rho) = \rho\hat{n} + \vec{x}_1$, $\rho > 0$ is a parameter and $\hat{n}$ is the direction of the light ray.The path reaches us at $\vec{x} = 0$ for $\rho = |\vec{x}_1| = r_1$. Write $\hat{n} = -\hat{x}_1 + \vec{\varepsilon}$. Since $\hat{n}, \hat{x}_1$ have modulus 1, $\varepsilon = |\vec{\varepsilon}| << 1$ is precisely the angle between $-\vec{x}_1$ and $\hat{n}$ at the source.The impact parameter is the proper distance of the path from the origin, when $\rho = |\vec{x}_1|$. The proper distance is determined by the 3-dimensional metric (15). That is $b = \tilde{R}(t_0) r_1 \theta = \tilde{R}(t_0) r_1 \varepsilon$, i.e. $\varepsilon = \frac{b}{\tilde{R}(t_0) r_1}$.

Then the solid angle of the cone is $\pi\varepsilon^2 = \frac{A}{r_1^2 \tilde{R}(t_0)^2}$, where $A = \pi b^2$ is the proper area of the mirror.The fraction of all isotropically emitted photons that reach the mirror is $f = \frac{A}{4\pi r_1^2 \tilde{R}(t_0)^2}$. Each photon carries an energy $h\nu_1$ at the source and $h\nu_0$ at the mirror. Photons emitted at intervals δt_1 will arrive at intervals δt_0. We have $\frac{\nu_1}{\nu_0} = \frac{\tilde{R}(t_0)}{\tilde{R}(t_1)}, \frac{\delta t_0}{\delta t_1} = \frac{\tilde{R}(t_0)}{\tilde{R}(t_1)}$. Therefore the power at the mirror is $P_0 = L\frac{\tilde{R}(t_1)^2}{\tilde{R}(t_0)^2}f$, where L is the luminosity of the source. The apparent luminosity is $l = \frac{P_0}{A} = L\frac{\tilde{R}(t_1)^2}{\tilde{R}(t_0)^2}\frac{1}{4\pi r_1^2 \tilde{R}(t_0)^2}$. In Euclidean space, the luminosity decreases with distance d according to $l = \frac{L}{4\pi d^2}$.This permits to define the luminosity distance:$d_L = \sqrt{\frac{L}{4\pi l}} = \tilde{R}(t_0)^2\frac{r_1}{\tilde{R}(t_1)}$. Using (21) we can write this in terms of the red shift:

$$d_L = (1+z)\int_0^z \frac{dz'}{\tilde{H}(z')}, \tilde{H} = \frac{\dot{\tilde{R}}}{\tilde{R}} \tag{24}$$

9. Supernova Ia data

The supernova Ia data gives, m (apparent or effective magnitude) as a function of z. This is related to distance d_L by $m = M + 5log(\frac{d_L}{10pc})$. Here M is common to all supernova and m changes with d_L alone.

We compare δ gravity to General Relativity(GR) with a cosmological constant:

$$H^2 = H_0^2(\Omega_m(1+z)^3 + (1-\Omega_m)), \Omega_\Lambda = 1 - \Omega_m$$

Notice that $\tilde{A} = 0$ for $w = 0$ in (19). So, it seems that we cannot fit the supernova data. However $w = 0$ is not the only component of the Universe. The massless particles that decoupled earlier still remain. It means that the true w is between $0 \leqslant w < \frac{1}{3}$, but very close to $w = 0$. So, we will fit the data with $w = 0.1, 0.01, 0.001$ and see how sensitive the predictions are to the value of w.

Using data from Essence[28], we notice that R^2 test changes very little for the chosen sequence of w's. Each fit determines the best l_2 for a given w. In this way we see that l_2 scales like $l_2 \sim \frac{a}{3w}$, a being independent of w. As an approximation to the limit $w = 0$, we get:

$$\tilde{R}(t) = R(t)\frac{\sqrt{a}}{\sqrt{a-t}} \tag{25}$$

$\sqrt{\frac{1}{3w}}$renormalizes the derivative of $\tilde{R}$ at $t = 0$. It is not divergent, because for $t \to 0$, $w \to \frac{1}{3}$. a is a free parameter determined by the best fit to the data.

Of course, the complete model must include the contribution of normal matter($w = 0$) plus relativistic matter ($w = \frac{1}{3}$). But, at later times, the data should tend to (25).

Let us fit the data to the simple scaling model (25).

We get:

$\Omega_m = 0.22 \pm 0.03, M = 43.29 \pm 0.03$, $\chi^2(perpoint) = 1.0328$, General Relativity

$a = 2.21 \pm 0.12, M = 43.45 \pm 0.06$, $\chi^2(perpoint) = 1.0327$, Delta Gravity

δ-gravity with non-relativistic(NR) matter alone give a fit to the data as good as GR with NR matter plus a cosmological constant.

According to the fit to data, a Big Rip will happen at $t = 2.21049$ in unities of t_0(today). It is a similar scenario as in [23].

Finally, we want to point out that since for $t \to 0$, we have $w \to \frac{1}{3}$, then $\tilde{R}(t) = R(t)$. Therefore the accelerated expansion is slower than (25) when we include both matter and radiation in the model.

10. Conclusions and Open Problems

Delta Gravity agrees with GR when $T_{\mu\nu} = 0$, imposing same boundary conditions for both tensor fields. In particular, the causal structure of delta gravity in vacuum is the same as in GR, since in this case the action (5) is proportional to the geodesic action in GR.

We recover the Newtonian approximation.

In a homogeneous and isotropic universe, we get accelerated expansion without a cosmological constant or additional scalar fields.

The computation of PPN(Post Newtonian) parameters is in progress, but we do not expect large departures from general relativity, because the Newtonian limit is the right one, as explained in section 6. Moreover the interstellar space has very small matter densities, so δ- gravity must give GR values for the PPN parameters(See comments after equation(4)). Additionally, please notice that all $\tilde{g}$ contributions are multiplied by the small parameter κ_2' of the order of 10^{-5} or less, so they are very small in the solar system.

Stellar evolution will not be changed from its Newtonian description, unless density of matter becomes very large. Even at the densities of white dwarfs the Poisson equation for the gravitational potential suffices.(See, for instance [8], chapter 11.3). δ- gravity implies it, as it is shown in section 6. Higher densities which are present in neutron stars may provide new tests of δ-gravity, since there we have to use the whole non-linear Einstein equations and the corresponding δ- gravity equations. But for the inner regions of massive stars, data is very scarce.

Notice that equation (19) implies that $\tilde{R} = R$ at the beginning of the Universe, when $w = \frac{1}{3}$, corresponding to ultrarelativistic matter. That is, the accelerated expansion started at a later time, which is needed if we want to recover the observational data of density perturbations and growth of structures in the Universe. An earlier acceleration of the expansion would prevent the growth of density perturbations.

Work is in progress to compute the growth of density perturbations and the anisotropies in the CMBR. The comparison of these calculations with the considerable amount of astronomical data that will be available in the near future will be a very stringent test of the present gravitational model.

It was noticed in [20] that the Hamiltonian of delta models is not bounded from below. Phantoms cosmological models [22], [23] also have this property. Although it is not clear whether this problem will subsist in a diffeomorphism invariant model as delta gravity or not, we want to mention some ways out of the difficulty.

a) Delta gravity is a gauge theory. Moreover it is diffeomorphism invariant. Thus the canonical Hamiltonian vanishes identically. It may be possible to truncate the Hilbert space, using the BRST formalism, to define a model with a Hamiltonian bounded from below. This is a difficult task that goes far beyond the present paper, but should be pursued in a future work.

b) In a supersymmetric model we have $H = Q^2$, where H is the Hamiltonian and Q is the hermitian supersymmetry charge. Thus the Hamiltonian is bounded from below. So, we expect that a delta supergravity model has a Hamiltonian bounded from below.

c) A Delta model has more symmetries than the original model. This permits to bound the Hamiltonian for a given value of a conserved quantity. We explain this in Appendix B.

Acknowledgements JA wants to thank the organizers of DSU2011 for a very stimulating experience at Beijing. The work of JA is partially supported by VRAID/DID/46/2010 and Fondecyt 1110378. He wants to thank R. Avila and P. González for several useful remarks; The author acknowledges interesting conversations with L. Infante, G. Palma, M. Bañados and A. Clocchiatti. In particular, JA wants to thank A. Clocchiatti for pointing out the data in [28]. Finally, JA wants to thank J. Gamboa for a careful reading of the manuscript.

[1] Clifford M. Will, "The Confrontation between General Relativity and Experiment", Living Rev. Relativity 9, (2006), http://www.livingreviews.org/lrr-2006-3;Slava G. Turyshev, Annual Review of Nuclear and Particle Science, Vol. 58: 207-248 (Volume publication date November 2008)

[2] For a modern review of string model see: M.B. Green, J.H. Schwarz and E. Witten, "Superstring Theory ", vols. 1, 2, Cambridge University Press 1987. J. Polchinski, "String Theory", vols. 1,2, Cambridge University Press 1998.
[3] for a review of Dark Matter and its detection , see S. Weinberg, Cosmology, Oxford University Press 2008; Hooper, D. and Baltz, Annu. Rev. Nucl. Part. Sci. 58, 293314(2008).
[4] A. G. Riess et al. (Supernova Search Team), Astron. J. 116, 1009 (1998),S. Perlmutter et al. (Supernova Cosmology Project), Astrophys. J. 517, 565 (1999). For a recent review, see R. R. Caldwell and M. Kamionkowski, The Physics of Cosmic Acceleration,astro-ph 0903.0866.
[5] Frieman, J. A., Turner, M. S. & Huterer, D. Dark energy and the accelerating Universe. Annu. Rev. Astron. Astrophys. 46, 385432 (2008).
[6] Milgrom, M.,A modification of the Newtonian dynamics as a possible alternative to the hidden mass hypothesis, ApJ, 270, 365(1983); Bekenstein, J. Relativistic gravitation theory for the MOND paradigm, Phys. Rev. D 70, 083509 (2004)
[7] See, for instance, S. Tsujikawa, Lect.Notes Phys.800:99-145,2010
[8] S. Weinberg, Gravitation and Cosmology(Wiley, New York, 1972)
[9] S. Weinberg, in General Relativity: An Einstein centenary survey, edited by S. W.Hawking and W.Israel (Cambridge University Press, 1979), chapter 16, p. 790.
[10] Ya.B. Zeldovich, JETP Lett., 6, 316 (1967); A. Sakharov, SOv. Phys. Dokl., 12, 1040 (1968); O. Klein, Phys. Scr. 9, 69 (1974); S. Adler, Rev. Mod. Phys., 54, 729 (1982).
[11] D.F. Litim, Phys.Rev.Lett.92:201301,2004; AIP Conf. Proc. 841, 322 (2006); e-Print: arXiv:0810.367; A. Codello, R. Percacci and C. Rahmede, Annals Phys.324:414-469,2009; M. Reuter and F. Saueressig, Lectures given at First Quantum Geometry and Quantum Gravity School, Zakopane, Poland (2007),arXiv:0708.1317
[12] J. Ambjorn, J. Jurkiewicz and R. Loll, Phys.Rev.Lett.85:924-927,2000.
[13] J. Alfaro, D. Espriu and D. Puigdomenech, Phys. Rev. D82:045018,2010.
[14] C.J. Isham, A. Salam and J.A. Strathdee, Annals Phys.62:98-119,1971.
[15] A.B. Borisov and V.I. Ogievetsky, Theor.Math.Phys.21:1179,1975; E.A. Ivanov and V.I. Ogievetsky, Lett.Math.Phys.1:309-313,1976.
[16] D. Amati and J. Russo, Phys.Lett. B 248, 44 (1990); J. Russo, Phys.Lett. B 254, 61 (1991); A. Hebecker, C. Wetterich, Phys.Lett.B574:269-275,2003; C. Wetterich, Phys. Rev. D 70: 105004, 2004.
[17] J. Alfaro, "Delta Gravity and Dark Energy", arXiv:1006.5765 [gr-qc]
[18] G. 't Hooft and M. Veltman, Ann. Inst. Henri Poincar, 20 (1974) 69
[19] J. Alfaro, bv gauge theories, hep-th 9702060.
[20] J. Alfaro and P. Labraña,Phys. Rev. D 65, 045002 (2002).
[21] J. Alfaro, P. Gonzalez, R. Avila, Class.Quant.Grav. 28 (2011) 215020.
[22] R.R. Caldwell, Physics Letters B 545 (2002) 2329.
[23] R. R. Caldwell, M. Kamionkowski, and N. N.Weinberg, Phys. Rev. Lett.91(2003)071301.
[24] W. Siegel, Fields,page 180.http://arxiv.org/abs/hep-th/9912205v3
[25] L. Landau and L.M. Lifshitz, The Classical Theory of Fields, Pergamon Press 1980.
[26] R. F. C. Vessot, et al, Phys. Rev. Lett. 45, 2081 (1980).
[27] Mohr, Peter J.; Taylor, Barry N.; Newell, David B. , Rev. Mod. Phys. 80: 633Ű730.
[28] W. M. Wood-Vasey et al.,Astrophys.J.666:694-715,2007.

11. Appendix A Review of δ-symmetries

Assume we have a group of transformations acting on the variables y with infinitesimal parameters ϵ. That is:

$$\delta y^i = \Lambda^i_\alpha(y)\epsilon^\alpha \tag{26}$$

We define the δ transformation by:

$$\delta \bar{y}^i = \Lambda^i_\alpha(y)_{,j}\bar{y}^j\epsilon^\alpha + \Lambda^i_\alpha(y)\bar{\epsilon}^\alpha \tag{27}$$

$k_{,i} = \frac{\partial k}{\partial y^i}$.

Notice that, we have introduced a new field $\bar{y}^i$ and a new transformation with parameter $\bar{\epsilon}^\alpha$. It is easy to see that (26, 27) form a closed algebra.

An invariant action under the extended symmetry is built in the same way. We assume that

$S(y)$ is invariant under (26):

$$\frac{\delta S}{\delta y^i}\Lambda^i_\alpha(y) = 0, \forall y, \text{all}\,\alpha \tag{28}$$

Then:

$$\bar{S}(y,\bar{y}) = S(y) + \frac{\delta S}{\delta y^i}\bar{y}^i$$

is invariant under (26, 27).

Proof:

$$\delta\bar{S}(y,\bar{y}) = \frac{\delta S}{\delta y^i}\Lambda^i_\alpha(y)\epsilon^\alpha + \frac{\delta^2 S}{\delta y^i \delta y^j}\Lambda^j_\alpha(y)\epsilon^\alpha\bar{y}^i + \frac{\delta S}{\delta y^i}(\Lambda^i_\alpha(y)_{,j}\bar{y}^j\epsilon^\alpha + \Lambda^i_\alpha(y)\bar{\epsilon}^\alpha) =$$
$$0 + (\frac{\delta^2 S}{\delta y^i \delta y^j}\Lambda^j_\alpha(y)\bar{y}^i + \frac{\delta S}{\delta y^i}\Lambda^i_\alpha(y)_{,j}\bar{y}^j)\epsilon^\alpha + 0\bar{\epsilon}^\alpha =$$
$$(\frac{\delta^2 S}{\delta y^i \delta y^j}\Lambda^i_\alpha(y) + \frac{\delta S}{\delta y^i}\Lambda^i_\alpha(y)_{,j})\epsilon^\alpha\bar{y}^j = \left\{\frac{\delta}{\delta y^j}(\frac{\delta S}{\delta y^i}\Lambda^i_\alpha(y))\right\}\epsilon^\alpha\bar{y}^j = 0$$

Last equality follows from equation (28).

Being careful with signs of permutations, these results are true for anti-commuting y, ϵ as well. In particular, super-symmetric transformations can be generalized to a δ- symmetry.

Other generalizations are possible. Suppose we have canonical transformations generated by $\epsilon(x,p)$:

$$\delta F = (\epsilon, F) \quad ,$$
$$\delta\bar{F} = (\epsilon,\bar{F}) + (\bar{\epsilon}, F) \tag{29}$$

equations (26, 27) are particular cases of (29). (A, B) is the Poisson bracket. Now we can prove the closure of the algebra in a more general context.

$$[\delta_\beta,\delta_\alpha]\,F = (\delta_\beta(\alpha,F) - \alpha\leftrightarrow\beta) = (\alpha,(\beta,F)) - (\beta,(\alpha,F)) = (F,(\beta,\alpha)) = ((\alpha,\beta),F) = \delta_{(\alpha,\beta)}F$$
$$[\delta_\beta,\delta_\alpha]\,\bar{F} = (\delta_\beta(\alpha,\bar{F}) - \alpha\leftrightarrow\beta) = (\alpha,(\beta,\bar{F})) - (\beta,(\alpha,\bar{F})) = (\bar{F},(\beta,\alpha)) = ((\alpha,\beta),\bar{F}) = \delta_{(\alpha,\beta)}\bar{F}$$
$$\left[\delta_\alpha,\delta_{\bar{\beta}}\right]F = 0$$
$$\left[\delta_\alpha,\delta_{\bar{\beta}}\right]\bar{F} = (\delta_\alpha(\bar{\beta},F) - \delta_{\bar{\beta}}(\alpha,\bar{F}) = (\bar{\beta},(\alpha,F)) - (\alpha,(\bar{\beta},F)) = (F,(\alpha,\bar{\beta})) = \delta_{(\bar{\beta},\alpha)}F$$
$$\left[\delta_{\bar{\alpha}},\delta_{\bar{\beta}}\right]\bar{F} = \delta_{\bar{\alpha}}(\bar{\beta},F) - \bar{\alpha}\leftrightarrow\bar{\beta} = 0$$

Replacing Poisson bracket by commutators is the realization of the algebra we used in [20].

12. Appendix B: The Delta Harmonic Oscillator

$$L = \frac{1}{2}\dot{x}^2 - \frac{1}{2}\omega^2x^2 + \dot{x}\dot{y} - \omega^2 xy$$

The canonical variables are:

$$\begin{aligned} p_x =& \quad \dot{x} + \dot{y} \\ p_y =& \quad \dot{x} \\ H =& \quad \dot{x}(\dot{x}+\dot{y}) + \quad \dot{y}\dot{x} - L = H_0 + \dot{x}\dot{y} + \omega^2 xy \end{aligned}$$

$$H_0 = \frac{1}{2}p_y^2 + \frac{1}{2}\omega^2x^2 \tag{30}$$

We know that H_0 is conserved, because the model satisfies the equations of motion for x. Since H is conserved, we have that Q is conserved.

Journal of Physics: Conference Series **384** (2012) 012027 doi:10.1088/1742-6596/384/1/012027

$$Q = \dot{x}\dot{y} + \omega^2 xy$$

Since H_0 is greater or equal than zero. H is bounded from below by the value of the conserved quantity Q.

This is a generic feature of delta models. The Hamiltonian is bounded from below, for a given value of a conserved quantity Q, if the Hamiltonian of the original model is bounded from below.

Casimir Energy of the Universe and the Dark Energy Problem

Shoichi Ichinose
Laboratory of Physics, School of Food and Nutritional Sciences, University of Shizuoka
Yada 52-1, Shizuoka 422-8526, Japan
E-mail: ichinose@u-shizuoka-ken.ac.jp

Abstract. We regard the Casimir energy of the universe as the main contribution to the cosmological constant. Using 5 dimensional models of the universe, the flat model and the warped one, we calculate Casimir energy. Introducing the new regularization, called *sphere lattice regularization*, we solve the divergence problem. The regularization utilizes the closed-string configuration. We consider 4 different approaches: 1) restriction of the integral region (Randall-Schwartz), 2) method of 1) using the minimal area surfaces, 3) introducing the weight function, 4) *generalized path-integral*. We claim the 5 dimensional field theories are quantized properly and all divergences are renormalized. At present, it is explicitly demonstrated in the numerical way, not in the analytical way. The renormalization-group function (β-function) is explicitly obtained. The renormalization-group flow of the cosmological constant is concretely obtained.

1. Introduction

At present, the energy budget of the universe is not explained theoretically. We do not understand the dark matter and the dark energy. The latter one, which is the present main topic, is regarded as the same problem as the cosmological constant one, which is the long-lasting problem[1, 2]. As for the problem, Polyakov[3, 4] made a notable conjecture that the cosmological constant, just like the QED coupling, flows to a small value in the IR region (screening phenomena). He made another comment[5] that the dark energy, like the black body radiation 150 years ago, hides secrets of fundamental physics. These views about the problem look to be revived in the recent strong trend of the AdS/CFT (holography) approach to the condensed matter physics or to the viscous fluid dynamics(Navier-Stokes equation)[6, 7].

The difficulty of the problem is strongly related to the unsuccessful situation of the quantum gravity. It also has the long research history since Feynman initiated in 1963[8] until the present string or D-brane research. In the microscopic side, the gravitational interaction has, when quantized, serious UV-divergences[9], while , in the macroscopic one, it has serious IR-divergences around the horizon (boundaries)[10, 4].

About 2 years ago (2010 January), Verlinde[11] made a shocking view about the role of the gravitational force. He made a very thoughtful analysis, including some Gedanken-experiments, about the gravitational force and finally reached the conclusion that the gravitational interaction (force) is *not* fundamental but is *emergent* (by some still-obscure mechanism). He referred some (not few) past literature about similar stand points such as Jacobson's[12] and Padmanabhan's[13]. The analysis demands very delicate treatment of the horizon which is the boundary of the theory. The physical quantities (IR-)diverge at the horizon. It strongly indicates the regularization problem in the near-horizon treatment where Hawking radiation (thermalization) occurs.

The biggest discrepancy, in all physical quantities, between observation and theory appears in the cosmological constant λ.

$$S = \int d^4x\sqrt{-g}\{\frac{1}{16\pi G_N}(R-2\lambda)+\mathcal{L}_{matter}\} \quad ,$$
$$\frac{\lambda_{th}}{\lambda_{obs}} \sim N_{DL}^2 \quad , \quad N_{DL} \equiv M_{pl}R_{cos} \sim 6.\times 10^{60} \quad , \tag{1}$$

where M_{pl} ($\equiv 1/\sqrt{G_N}$) and R_{cos} ($\equiv 1/H_0, H_0$: Hubble constant) are Planck mass and the size of the universe respectively. N_{DL} is an analog of Dirac's large number[14]. [1] The cause of the theoretical difficulty lies in the two extreme-ends of mysterious branches of physics: the quantum gravity and the cosmology.

Casimir energy is generally the vacuum energy of the *free*-part of the system dynamics. For the harmonic oscillator, it is the energy of the zero-point oscillation. By definition, it is independent of the coupling. It depends, however, on the *boundaries* and the system *topology*. It is caused by the quantum *fluctuation*. Because we have to deal with serious UV and IR divergences, highly-delicate *regularization* is required. The well-known one is that for the 4D electromagnetism (free wave theory). The system of two parallely-placed metalic plates separated by the length l, has Casimir energy as follows.

$$\text{4D Space-time}: \ V_{Cas}(l) = \frac{B_4}{l^3}, \quad F_{Cas}(l) = -\frac{\partial V}{\partial l} = 3\frac{B_4}{l^4}, \quad B_4(\text{4-th Bernoulli No}) = -\frac{1}{30},$$
$$\text{5D Space-time}: \ V_{Cas}(l) = -\frac{3}{32\pi^2}\frac{\zeta(5)}{l^4}, \quad F_{Cas}(l) = -\frac{\partial V}{\partial l} = -\frac{3}{8\pi^2}\frac{\zeta(5)}{l^5}, \quad \zeta(5) = 1.03693\cdots. \tag{2}$$

2. Casimir Energy of 5 Dimensional Electromagnetism

For the electromagnetism in the flat 5D space-time ($ds^2 = \eta_{\mu\nu}dx^\mu dx^\nu + dy^2$, $\eta_{\mu\nu} = \mathrm{diag}(-1,1,1,1)$), Casimir energy is given by [15]

$$E_{Cas}(\Lambda, l) = \int_{\tilde{p}\leq\Lambda}\frac{d^4p}{(2\pi)^4}\int_0^l dy(F_f^-(\tilde{p},y)+4F_f^+(\tilde{p},y)) \quad ,$$
$$F_f^{\mp}(\tilde{p},y) = -\int_{\tilde{p}}^{\infty} d\tilde{k}\frac{\mp\cosh\tilde{k}(2y-l)+\cosh\tilde{k}l}{2\sinh(\tilde{k}l)} \quad , \tag{3}$$

where $\mp$ is the y-parity ($y \leftrightarrow -y$), l is the periodicity $y \to y+2l$, and Λ is the momentum cut-off. $\tilde{p}$ is the magnitude of the 4D Euclidean momentum ($p_1, p_2, p_3, p_4 = ip_0$). This is consistent with Ref.[16].

As for the warped (AdS_5) geometry, $ds^2 = (\eta_{\mu\nu}dx^\mu dx^\nu + dz^2)/\omega^2 z^2$ (ω: 5D curvature), the corresponding one is given by [17]

$$-E_{Cas}^{\Lambda,\mp}(\omega,T) = \int\frac{d^4p_E}{(2\pi)^4}\Big|_{\tilde{p}\leq\Lambda}\int_{1/\omega}^{1/T} dz\ F_\omega^{\mp}(\tilde{p},z), F_\omega^{\mp}(\tilde{p},z) = \frac{1}{(\omega z)^3}\int_{\tilde{p}^2}^{\infty}\{G_k^{\mp}(z,z)\}dk^2,$$
$$G_p^{\mp}(z,z') = \mp\frac{\omega^3}{2}z^2z'^2\frac{\{\mathbf{I}_0(\frac{\tilde{p}}{\omega})\mathbf{K}_0(\tilde{p}z)\mp\mathbf{K}_0(\frac{\tilde{p}}{\omega})\mathbf{I}_0(\tilde{p}z)\}\{\mathbf{I}_0(\frac{\tilde{p}}{T})\mathbf{K}_0(\tilde{p}z')\mp\mathbf{K}_0(\frac{\tilde{p}}{T})\mathbf{I}_0(\tilde{p}z')\}}{\mathbf{I}_0(\frac{\tilde{p}}{T})\mathbf{K}_0(\frac{\tilde{p}}{\omega})-\mathbf{K}_0(\frac{\tilde{p}}{T})\mathbf{I}_0(\frac{\tilde{p}}{\omega})}, \tag{4}$$

[1] The original Dirac's definition is [electromagnetic force]/[gravitational force] $=\frac{\alpha\hbar c}{G_N m_e m_p} \sim 2.3\times 10^{39}$ ($\alpha = e^2/4\pi\varepsilon_0\hbar c$) where the force is that working between the electron and the proton in a H-atom. The substitute, in the present case, is naturally [size of the universe]/[Planck length]. Note that Dirac's value is approximately equal to ([Planck mass]/[mass of nucleon])^2.

where $\mathbf{I}_0$ and $\mathbf{K}_0$ are 0-th modified Bessel functions. For simplicity, we take the bulk particle mass M in such a way that $M^2 = -4\omega^2 < 0$. T is defined by $T = \omega \exp(-l\omega)$.

3. Behavior of Casimir Energy

The integral region, for the flat case (3), is shown in Fig.1. μ and ϵ are the IR cutoff of $\tilde{p}$-axis and the UV cutoff of y-axis respectively. Fig.2 shows the integrand of (3): $\tilde{p}^3(F_f^- + 4F_f^+) \equiv \tilde{p}^3 F(\tilde{p}, y)$. In Fig.3 the behavior of the integrand of F^- (3) is shown. The (inverse) table shape says the

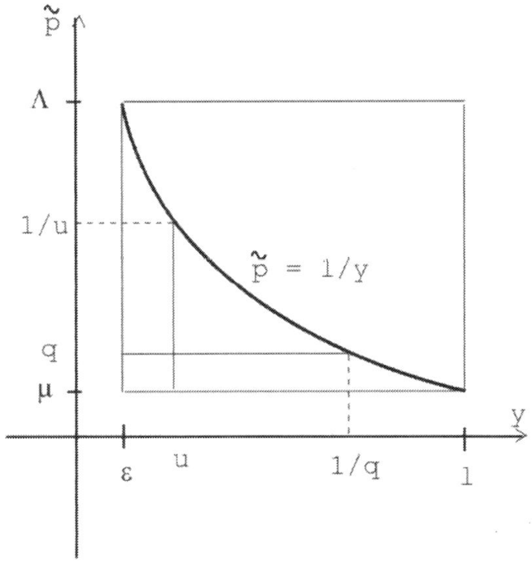

Figure 1. Space of (y,$\tilde{p}$) for the integration (3). The hyperbolic curve is used in (6).

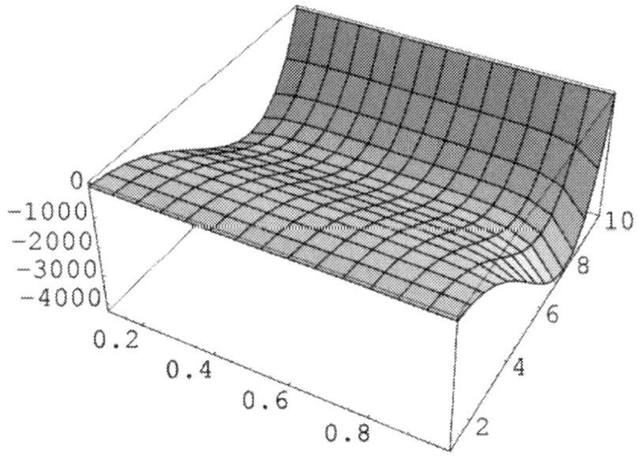

Figure 2. Behaviour of $\tilde{p}^3 F(\tilde{p}, y)$ in (3). $l = 1$, $\Lambda = 10$, $0.1 \le y < 1$, $1 \le \tilde{p} \le 10$.

"Rayleigh-Jeans" dominance because Casimir energy density is proportinal to the cubic power of $\tilde{p}$ in the region $\tilde{p} \ll \Lambda$. Numerically $E_{Cas}(\Lambda, l)$ is obtained as

$$E_{Cas}(\Lambda, l) = \frac{2\pi^2}{(2\pi)^4}\left[-0.1249 l\Lambda^5 - (1.41, 0.706, 0.353) \times 10^{-5}\, l\Lambda^5 \ln(l\Lambda)\right] \quad . \tag{5}$$

In Fig.1, the region below the hyperbolic curve is that of Randall-Schwartz(RS)[18]. They claimed the rectangular region should be replaced by this restricted one for the purpose of reducing the divergences.

$$E_{Cas}^{RS} = \frac{2\pi^2}{(2\pi)^4}\int_{1/l}^{\Lambda} dq \int_{1/\Lambda}^{1/q} dy\ q^3 F(q, y) = \frac{2\pi^2}{(2\pi)^4}[-8.93814 \times 10^{-2}\ \Lambda^4] \quad , \tag{6}$$

which shows slightly milder than (5). $\ln(l\Lambda)$-term does not appear.

As for the Warped model, the situation is similar. Fig.4 shows the integral region. The z-axis range is $\omega^{-1} \le z \le T^{-1}$. We take IR-regularization-point of $\tilde{p}$ as $\mu = \Lambda T/\omega$. In Fig.5 the behavior of the integrand of E_{Cas}^- (4) is shown. Fig.6 shows the warped version of Fig.3. E_{Cas}^- (4) is numerically obtained as

$$E_{Cas}^{\Lambda,-}(\omega, T) = \frac{2\pi^2}{(2\pi)^4} \times \left[-0.0250\frac{\Lambda^5}{T}\right] \quad , \tag{7}$$

Figure 3. Behaviour of the integrand of F^-,(3). $l = 1, \Lambda = 100, 0 \leq y \leq l = 1, 1 \leq \tilde{k} \leq \Lambda = 100$. The flat plane locates at the height -0.5.

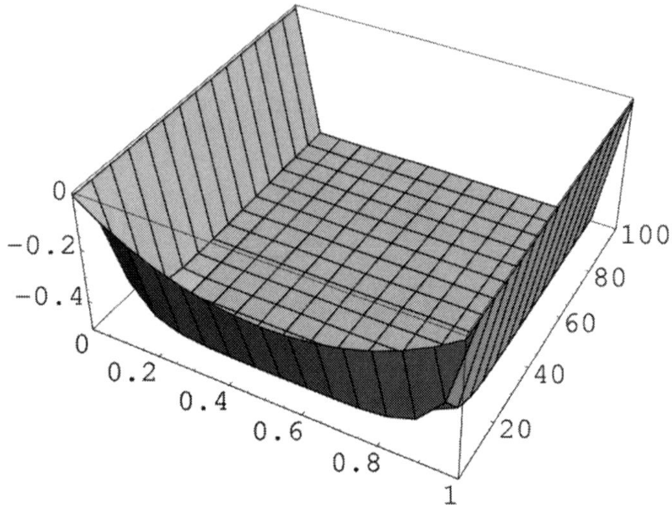

which does not depend on ω and has no $\ln(\Lambda/T)$ term. When restricted to RS-region, the above value changes to

$$E_{Cas}^{-RS}(\omega, T) = \frac{2\pi^2}{(2\pi)^4}\int_{\mu}^{\Lambda} dq \int_{1/\omega}^{\Lambda/\omega q} dz\; q^3 F^-(q, z)$$
$$= \frac{2\pi^2}{(2\pi)^4}\frac{\Lambda^5}{\omega}\left\{-1.58 \times 10^{-2} - 1.69 \times 10^{-4} \ln \frac{\Lambda}{\omega}\right\} \quad , \qquad (8)$$

which is independent of T and has the $\ln(\Lambda/\omega)$ term. Divergence behavior is the same as the unrestricted case (7).

4. Sphere Lattice Regularization

The Randall-Schwartz's way of restricting the integral region does not sufficiently work to reduce the divergences. In Ref.[19], we proposed a new way of the restriction based on the isotropy property of the system and the minimal area principle. Let us introduce two 4D hyper-surfaces, B_{UV} and B_{IR}, in the 5D bulk space-time.

$$B_{UV} \quad : \quad \tilde{p}^{-1} = \sqrt{(x^1)^2 + (x^2)^2 + (x^3)^2 + (x^4)^2} = r_{UV}(y) \quad , \quad \epsilon = \frac{1}{\Lambda} < y < l \quad ,$$
$$B_{IR} \quad : \quad \tilde{p}^{-1} = \sqrt{(x^1)^2 + (x^2)^2 + (x^3)^2 + (x^4)^2} = r_{IR}(y) \quad , \quad \epsilon = \frac{1}{\Lambda} < y < l \quad , \qquad (9)$$

where x^a $(a = 1, 2, 3, 4)$ is Euclidian coordinate. See Fig.7 and Fig.8. The surfaces, in the "brane" located at y, are 3D sphere with the radius $r(y)$. The radius changes along the extra axis y. The function form of $r(y)$ is given by the minimal area (of the hyper surface) principle. We integrate the region bounded by B_{IR} from below and by B_{UV} from above as shown in Fig.7. Some regularization parameters $(\Lambda, \mu, \Lambda', \mu', \varepsilon)$ are defined there. The renormalization group interpretation is shown in Fig.8. The surface B_{UV} is stereo-graphically shown in Fig.9. It shows the present regularization utilizes the *closed-string* configuration. The integral region for the warped case is shown in Fig.10.

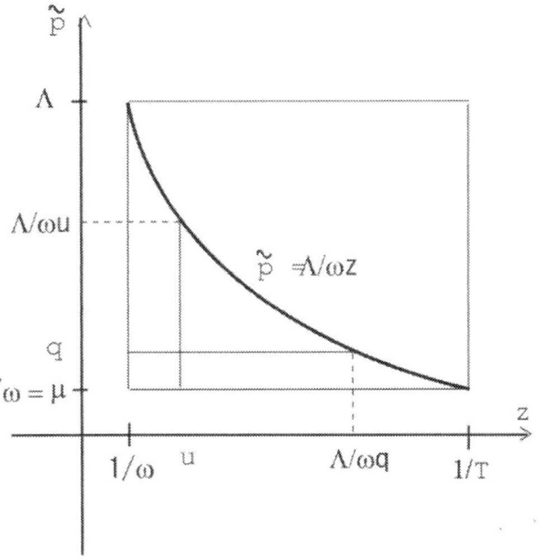

Figure 4. Space of (z,$\tilde{p}$) for the integration (4). The hyperbolic curve is used in (8).

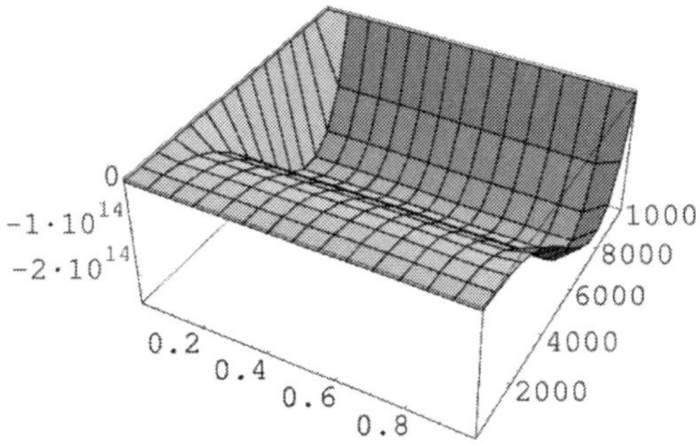

Figure 5. Behaviour of $(-1/2)\tilde{p}^3 F^-(\tilde{p},z)$ in (4). $T = 1, \omega = 10^4, \Lambda = 10^4$. $1.0001/\omega \leq z < 0.9999/T$, $\Lambda T/\omega \leq \tilde{p} \leq \Lambda$.

Figure 6. Behavior of $\ln|\frac{1}{2}\mathcal{F}^-(\tilde{k},z)| = \ln|\tilde{k}\ G_{\tilde{k}}^-(z,z)/(\omega z)^3|$. $\omega = 10^4, T = 1, \Lambda = 2\times 10^4$. $1.0001/\omega \leq z \leq 0.9999/T$. $\Lambda T/\omega \leq \tilde{k} \leq \Lambda$. Note $\ln|(1/2)\times(1/2)| \approx -1.39$.

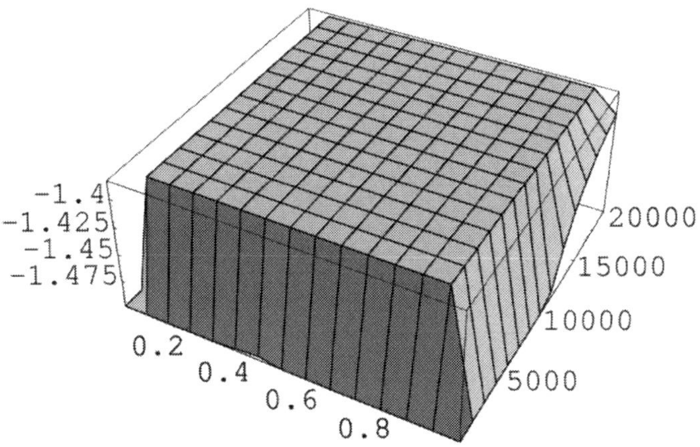

5. Introduction of the Weight Function $W(\tilde{p}, y)$ or $W(\tilde{p}, z)$

Another way to reduce the divergences is, instead of restriction, to introduce the *weight function* $W(\tilde{p}, y)$ or $W(\tilde{p}, z)$ in the integration over 5D space-time. The determination of the function form is explained later using the minimal area principle. Now we assume some forms as sample examples, and check whether Casimir energy is finitely obtained. For the flat case, we take the following forms.

$$W(\tilde{p}, y) =$$

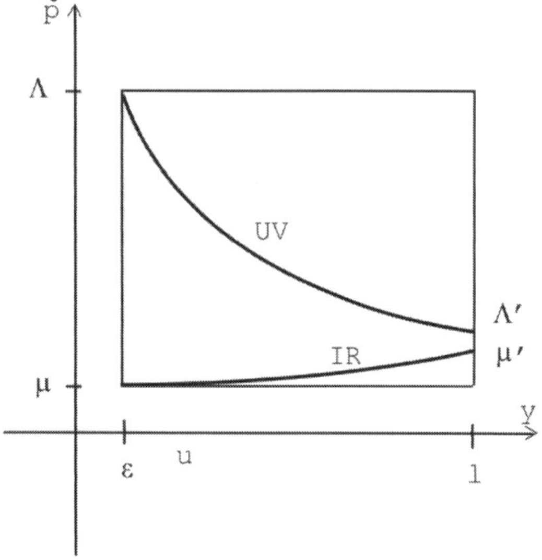

Figure 7. Space of ($\tilde{p}$,y) for the integration (present proposal as the substitute of Fig.1).

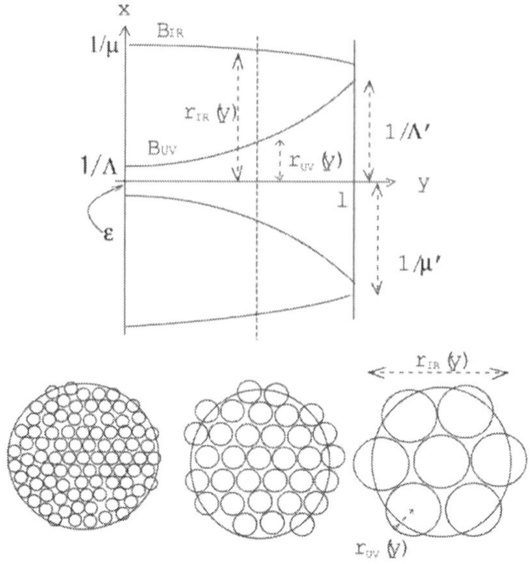

Figure 8. Regularization Surface B_{IR} and B_{UV} in the 5D coordinate space (x^μ, y), Flow of Coarse Graining (Renormalization) and Sphere Lattice Regularization.

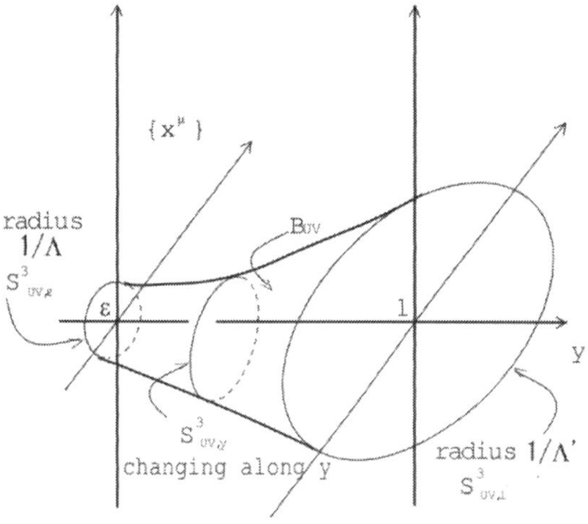

Figure 9. UV regularization surface in 5D coordinate space.

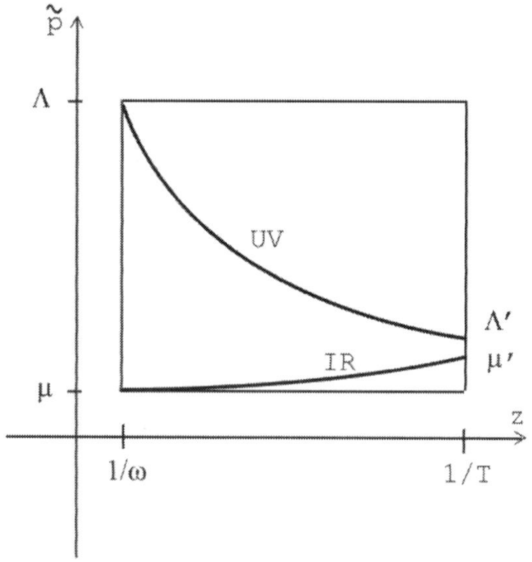

Figure 10. Space of ($\tilde{p}$,z) for the integration (present proposal as the substitute of Fig.4).

$$\left\{ \begin{array}{lll} (N_1)^{-1}e^{-(1/2)l^2\tilde{p}^2-(1/2)y^2/l^2} \equiv W_1(\tilde{p},y),\ N_1 = 1.557/8\pi^2 & \text{elliptic suppression} \\ (N_2)^{-1}e^{-\tilde{p}y} \equiv W_2(\tilde{p},y),\ N_2 = 2(l\Lambda)^3/8\pi^2 & \text{hyperbolic suppression1} \end{array} \right. \quad (10)$$

W_2 is considered to imitate the Randall-Schwartz restriction. Fig.11 shows W_1-weighted case of Fig.2. We notice, in the "valley", the depth, the location and the shape change. The Casimir

Figure 11. Behaviour of $\tilde{p}^3 W_1(\tilde{p}, y)F(\tilde{p}, y)$(elliptic suppression). $\Lambda = 10,\ l = 1$. $1/\Lambda \le y \le 0.99999l,\ 1/l \le \tilde{p} \le \Lambda$.

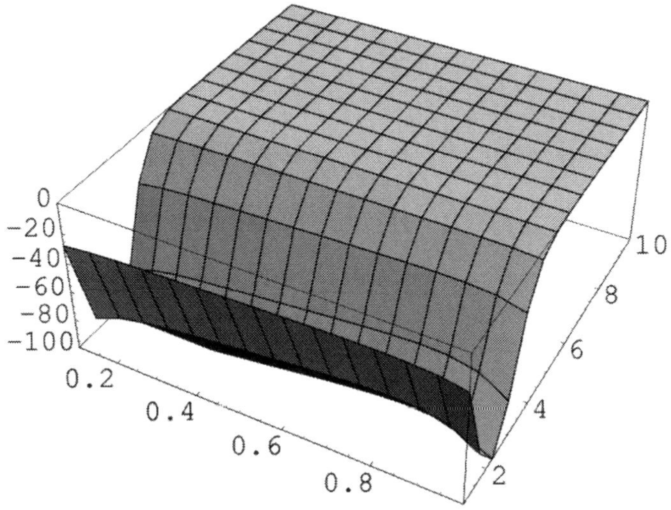

energy is numerically obtained as

$$E^W_{Cas} = \left\{ \begin{array}{ll} -(2.500, 2.501, 2.501)\frac{\Lambda}{l^3} + (-0.142, 1.09, 1.13) \times 10^{-4}\frac{\Lambda \ln(l\Lambda)}{l^3} & \text{for} \quad W_1 \\ -(6.0392, 6.0394, 6.03945) \times 10^{-2}\frac{\Lambda}{l^3} - (24.7, 2.79, 1.60) \times 10^{-8}\frac{\Lambda \ln(l\Lambda)}{l^3} & \text{for} \quad W_2 \end{array} \right. \tag{11}$$

Triplet data show unstableness of the value due to insufficient range of l and Λ. $\ln(l\Lambda)$ term vanishes for W_2 in the present calculational precision.

For the warped case. the situation goes similarly. We take the following weights.

$$W(\tilde{p}, z) = \left\{ \begin{array}{ll} (N_1)^{-1}e^{-(1/2)\tilde{p}^2/\omega^2 - (1/2)z^2T^2} \equiv W_1(\tilde{p}, z),\ N_1 = 1.711/8\pi^2 & \text{elliptic suppr.} \\ (N_2)^{-1}e^{-\tilde{p}zT/\omega} \equiv W_2(\tilde{p}, z),\ N_2 = 2\frac{\omega^3}{T^3}/8\pi^2 & \text{hyperbolic suppr.1} \end{array} \right. \tag{12}$$

Fig.12 is W$_1$-weighted case of Fig.5. The "valley" changes in its depth, shape and location. Casimir energy is given by

$$-E^W_{Cas} = \left\{ \begin{array}{ll} \frac{\omega^4}{T}\Lambda \times 1.2\left\{1 + 0.11\ \ln\frac{\Lambda}{\omega} - 0.10\ \ln\frac{\Lambda}{T}\right\} & \text{for} \quad W_1 \\ \frac{T^2}{\omega^2}\Lambda^4 \times 0.062\left\{1 + 0.03\ \ln\frac{\Lambda}{\omega} - 0.08\ \ln\frac{\Lambda}{T}\right\} & \text{for} \quad W_2 \end{array} \right. \tag{13}$$

For the hyperbolic case (W_2), the divergence is slightly milder than the case of RS-restriction (8) ($\Lambda^5 \to \Lambda^4$). It is new that $\ln(\Lambda/T)$ appears.

After calculating for 13 different weights[15, 17], we conclude Casimir energy can be expressed as follows (except the hyperbolic weights). For the flat case,

$$E^W_{Cas}/\Lambda l = -\frac{\alpha}{l^4}\left(1 - 4c\ln(l\Lambda)\right) \quad . \tag{14}$$

Figure 12. Behavior of $(-N_1/2)\tilde{p}^3 W_1(\tilde{p}, z)F^-(\tilde{p}, z)$(elliptic suppression). $\Lambda = 20000$, $\omega = 5000$, $T = 1$. $1.0001/\omega \leq z \leq 0.9999/T$, $\mu = \Lambda T/\omega \leq \tilde{p} \leq \Lambda$.

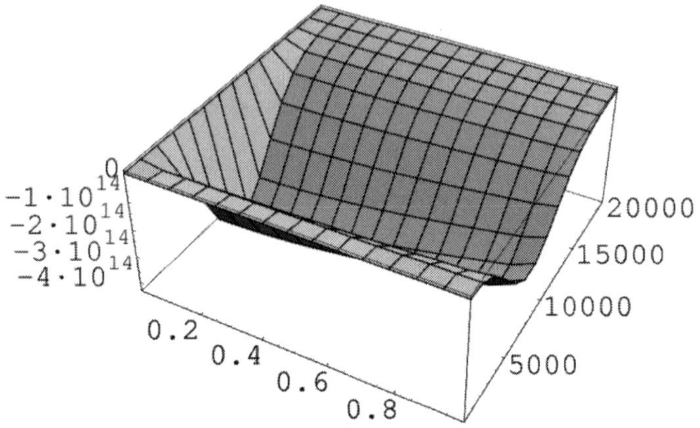

For the warped case

$$E^W_{Cas}/\Lambda T^{-1} = -\alpha\omega^4 \left(1 - 4c\ln(\Lambda/\omega) - 4c'\ln(\Lambda/T)\right) \quad . \tag{15}$$

The parameters α, c, c' depend on the form of the chosen weight. The factors Λl and ΛT^{-1} are the area of the rectangular regions (normalization constants). This result says the divergences of the 5D Casimir energy reduces to the log-divergence if we take into accout the weight properly. The final log-divergence can be renormalized into the boundary parameters l (or T) and ω. See Sec.8.

6. Meaning of Introduction of the Weight Function (1): Minimal Area Principle

In Ref.[19], we presented the following way to determine the form of W. Let us explain it in the warped case. $W(\tilde{p}, z)$ appears as

$$\begin{aligned} -E^W_{Cas}(\omega, T) &= \int \frac{d^4p_E}{(2\pi)^4} \int_{1/\omega}^{1/T} dz\ W(\tilde{p}, z)F^{\mp}(\tilde{p}, z) \\ &= \frac{2\pi^2}{(2\pi)^4} \int d\tilde{p} \int_{1/\omega}^{1/T} dz\ \tilde{p}^3 W(\tilde{p}, z)F^{\mp}(\tilde{p}, z) \quad . \end{aligned} \tag{16}$$

The integral region is the rectangular of Fig.10. We express the above expression by the following path-integral.

$$\begin{aligned} -E^W_{Cas}(\omega, T) &= \int \mathcal{D}\tilde{p}(z) \int_{1/\omega}^{1/T} dz\ S[\tilde{p}(z), z] \ , \\ S[\tilde{p}(z), z] &= \frac{2\pi^2}{(2\pi)^4} \tilde{p}(z)^3 W(\tilde{p}(z), z)F^{\mp}(\tilde{p}(z), z). \end{aligned} \tag{17}$$

All possible paths $\{\tilde{p}(z)|\omega^{-1} < z < T^{-1}\}$ are summed. The dominant contribution, in the above path-integral, is given by $\delta S = 0$.

$$\text{Dominant Path } \tilde{p}_W(z) \ : \quad \frac{d\tilde{p}}{dz} = \frac{-\frac{\partial \ln(WF)}{\partial z}}{\frac{3}{\tilde{p}} + \frac{\partial \ln(WF)}{\partial \tilde{p}}} \quad . \tag{18}$$

This result says the dominant path $\tilde{p}_W(z)$ is determined by $W(\tilde{p}, z)$. A concrete example is the valley bottom line in Fig.12.

On the other hand, there exists another path which is determined independently of the above dominant path. That is the minimal area curve $r_g(z)$ which is given by

$$\text{Minimal Surface Curve } r_g(z) \ : \quad 3 + \frac{4}{z} r' r - \frac{r'' r}{r'^2 + 1} = 0 \quad , \quad \frac{1}{\omega} \leq z \leq \frac{1}{T} \quad , \tag{19}$$

where $r' = \frac{dr}{dz}, r'' = \frac{d^2 r}{dz^2}$. This differential equation is obtained by minimizing the area A of the hyper-surface (9): $\delta A = 0$ where A is given by

$$ds^2 = (\delta_{ab} + \frac{x^a x^b}{(rr')^2}) \frac{dx^a dx^b}{\omega^2 z^2} \equiv g_{ab}(x) dx^a dx^b,$$
$$A = \int \sqrt{\det g_{ab}} \ d^4x = \int_{1/\omega}^{1/T} \frac{1}{\omega^4 z^4} \sqrt{r'^2 + 1} \ r^3 dz. \tag{20}$$

Hence this path $r_g(z)$ is determined by the *induced geometry* $g_{ab}(x)$. We require [15] the following relation in order to define $W(\tilde{p}, z)$.

$$\tilde{p}_W(z) = \tilde{p}_g(z) \ (= \frac{1}{r_g(z)}) \quad . \tag{21}$$

In the above procedure, we have defined the integral measure $d^4 p_E dz W(\tilde{p}, z)$ by the 5D bulk geometry or by the 4D induced geometry.

7. Meaning of Introduction of the Weight Function (2): Fluctuation of Space-Time

For the purpose of naturally introducing the idea of the previous sections, we newly define Casimir energy in the higher dimension by the following *generalized path-integral.* (warped case).

$$-\mathcal{E}_{Cas}(\omega, T, \Lambda) \equiv \int_{1/\Lambda}^{1/\mu} d\rho \int_{\substack{r(1/\omega) \\ = r(1/T) \\ = \rho}} \prod_{a,z} \mathcal{D}x^a(z) \times$$
$$F(\frac{1}{r}, z) \ \exp\left[-\frac{1}{2\alpha'} \int_{1/\omega}^{1/T} \frac{1}{\omega^4 z^4} \sqrt{r'^2 + 1} \ r^3 dz\right], \tag{22}$$

where $\mu = \Lambda T/\omega$ is the IR-cutoff parameter, and we take the limit $\Lambda T^{-1} \to \infty$ at the final stage. $1/2\alpha'$ is the string (surface) tension parameter (α' has the dimension of $[\text{Length}]^4$). $F(\tilde{p}, z)$ comes from the quantum fluctuation of the bulk matter field.

The above path-integral expression says the 4D *coordinates* x^a play the role of *operators* of the quantum statistical mechanics [20]. The extra coordinate z is the parameter of the *inverse temperature*. It looks that *the space-time coordinates are fluctuating.*

We expect the numerical or analytical evaluation of Casimir energy (22) directly gives similar results explained so far.

Figure 13. Graph of Planck's radiation formula. $\mathcal{P}(\beta,k) = \frac{1}{(c\hbar)^3}\frac{1}{\pi^2}k^3/(e^{\beta k}-1)$ $(1 \leq \beta \leq 2,\ 0.01 \leq k \leq 10)$.

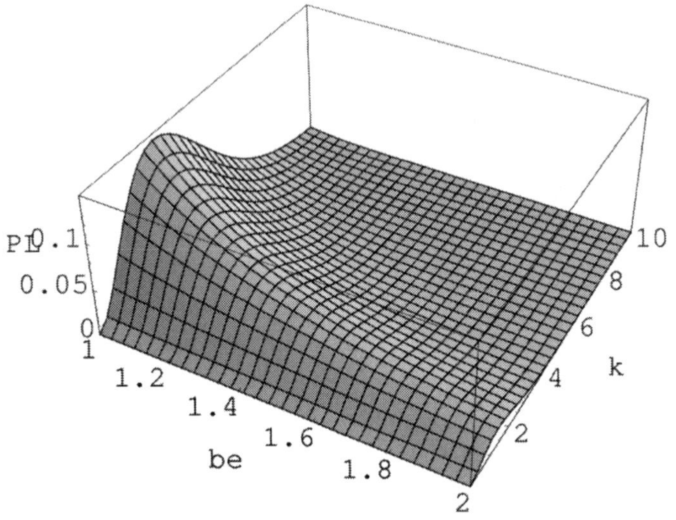

8. Conclusion

We show, in Fig.13, the Planck's radiation distribution in the stereo-graphical way by adding the inverse temperature axis. The behavior looks similar to Fig.12 except the sign. It says the extra coordinate corresponds to the inverse temperature.

The log-divergences in Casimir energy (14) and (15) are familiar in the quantum field theory. For the warped case, they are renormalized into the boundary parameter ω.

$$\frac{E^W_{Cas}}{\Lambda T^{-1}} = -\alpha\omega^4\left(1-4c\ln(\frac{\Lambda}{\omega})-4c'\ln(\frac{\Lambda}{T})\right) = -\alpha(\omega_r)^4 \quad ,$$
$$\omega_r = \omega\sqrt[4]{1-4c\ln(\frac{\Lambda}{\omega})-4c'\ln(\frac{\Lambda}{T})} \quad . \qquad (23)$$

Local counterterms are unnecessary. Divergences are directly absorbed into the boundary parameter. Note that c and c' are pure numbers and represent the interaction between the bounaries and the (free) field. Compare this with the ordinary renormalization, like QED, where the β-function depends on the coupling. When c and c' are sufficiently small, the β-function is given as

$$|c| \ll 1\ ,\ |c'| \ll 1 \quad , \quad \omega_r = \omega(1-c\ln(\Lambda/\omega)-c'\ln(\Lambda/T)) \quad ,$$
$$\beta \equiv \frac{\partial}{\partial(\ln\Lambda)}\ln\frac{\omega_r}{\omega} = -c-c' \quad . \qquad (24)$$

The scaling behavior of ω is determined by the sign of $c+c'$. Because we identify Casimir energy (23) with the cosmological constant, the sign also determines the scaling behavior of the constant. Note that the direction of flow, which determines the attractive or repulsive force, is not given by the derivative (w.r.t. the boundary parameter) of Casimir energy.

For the flat case (14), the other boundary parameter l is renormalized[15].

Parameters, which appears in the 5D warped model, can be fitted in the way consistent with the present observation (except the sign). First we take the 4D momentum cut-off $\Lambda = M_{pl} \sim$

1.2×10^{19}GeV. The present final result (23) says $|S(\text{Euclidian Action})| \sim \int d^4x\sqrt{-g}\lambda_{obs}/G_N \propto \omega^4$. The observational data says $|S(\text{Euclidian Action})| \sim \int d^4x\sqrt{-g}\lambda_{obs}/G_N \sim (R_{cos})^4(10^{-3}\text{eV})^4$ where $R_{cos} \sim 5. \times 10^{41}\text{GeV}^{-1}$ is the size of the present universe. The experimental result about the Newton's gravitational force tells us the warped parameter ω is taken as $\omega \sim 10^{-3}$eV. Note that $\omega \sim \sqrt{M_{pl}/R_{cos}}$, $R_{cos} \cdot \omega \sim \sqrt{M_{pl} \cdot R_{cos}} = \sqrt{N_{DL}}$. Hence $|S(\text{Euclidian Action})| \sim {N_{DL}}^2 = 4. \times 10^{121}$. From (23), $|E_{Cas}| \sim \Lambda T^{-1}\omega^4$. We can identify $T^{-4}|E_{Cas}| \sim \Lambda T^{-5}\omega^4$ with $|S(\text{Euclidian Action})| \sim (R_{cos})^4\omega^4$. Hence $T \sim {R_{cos}}^{-1}{N_{DL}}^{1/5} \sim 3. \times 10^{-30}$GeV. $\mu = \Lambda T/\omega \sim M_{pl}{N_{DL}}^{-3/10} \sim 7.$ GeV. The IR cutoff μ is near to the nucleon mass or the weak boson mass. If we interpret, in Fig.8, the number of small 4D balls within the S^3 boundary is the degree of freedom of the present system, then it is grossly given by $\Lambda^4/\mu^4 = \omega^4/T^4 \sim {N_{DL}}^{6/5} \sim 9. \times 10^{72}$. Finally we note the neutrino mass, m_ν, is similar to the warp parameter (5D bulk curvature) ω.

As stated in Sec.7, the present formulation of the higher dimensional quantum field theory gives the picture of the fluctuating space-time. It is known, in the string theory, the uncertainty principle appears in the space-time coordinates[21].

9. Acknowledgment

The past development is given in the proceedings of some conferences[22, 23, 24, 25, 26]. The author thanks the audience.

[1] S. Weinberg, Rev.Mod.Phys.**61**(1989)1

[2] T. Padmanabhan, Phys.Rept.**380**(2003)235

[3] A.M. Polyakov, Sov.Phys.Usp.**25**(1982)187 [Usp.Fiz.Nauk **136**(1982)538]

[4] A.M. Polyakov, Nucl.Phys.**B797**(2008)199, arXiv:0709.2899(hep-th)

[5] A.M. Polyakov, Nucl.Phys.**B834**(2010)316, arXiv:0912.5503

[6] I. Bredberg, C. Keeler, V. Lysov and A. Strominger, "From Navier-Stokes To Einstein", arXiv: 1101.2451(hep-th)

[7] V. Lysov and A. Strominger, "From Petrov-Einstein to Navier-Stokes", arXiv: 1104.5502(hep-th), 2011

[8] R.P. Feynman, Acta Phys. Polonica**24**(1963)697

[9] G. 'tHooft and M. Veltman, Ann. Inst. Henri Poincaŕe **20**(1974)69

[10] E. Witten, "Quantum Gravity IN De Sitter Space", arXiv: hep-th/0106109, 2001

[11] Erik Verlinde, JHEP**1104**:029, 2011, arXiv:1001.0785

[12] T. Jacobson, Phys.Rev.Lett.**75**(1995)1260, arXiv:gr-qc/9504004

[13] T. Padmanabhan, Rep.Prog.Phys.**73**(2010)046901, arXiv:0911.5004

[14] P.A.M. Dirac, Nature **139**(1937)323; Proc.Roy.Soc.**A165**(1938)199; "Directions in Physics", John Wiley & Sons, Inc., New York, 1978

[15] S. Ichinose,*Prog.Theor.Phys.***121**(2009)727, ArXiv:0801.3064v8[hep-th].

[16] T. Appelquist and A. Chodos, *Phys. Rev.* **D28**(1983)772
T. Appelquist and A. Chodos, *Phys. Rev. Lett.* **50**(1983)141

[17] S. Ichinose, "Casimir Energy of 5D Warped System and Sphere Lattice Regularization", ArXiv:0812.1263[hep-th], US-08-03.

[18] L. Randall and M.D. Schwartz, *JHEP* **0111** (2001) 003, hep-th/0108114

[19] S. Ichinose and A. Murayama, *Phys. Rev.* **D76**(2007)065008, hep-th/0703228

[20] R.P. Feynman, "Statistical Mechanics", W.A. Benjamin, Inc., Massachusetts, 1972

[21] T. Yoneya, *Duality and Indeterminacy Principle in String Theory* in "Wandering in the Fields", eds. K. Kawarabayashi and A. Ukawa (World Scientific,1987), p.419
T. Yoneya, *String Theory and Quantum Gravity* in "Quantum String Theory", eds. N. Kawamoto and T. Kugo (Springer,1988), p.23
T. Yoneya, *Prog.Theor.Phys.***103**(2000)1081

[22] S. Ichinose, Proc. of VIII Asia-Pacific Int. Conf. on Gravitation and Astrophysics (ICGA8,Aug.29-Sep.1,2007,Nara Women's Univ.,Japan),Press Section p36-39, arXiv:/0712.4043

[23] S. Ichinose, *Int.Jour.Mod.Phys.***23A**(2008)2245-2248, Proc. of Int. Conf. on Prog. of String Theory and Quantum Field Theory (Dec.7-10,2007,Osaka City Univ.,Japan), arXiv:/0804.0945

[24] S. Ichinose, Int.Jour.Mod.Phys.24A(2009)3620, Proc. of Int. Conf. on Particle Physics, Astrophysics and Quantum Field Theory: 75 Years since Solvay (Nov.27-29, 2008, Nanyang Executive Centre, Singapore), arXiv:0903.4971

[25] S. Ichinose, Jour.Phys.:Conf.Scr.**222**(2010)012048. ArXiv:1001.0222[hep-th], Proc. of First Mediterranean Conference on Classical and Quantum Gravity (09.9.14-18, Kolymbari, Crete, Greece),

[26] S. Ichinose, Proc. of Int. Workshop on 'Strong Coupling Gauge Theories in LHC Era'(09.12.8-11, Nagoya Univ., Nagoya, Japan), editted by H. Fukaya et al, p407 (World Scientific). arXiv:1003.5041(hep-th).

Journal of Physics: Conference Series **384** (2012) 012029 doi:10.1088/1742-6596/384/1/012029

Scalar-tensor cosmological models converging to general relativity: potential dominated and matter dominated cases

Margus Saal, Laur Järv and Piret Kuusk

Institute of Physics, University of Tartu, Riia 142, Tartu 51014, Estonia

E-mail: margus@fi.tartu.ee

Abstract. We consider Friedmann-Lemaître-Robertson-Walker flat cosmological models in the framework of general Jordan frame scalar-tensor theories of gravity in two different cases: in the dust matter dominated era and in the potential dominated era. Motivated by the local weak field constraints and by cosmological observations, we develop and use an approximation scheme for the regime close to the so-called limit of general relativity. The ensuing nonlinear approximate equations for the scalar field and the Hubble parameter can be solved analytically in cosmological time in both cases. We find criteria for the functions ω and V characterizing a scalar-tensor theory, to determine whether the theory does or does not possess solutions converging to general relativity asymptotically in time. The converging solutions can be subsumed under two principal classes: exponential or polynomial convergence, and damped oscillations around general relativity. The classes of scalar-tensor theories of gravity which contain these types of solutions and satisfy observational constraints, are candidates to explain possible deviations from the standard ΛCDM model. Finally, the effective equation of state parameter $w_{\rm eff}$ is used to illustrate possible asymptotic cosmological dynamics.

1. Introduction

Various cosmological observations of our Universe can be fairly well accommodated within the ΛCDM concordance model [1, 2] based on the theory of general relativity (GR). However, there is a number of viable alternative theories of gravity which also manage to conform sufficiently well with observational data [3, 4, 5, 6]. One such family of theories is provided by scalar-tensor gravity (STG) which employ a scalar field Ψ besides the usual spacetime metric tensor $g_{\mu\nu}$ to describe the gravitational interaction. In the so-called Jordan frame and Brans-Dicke like parametrization STG form a collection of theories which contain two functional degrees of freedom, a coupling function $\omega(\Psi)$ and a scalar potential $V(\Psi)$. As has been discussed by many authors previously [7, 8, 9, 10, 11, 12, 13, 14, 15, 16], for a range of choices of ω and V the cosmological evolution of dust and potential dominated STG models naturally converges close to the one expected from GR. On the other hand, cosmological observations and data from the local weak field tests permit to suppose that only these STG models are viable which involve this property. Namely, limits of observed values of parametrized post-Newtonian (PPN) parameters $|\beta-1| < 10^{-4}$, $|\gamma-1| < 10^{-5}$ and the time variation of the gravitational constant $|\dot{G}/G| < 10^{-13}$ yr^{-1} strongly constrain the parameters of the theory at present [17, 18]. Constraints from the cosmological dynamics and from the Big-Bang nucleosynthesis (BBN) indicate that the Universe

Journal of Physics: Conference Series **384** (2012) 012029 doi:10.1088/1742-6596/384/1/012029

was very close to GR already at the beginning of matter dominated era [19]. Yet, at the same time STG models may also offer a possibility to explain small observational differences from pure GR ΛCDM behaviour, e.g. recent data are not excluding the possibility that the effective equation of state (EoS) of dark energy is changing in the cosmological time [20, 21].

In this paper we consider the flat Friedmann-Lemaître-Robertson-Walker (FLRW) cosmological models in the framework of general STG, close to GR (referred to as 'the limit of GR' or 'GR point'). We apply the approximation scheme for the field equations to capture the dynamics near the GR limit. We follow our recent papers [22, 23, 24, 25] where we presented and justified an approximation scheme, as well as derived the non-linear equations for deviations of the scalar field and for the Hubble parameter. The resulting equations explicitly contain cosmological time and can be solved analytically in the potential dominated and in the matter dominated epoch. As an example, we use the effective barotropic index $\mathrm{w}_{\mathrm{eff}}$ to illustrate the cosmological dynamics near GR point. For some related recent studies see Refs. [26, 27, 28, 29, 30, 31, 32, 33, 34, 35].

In section 2 we recall STG FLRW equations and outline the approximation method worked out in Ref. [22]. In section 3 we review the solutions of resulting nonlinear equations in the regime when the density of scalar potential $V(\Psi)$ dominates over the energy density of matter ρ at cosmological scales (i.e. $V \neq 0$, $\rho = 0$) as expected for the dark energy era [22, 23, 24]. Section 4 is devoted to the dust matter dominated era, introducing the results from a recent publication [25]. Some characteristic examples and implications for selecting a model of STG viable for realistic cosmological scenario is also briefly discussed in sections 3 and 4. Finally, section 5 provides a brief summary.

2. Equations and Approximation Method

We consider a general scalar-tensor theory in the Jordan frame given by the action

$$S = \frac{1}{2\kappa^2}\int d^4x\sqrt{-g}\left[\Psi R(g) - \frac{\omega(\Psi)}{\Psi}\nabla^\rho\Psi\nabla_\rho\Psi - 2\kappa^2 V(\Psi)\right] + S_m(g_{\mu\nu},\chi_m)\,. \tag{1}$$

Here $\omega(\Psi)$ is a coupling function (we assume $2\omega(\Psi)+3 \geq 0$ to avoid ghosts in the Einstein frame), $V(\Psi) \geq 0$ is a scalar potential, κ^2 is the non-variable part of the effective gravitational constant $\frac{\kappa^2}{\Psi}$, and S_m is the matter contribution to the action as all other fields are included in χ_m.

The field equations for the flat Friedmann-Lemaître-Robertson-Walker (FLRW) line element $ds^2 = -dt^2 + a(t)^2\left(dr^2 + r^2(d\theta^2 + \sin^2\theta\, d\varphi^2)\right)$ and perfect barotropic fluid matter, $p = \mathrm{w}\rho$, read

$$\begin{aligned}
H^2 &= -H\frac{\dot\Psi}{\Psi} + \frac{1}{6}\frac{\dot\Psi^2}{\Psi^2}\,\omega(\Psi) + \frac{\kappa^2}{3}\frac{\rho}{\Psi} + \frac{\kappa^2}{3}\frac{V(\Psi)}{\Psi}\,, &(2)\\
2\dot H + 3H^2 &= -2H\frac{\dot\Psi}{\Psi} - \frac{1}{2}\frac{\dot\Psi^2}{\Psi^2}\,\omega(\Psi) - \frac{\ddot\Psi}{\Psi} - \frac{\kappa^2}{\Psi}\mathrm{w}\rho + \frac{\kappa^2}{\Psi}\,V(\Psi)\,, &(3)\\
\ddot\Psi &= -3H\dot\Psi - \frac{1}{2\omega(\Psi)+3}\,\frac{d\omega(\Psi)}{d\Psi}\,\dot\Psi^2 + \frac{\kappa^2}{2\omega(\Psi)+3}\,(1-3\mathrm{w})\,\rho \\
&\quad + \frac{2\kappa^2}{2\omega(\Psi)+3}\left[2V(\Psi) - \Psi\,\frac{dV(\Psi)}{d\Psi}\right], &(4)\\
\dot\rho &= -3H\,(\mathrm{w}+1)\,\rho &(5)
\end{aligned}$$

where $H \equiv \dot a/a$ and we will assume that $\rho \geq 0$.

Upon introducing the notation

$$A(\Psi) \equiv \frac{d}{d\Psi}\left(\frac{1}{2\omega(\Psi)+3}\right), \qquad W(\Psi) \equiv 2\kappa^2\left(2V(\Psi) - \frac{dV(\Psi)}{d\Psi}\Psi\right) \tag{6}$$

Journal of Physics: Conference Series **384** (2012) 012029 doi:10.1088/1742-6596/384/1/012029

and substituting H from Eq. (2) into Eq. (4), we get

$$\begin{aligned}\ddot{\Psi} &= \left(\frac{3}{2\Psi}+\frac{1}{2}A(\Psi)(2\omega(\Psi)+3)\right)\dot{\Psi}^2+\frac{W(\Psi)}{2\omega(\Psi)+3}+\frac{\kappa^2}{2\omega(\Psi)+3}(1-3\mathrm{w})\rho \\ &\pm\frac{1}{2\Psi}\sqrt{3(2\omega(\Psi)+3)\dot{\Psi}^2+12\kappa^2\Psi(\rho+V(\Psi))}\;\dot{\Psi}\,. \end{aligned}\tag{7}$$

In the limit $\frac{1}{(2\omega(\Psi)+3)}\to 0$, $\dot{\Psi}\neq 0$ the system faces a spacetime curvature singularity, since H diverges, and likewise behaves $\ddot{\Psi}$. At first the limit (a) $\frac{1}{(2\omega(\Psi)+3)}\to 0$ and (b) $\dot{\Psi}\to 0$ seems only slightly less mathematically precarious for the equations are left just indeterminate (contain terms $\frac{0}{0}$). Yet the latter situation is of particular physical importance, as the experiments in the Solar System (where local matter density dominates over the scalar potential), i.e. the limits of observed values of PPN parameters and the time variation of the gravitational constant suggest the present cosmological background value of the scalar field to be very close to the limit (a)-(b) [18].

Let us define $\Psi_\star$ by $\frac{1}{2\omega(\Psi_\star)+3}=0$ and and focus upon the solutions near this point,

$$\Psi=\Psi_\star+x(t)\,,\qquad \dot{\Psi}=\dot{\Psi}_\star+\dot{x}(t)=\dot{x}(t)\,,\tag{8}$$

where x and $\dot{x}$ span the neighbourhood of first order small distance from $(\Psi_\star,\,\dot{\Psi}_\star)$. As phase space variables x and $\dot{x}$ are independent from each other, their ratio $\dot{x}/x$ is indeterminate at $(x=0,\,\dot{x}=0)$.

Under additional mathematical assumptions, (c) $A_\star\equiv A(\Psi_\star)\neq 0$ and (d) $\frac{1}{2\omega+3}$ is differentiable at $\Psi_\star$, we can expand in series

$$\frac{1}{2\omega(\Psi)+3} = \frac{1}{2\omega(\Psi_\star)+3}+A_\star x+\ldots\approx A_\star x\,,\tag{9}$$

$$(2\omega(\Psi)+3)\dot{\Psi}^2 = \frac{\dot{x}^2}{0+A_\star x+\ldots}=\frac{\dot{x}^2}{A_\star x}(1+O(x))\approx\frac{\dot{x}^2}{A_\star x}\,.\tag{10}$$

The latter result actually informs us that in order to avoid a spacetime singularity $\frac{\dot{x}^2}{x}$ must not diverge, hence we should treat $x(t)$ and $\dot{x}(t)$ as the same order (small) quantities. In passing let us remark that in our previous papers [22, 23, 24, 25] we have tentatively called (a)-(d) 'the limit of general relativity' since under these conditions the set of STG cosmological equations (2)-(5) reduces to those of pure GR (with a cosmological constant if $V(\Psi_\star)\neq 0$).

3. Potential Dominated Regime

In this section we focus upon the case when matter density can be neglected in favor of the potential (i.e. take $V\neq 0$, $\rho=0$) and consider the limit (a)-(d). In approximate equations we must recognize the term $\frac{\dot{x}^2}{x}$ as being of the same order as x and $\dot{x}$. Thus, keeping the term $\frac{\dot{x}^2}{x}$ in the approximation of Eq. (7) we obtain a second order nonlinear differential equation [22]

$$\ddot{x}+C_1\,\dot{x}-C_2\,x=\frac{\dot{x}^2}{2x}\,.\tag{11}$$

Here we have defined the values of some parameters at $(\Psi_\star,\,\dot{\Psi}_\star)$ as

$$C_1\equiv\pm\sqrt{\frac{3\kappa^2V(\Psi_\star)}{\Psi_\star}}\,,\qquad C_2\equiv A_\star\,W_\star\,,\tag{12}$$

where $W_\star \equiv W(\Psi_\star)$ and $V(\Psi_\star) \geq 0$. The three parameters $A_\star$, $W_\star$, C_1 determine the leading terms in expansions of the two functions $\omega(\Psi)$, $V(\Psi)$ which specify a STG. The phase trajectories for the nonlinear approximate system correspondng to equation (11) near GR point are classified and analysed in Ref. [22]. The fixed points and the corresponding eigenvalues in the case of linear system (neglecting the term $\dot{x}^2/2x$) are found in Refs. [36, 37].

The general solution of Eq. (11) reads [24]

$$\pm x(t) = \begin{cases} e^{-C_1 t}\left[M_1 e^{\frac{1}{2}t\sqrt{C}} - M_2 e^{-\frac{1}{2}t\sqrt{C}}\right]^2 , & C > 0 , \\ e^{-C_1 t}\left[\tilde{M}_1 t - \tilde{M}_2\right]^2 , & C = 0 , \\ e^{-C_1 t}\left[N_1 \sin(\frac{1}{2}t\sqrt{|C|}) - N_2 \cos(\frac{1}{2}t\sqrt{|C|})\right]^2 , & C < 0 , \end{cases} \tag{13}$$

where $C = \sqrt{C_1^2 + 2C_2}$ and M_1, M_2, $\tilde{M}_1$ $\tilde{M}_2$, N_1, N_2 are constants of integration determined by initial conditions. In order to successfully meet the observational constraints, let us now focus upon solutions which approach the GR limit asymptotically in time (PPN parameters approach the GR values $\beta = 1$ and $\gamma = 1$ as $t \to \infty$). Our results [22, 24] allow one to immediately decide whether any STG with particular $\omega(\Psi)$ and $V(\Psi)$ is viable or not. Furthermore, the behavior of solutions which approach GR can be classified under two characteristic types: (i) exponential or linear exponential convergence ($C > 0$ or $C = 0$), (ii) damped oscillations around GR ($C < 0$).

For the evolution of the universe in scalar-tensor cosmology we may envisage a realistic scenario where during the matter domination era the scalar field has already dynamically relaxed sufficiently close to the GR limit. Later when the cosmological energy density of the potential becomes more significant, the solutions given here can be taken to provide a rough description. As an illustration for dynamics given by the solution (13), we use the effective barotropic index which determines the behaviour of dark energy. The same approximation scheme applied to the scalar field equation (7) allows to expand the effective barotropic index as (the expansion of H, $\dot{H}$ and PPN parameters, see Ref. [24])

$$\mathrm{w}_{\mathrm{eff}} \equiv -1 - \frac{2\dot{H}}{3H^2} \approx -1 + \frac{1}{C_1^2 \Psi_\star}\left[\frac{3}{2}\left(1 + \frac{1}{\Psi_\star A_\star}\right)\frac{\dot{x}^2}{x} - 4C_1\dot{x} + 3C_2 x\right] + \ldots . \tag{14}$$

Among the models for which the 'GR point' works as an attractor, we can determine whether a model in the theory characterized by distinct parameters $(C_1, C_2, A_\star)$ approaches the de Sitter spacetime from the quintessence side ($\mathrm{w}_{\mathrm{eff}} > -1$) or from the phantom side ($\mathrm{w}_{\mathrm{eff}} < -1$). Note that a necessary condition for crossing the so-called phantom divide, $\mathrm{w}_{\mathrm{eff}} = -1$, is vanishing of the second term in Eq. (14). Depending on the model, exponential solutions may cross the phantom divide line at most twice before approaching $\mathrm{w}_{\mathrm{eff}} = -1$ from either above or below. In the oscillating type of solutions the dark energy effective barotropic index oscillates either in the quintessence regime ($\mathrm{w}_{\mathrm{eff}} > -1$), phantom regime ($\mathrm{w}_{\mathrm{eff}} < -1$), or crossing the phantom divide line once or twice during each period.

As an illustration, Figure 1 depicts the dynamics of $\mathrm{w}_{\mathrm{eff}}$ for three sample solutions in different STG models:

$$\omega(\Psi) = \frac{3\Psi}{2(1-\Psi)} , \qquad \kappa^2 V(\Psi) = \frac{2}{3}\left[1 + (1-\Psi)^2\right] , \tag{15}$$

$$\omega(\Psi) = \frac{5\Psi}{7(1-\Psi)} , \qquad \kappa^2 V(\Psi) = 3e^{3(1-\Psi)} , \tag{16}$$

$$\omega(\Psi) = \frac{\Psi}{2(1-\Psi)} , \qquad \kappa^2 V(\Psi) = 3e^{3(1-\Psi)} . \tag{17}$$

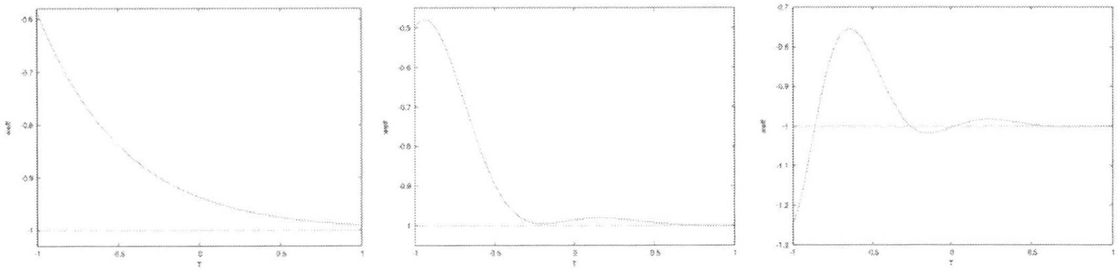

Figure 1. Examples of the time evolution of $w_{\rm eff}$ for different STG models: (15) left, (16) middle, (17) right. The evolution is measured in the units of the analogue of Hubble time, $T = H_\star t = \frac{C_1}{3} t$, and present moment is chosen to be at $T = 0$.

The first model belongs to the class with $C > 0$, $C_1 > 0$, $C_2 < 0$ and the sample solution shows a monotonic quintessence type convergence towards de Sitter. The second model belongs to the class with $C < 0$, $C_1 > 0$, $C_2 < 0$ and is characterized by damped oscillations in the quintessence regime. The third model also belongs to the same class as previous example but exhibits oscillations through the phantom divide line. The initial conditions of these solutions have been chosen such that the corresponding PPN parameters are within observationally allowed limits. We may notice that it is possible to have the period of oscillations to be about the same order of magnitude as the age of the Universe.

4. Matter Dominated Regime

In what follows we focus to the era when the dust matter density dominates over potential (i.e. take $\rho \neq 0$, $\mathrm{w} = 0$, $V = 0$) and consider the limit (a)-(d). The system (2)-(5) contains three variables $\{\Psi, H, \rho\}$, but one of them is related to the others via the Friedmann constraint (2). Upon eliminating ρ we have two equations [25]

$$\ddot{\Psi} = -3H\dot{\Psi} + \frac{1}{2}(2\omega+3)A(\Psi)\dot{\Psi}^2 + \frac{1}{(2\omega+3)}\left(3\Psi H^2 + 3H\dot{\Psi} - \frac{\dot{\Psi}^2}{2\Psi}\omega\right), \tag{18}$$

$$\dot{H} = -\frac{3}{2}H^2 + H\frac{\dot{\Psi}}{2\Psi} - \frac{\dot{\Psi}^2}{4\Psi^2}\omega - \frac{1}{4}(2\omega+3)A(\Psi)\frac{\dot{\Psi}^2}{\Psi} - \frac{1}{2(2\omega+3)}\left(3H^2 + 3H\frac{\dot{\Psi}}{\Psi} - \frac{\dot{\Psi}^2}{2\Psi^2}\omega\right), \tag{19}$$

where we have introduced notation (6).

Let us Taylor expand the functions in Eqs. (18), (19) using approximations (8), (9) and an additional approximation for the Hubble parameter

$$H(t) = H_\star(t) + h(t). \tag{20}$$

We assume that $H_\star(t)$ satisfies the equation $\dot{H}_\star = -\frac{3}{2}H_\star^2$ as in the Friedmann solution of the dust dominated pure GR model. It determines the time evolution of $H_\star$ to be

$$H_\star = \frac{2}{3(t-t_s)}. \tag{21}$$

Here t_s is a constant of integration which fixes the beginning of time scale; in what follows we choose $t_s = 0$, $t > 0$.

We take $x, \dot{x}, h, \dot{x}^2/x$ to be of the same order small quantities. Approximate first order equations now read with $H_\star$ given by Eq. (21) as

$$\ddot{x} = \frac{\dot{x}^2}{2x} - 3H_\star \dot{x} + 3A_\star \Psi_\star H_\star^2 x \,, \tag{22}$$

$$\dot{h} + 3H_\star h = -\frac{1}{4\Psi_\star}\left(1 + \frac{1}{2A_\star \Psi_\star}\right)\frac{\dot{x}^2}{x} + \frac{1}{2\Psi_\star} H_\star \dot{x} - \frac{3}{2} A_\star H_\star^2 x \,. \tag{23}$$

The effective barotropic index is

$$\mathrm{w_{eff}} \equiv -1 - \frac{2\dot{H}}{3H^2} \approx -\frac{2}{3H_\star^2}\left(\dot{h} + 3H_\star h\right) . \tag{24}$$

Notice that Eqs. (22), (23) contain time t explicitly due to $H_\star$ (21). This means that the corresponding system of first order equations is not autonomous and the standard phase space analysis is not applicable. However, we can straightforwardly integrate Eqs. (22), (23) in cosmological time and analyse the behaviour of solutions in the neighbourhood of the limit of general relativity. It turns out that the type of solution $x(t)$ of Eq. (22) depends on the constant

$$D \equiv 1 + \frac{8}{3} A_\star \Psi_\star \tag{25}$$

which characterizes the underlying STG. Knowing solutions $x(t)$ we can also find the solutions $h(t)$ of Eq. (23) and determine the effective barotropic index $\mathrm{w_{eff}}$ from Eq. (24).

The general solution of Eq. (22) reads [25]

$$\pm x(t) = \begin{cases} \frac{1}{t}\left[M_1 t^{\frac{\sqrt{D}}{2}} - M_2 t^{-\frac{\sqrt{D}}{2}}\right]^2 , & D > 0 \,, \\ \frac{1}{t}\left[\tilde{M}_1 \ln t - \tilde{M}_2\right]^2 , & D = 0 \,, \\ \frac{1}{t}\left[N_1 \sin(\frac{1}{2}\sqrt{|D|}\ln t) - N_2 \cos(\frac{1}{2}\sqrt{|D|}\ln t)\right]^2 , & D < 0 \,. \end{cases} \tag{26}$$

Here M_1, M_2, $\tilde{M}_1$ $\tilde{M}_2$, N_1, N_2 are constants of integration and are related to the initial data $x_* = x(t_*)$, $\dot{x}_* = \dot{x}(t_*)$ at some arbitrary time t_*. The $\pm$ follows from an invariance property of Eq. (22) under reflection $x \to -x$, i.e. it allows solutions which can lie in the region $\Psi > \Psi_\star$ ($x > 0$) or $\Psi < \Psi_\star$ ($x < 0$).

Let us make some general comments about the solutions, while the detailed analysis as well as solutions for $h(t)$ is presented in the recent paper [25]. For the first class of solutions, $D > 0$, we can conclude that asymptotically at $t \to \infty$ there are two distinctive behaviors. For STGs with $\sqrt{D} < 1$ (i.e. $A_\star \Psi_\star < 0$) all cosmological solutions irrespective of their initial conditions monotonically approach the general relativistic dust matter FLRW cosmology, $\Psi(t) \to \Psi_\star = \text{const.}$, $H(t) \to H_\star(t) = 2/(3t)$, $\mathrm{w_{eff}}(t) \to 0$, since all first order corrections vanish at this limit. On the other hand STGs with $\sqrt{D} > 1$ (i.e. $A_\star \Psi_\star > 0$) allow only solutions that will diverge, $x(t) \to \infty$, $h(t) \to \infty$, $\mathrm{w_{eff}}(t) \to \infty$, meaning that solutions in these theories can linger near general relativity only for a certain period, while as time evolves they will leave and the approximation scheme will break down eventually. The second class of solutions, $D = 0$, are of marginal interest due to the fine-tuning of the STG by the condition $A_\star \Psi_\star = -\frac{3}{8}$. For the third class of solutions, $D < 0$, the first order corrections to the GR FLRW dust cosmological model are proportional to negative powers of cosmological time t, so they vanish at $t \to \infty$ and solutions approach the general relativistic dust matter cosmology in the manner of damped oscillations. The amplitude of the deviations monotonically decreases while the period monotonically increases in time.

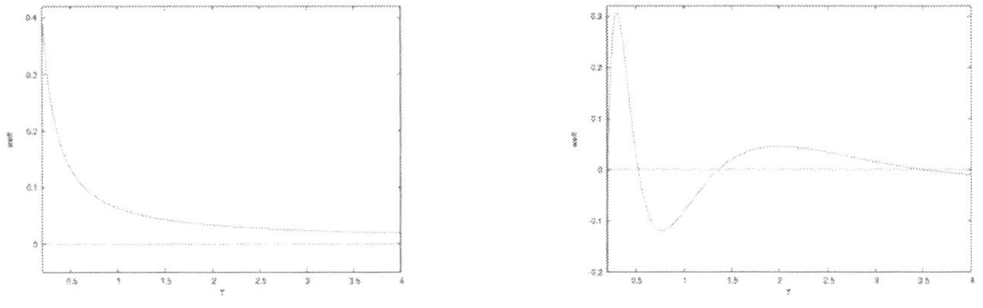

Figure 2. Examples of the time evolution of w_{eff} for different STG models: (27) left and (28) right. The evolution is measured in the units of the analogue of Hubble time, $T = H_* t$ and the initial conditions are fixed at $T_* = 1$.

As an illustration, Figure 2 depicts the dynamics of w_{eff} for two sample solutions in different STG models:

$$\omega(\Psi) = \frac{5\Psi}{3(1-\Psi)}, \tag{27}$$

$$\omega(\Psi) = \frac{\Psi}{9(1-\Psi)}. \tag{28}$$

The first model belongs to the class with $D > 0$ and the sample solution shows a monotonic convergence towards $w_{\mathrm{eff}} = 0$. The second model belongs to the class with $D < 0$ and is characterized by damped oscillations around $w_{\mathrm{eff}} = 0$.

Finally, let us once again summarize the conditions of the models with the aim to view different epochs as parts of a single cosmological scenario. In both cases there are general conditions for solutions to converge towards the GR value $\Psi_\star$ asymptotically in time: in the dust dominated model it reads

$$A_\star \Psi_\star \equiv \left[\frac{d}{d\Psi}\left(\frac{1}{2\omega(\Psi)+3}\right)\Psi\right]_\star < 0 \tag{29}$$

and in the potential dominated model [22]

$$V(\Psi_\star) > 0\,, \qquad \left[\frac{\Psi}{2V}\frac{dV}{d\Psi}\right]_\star < 1\,. \tag{30}$$

A realistic STG cosmological scenario compatible with observations would better need to have GR as an attractor in both dust dominated and matter dominated regimes. Therefore for a credible STG both conditions (29) and (30) must be satisfied, thus constraining the set of functions $\omega(\Psi)$ and $V(\Psi)$ one can consider for constructing a viable model.

5. Summary

We have derived and solved nonlinear approximate equations for small deviations of the scalar field in the framework of general STG FLRW flat cosmological models in two eras near the limit of general relativity as favored by various observational constraints. First we look the era when the energy density of the scalar potential dominates over the energy density of the ordinary matter and the Universe has evolved close to the limit of GR (which acts as an attractor for certain classes of STG). Secondly we consider the era of dust dominated matter close to the limit of GR. The behaviour of solutions can be used to analyse the cosmological expansion near the 'GR point' where the weak field constraints are satisfied. The solutions which approach GR can be classified under two characteristic types: (a) exponential or polynomial convergence, and (b) damped oscillations around general relativity. The classes of STGs which contain these solutions are of particular interest since there is a dynamical mechanism naturally driving the solutions to satisfy observational constraints. On the other hand, they are good candidates to explain possible deviations from the ΛCDM. Combining the results of the dust dominated epoch with results of the potential dominated regime, provides a reasonable viability filter for STG models in terms of the conditions (29),(30).

Acknowledgments

This work was supported by the Estonian Science Foundation Grant No. 8837 and by Estonian Ministry for Education and Science Support Grant No. SF0180013s07. MS also acknowledges the Estonian Science Foundation Post-doctoral research grant No. JD131.

References

[1] Komatsu E *et al.* [WMAP Collaboration] 2011 *Astrophys. J. Suppl.* **192** 18
[2] Kessler R *et al.* 2009 *Astrophys. J. Suppl.* **185** 32
[3] Sotiriou T P and Faraoni V 2010 *Rev. Mod. Phys.* **82** 451
[4] Tsujikawa S 2010 *Lect. Notes Phys.* **800** 99
[5] Clifton T, Ferreira P G, Padilla A and Skordis C 2011 Modified Gravity and Cosmology *Preprint* 1106.2476 [astro-ph.CO]
[6] Nojiri S I and Odintsov S D 2011 *Phys. Rept.* **505** 59
[7] Damour T and Nordtvedt K 1993 *Phys. Rev.* D **48** 3436
[8] Barrow J D and Mimoso J P 1994 *Phys. Rev.* D **50** 3746
[9] Gérard J M and Mahara I 1995 *Phys. Lett.* B **346** 35
[10] Serna A and Alimi J M 1996 *Phys. Rev.* D **53** 3074
[11] Barrow J D and Parsons P 1997 *Phys. Rev.* D **55** 1906
[12] Santiago D I, Kalligas D and Wagoner R V 1998 *Phys. Rev.* D **58** 124005
[13] Mimoso J P and Nunes A M 1998 *Phys. Lett.* A **248** 325
[14] Navarro A, Serna A and Alimi J M 1999 *Phys. Rev.* D **59** 124015
[15] Serna A, Alimi J M and Navarro A 2002 *Class. Quant. Grav.* **19** 857
[16] Barrow J D and Shaw D J 2008 *Class. Quant. Grav.* **25**, 085012
[17] Bertotti B, Iess L and Tortora P 2003 *Nature* **425** 374
[18] Will C M 2005 *Living Rev. Rel.* **9** 3
[19] Coc A, Olive K A, Uzan J P and Vangioni E 2006 *Phys. Rev.* D **73**, 083525
[20] Huang Q G, Li M, Li X D and Wang S 2009 *Phys. Rev.* D **80** 083515
[21] Cai R G, Su Q and Zhang H B 2010 *J. Cosmol. Astropart. Phys.* JCAP04(2010)012
[22] Järv L, Kuusk P and Saal M 2010 *Phys. Rev.* D **81** 104007
[23] Järv L, Kuusk P and Saal M 2011 *J. Phys. Conf. Ser.* **283** 012017
[24] Järv L, Kuusk P and Saal M 2010 *Phys. Lett.* B **694** 1
[25] Järv L, Kuusk P and Saal M 2011 Scalar-tensor cosmologies with dust matter in the general relativity limit *Preprint* 1112.5308 [gr-qc]
[26] Järv L, Kuusk P and Saal M 2007 *Phys. Rev.* D **75** 023505
[27] Nojiri S I, Odintsov S D and Štefančić H 2006 Phys. Rev. D **74** 086009
[28] Carloni S, Capozziello S, Leach J A and Dunsby P K S 2008 *Class. Quant. Grav.* **25** 035008
[29] Agarwal N and Bean R 2008 *Class. Quant. Grav.* **25** 165001
[30] Hrycyna O and M. Szydłowski M 2009 *J. Cosmol. Astropart. Phys.* JCAP04(2009)026
[31] Gupta G, Saridakis E N and Sen A A 2009 *Phys. Rev.* D **79** 123013
[32] Hrycyna O and M. Szydłowski M 2010 *Phys. Lett.* B **694** 191
[33] Batista C E M and Zimdahl W 2010 *Phys. Rev.* D **82** 023527
[34] Hrycyna O and Szydłowski M 2010 *J. Cosmol. Astropart. Phys.* JCAP12(2010)016
[35] Leon G, Silveira P and Fadragas F R 2010 Phase-space of flat Friedmann-Robertson-Walker models with both a scalar field coupled to matter and radiation *Preprint* 1009.0689 [gr-qc]
[36] Järv L, Kuusk P and Saal M 2008 *Phys. Rev.* D **78** 083530
[37] Järv L, Kuusk P and Saal M 2009 *Int. J. Mod. Phys.* A **24** 1631

Journal of Physics: Conference Series **384** (2012) 012030 doi:10.1088/1742-6596/384/1/012030

Gravitational Waves in Viable Modified Gravity Theories

C. Q. Geng

Department of Physics, National Tsing Hua University, Hsinchu 300, Taiwan
Physics Division, National Center for Theoretical Sciences, Hsinchu 300, Taiwan

E-mail: geng@phys.nthu.edu.tw

Abstract. We review our recent work [1] on gravitational waves in viable $f(R)$ models. We concentrate on the exponential gravity and Starobinsky models. We show that in both cases, the mass of the scalar mode is order of $10^{-33}eV$ when it propagates in vacuum. In the presence of matter density, such as galaxy, the scalar mode can be heavy. In particular, it becomes almost infinity so that the scalar mode of gravitational wave for the exponential model disappears like the ΛCDM, whereas it can be as low as $10^{-24}eV$ in the Starobinsky model, corresponding to the lowest frequency of 10^{-9} Hz, which may be detected by the current and future gravitational wave probes in space.

1. Introduction

It has been widely accepted that one simple way to modify general relativity is to promote the Ricci scalar R in the Einstein-Hilbert action into an $f(R)$ function, which is the so-called $f(R)$ theory [2, 3, 4]. A viable model of $f(R)$ can generate a late-time accelerating expansion of our universe, have the radiation-dominated stage followed by the matter-dominated one, and be consistent with the solar-system constraint under chameleon mechanism. The conditions for such a viable $f(R)$ model include (i) the positivity of the effective gravitational coupling; (ii) the stability of cosmological perturbations; (iii) the stability of the late-time de-Sitter point; (iv) the asymptotic behavior to ΛCDM at the high curvature regime; (v) the solar system constraint; and (vi) the constraint from the violation of the equivalence principle. The typical examples of the viable $f(R)$ models are Hu-Sawicki [5], Starobinsky [6], Tsujikawa [7] and exponential gravity models [8] as shown in Table 1.

In Ref. [9], Chiba showed that an $f(R)$ model will allow a new scalar degree of freedom. This corresponds to a new scalar mode of gravitational wave besides the ordinary tensor one of general relativity. This new scalar mode will be massive and propagate as a longitudinal polarization. Various discussions and predictions about this extra scalar mode of gravitational wave have been given in the literature [10, 11, 12, 13]. However, most of them were concentrated on either quadratic or inverse-curvature type of $f(R)$ models, which is highly restricted by the observational results [3]. In the talk, we will review our recent work [1] on gravitational waves in viable $f(R)$ models. To illustrate our results, we will concentrate on the exponential and Starobinsky models. Our study can be easily extended to other viable models.

Table 1. Viable $f(R)$ models

model	$f(R)$	Constant parameters
(a) Hu-Sawicki	$R - \frac{c_1 R_{\rm HS}(R/R_{\rm HS})^p}{c_2(R/R_{\rm HS})^p+1}$	c_1, c_2, $p(>0)$,$R_{\rm HS}(>0)$
(b) Starobinsky	$R + \lambda R_c \left[\left(1+\frac{R^2}{R_c^2}\right)^{-n} - 1\right]$	$\lambda(>0)$, $n(>0)$, R_c
(c) Tsujikawa	$R - \mu R_{\rm T} \tanh\left(\frac{R}{R_{\rm T}}\right)$	$\mu(>0)$, $R_{\rm T}(>0)$
(d) Exponential	$R - \beta R_{\rm E}\left(1 - e^{-R/R_{\rm E}}\right)$	β, $R_{\rm E}$

We use units of $k_B = c = \hbar = 1$ and the gravitational constant $G = M_{Pl}^{-2}$ with the Planck mass of $M_{Pl} = 1.22089 \times 10^{19} GeV$.

2. Gravitational Waves in Viable $f(R)$ Gravity

We start by considering a general Einstein-Hilbert action

$$S = \frac{1}{2\kappa^2}\int d^4x\sqrt{-g}f(R) + S_m(g_{\mu\nu}, \Psi_{\mu\nu}), \tag{1}$$

where $f(R)$ is an arbitrary function of the Ricci scalar R, S_m is the action of the matter part and $\kappa^2 \equiv 8\pi G$. In the metric formalism, we vary the action (1) with respect to $g_{\mu\nu}$, and the modified Einstein field equation can be obtained as

$$f'(R)R_{\mu\nu} - \frac{1}{2}f(R)g_{\mu\nu} + (g_{\mu\nu}\Box - \nabla_\mu\nabla_\nu)\,f'(R) = \kappa^2 T_{\mu\nu}, \tag{2}$$

where a prime denotes the derivative with respect to R, ∇_μ is the covariant derivative and $\Box = g^{\mu\nu}\nabla_\mu\nabla_\nu$ is the d'Alembert operator. The trace of the field equation (2) gives

$$f'(R)R - 2f(R) + 3\Box f'(R) = \kappa^2 T, \tag{3}$$

where $T = g^{\mu\nu}T_{\mu\nu} = -\rho + 3a^2 P$ is the trace of the matter energy-momentum tensor, and a is the scale factor.

For $f(R)$, the de Sitter stage is a vacuum solution with a positive constant background curvature R_d, which is assumed to be homogeneous and static. Consequently, one has

$$\nabla_\mu f'(R_d) = 0 \quad \text{and} \quad f'(R_d)R_d = 2f(R_d). \tag{4}$$

Moreover, from Eq. (2), the Ricci tensor satisfies $R_{\mu\nu}|_{R_d} = g_{\mu\nu}R_d/4$.

In order to investigate gravitational wave in $f(R)$ theories, we need to study the linearized theory of $f(R)$ gravity. Consider a small perturbation from the FRW metric:

$$g_{\mu\nu} = \overline{g}_{\mu\nu} + h_{\mu\nu}, \tag{5}$$

where $|h_{\mu\nu}| \ll 1$ is the perturbation and $\overline{g}_{\mu\nu} = diag(-1, a^2, a^2, a^2)$ is the FRW background metric. If the evolution of the system is much shorter than Hubble time, we can approximate the background spacetime to be nearly the Minkowski one with $\overline{g}_{\mu\nu} \approx \eta_{\mu\nu} = diag(-1, 1, 1, 1)$. We keep the theory to be the first order in $h_{\mu\nu}$ and neglect terms higher than $\mathcal{O}\left(h^2\right)$.

The different between gravitational waves in $f(R)$ and general relativity is that it contains an extra scalar degree of freedom in $f(R)$. This comes from the non-vanishing trace of the field equation. Eq. (3) can be viewed as equation of motion for a scalar field Φ. By the identifications [14, 15]

$$\Phi \to f'(R) \quad \text{and} \quad \frac{dV_{eff}}{d\Phi} \to \frac{2f(R) - f'(R)R - \kappa^2\rho}{3}, \tag{6}$$

we obtain the Klein-Gordon equation for the scalar field Φ:

$$\Box\Phi = \frac{dV_{eff}}{d\Phi}. \tag{7}$$

In order to have a stable perturbation of spacetime, we must require the background scalar Φ_0 to stay at the stable minimum of the effective potential V_{eff}, i.e., $dV_{eff}/d\Phi = 0$ and $d^2V_{eff}/d\Phi^2 > 0$, corresponding to the conditions for the de-Sitter point curvature (4) and the positivity of the scalar mass, respectively, in vacuum. Perturbing the trace of the field equation (3) with a nonzero constant background curvature R_0 yields

$$3\Box\delta f' + R_0\delta f' + f'(R_0)\delta R - 2\delta f = 0. \tag{8}$$

Using the relations $\delta f = f'(R_0)\delta R$ and $\delta f' = f''(R_0)\delta R$, we obtain the massive wave equation for the scalar mode [11]

$$\Box h_f = m_s^2 h_f, \tag{9}$$

where $h_f \equiv \delta f'/f'(R_0)$ is the field of the scalar mode and

$$m_s^2 = \frac{1}{3}\left(\frac{f'(R_0)}{f''(R_0)} - R_0\right) \tag{10}$$

is the mass squared of it. Note that $m_s^2 = V''_{eff}(\Phi)$ [5]. For any viable $f(R)$ model, the condition $m_s^2 > 0$ is needed for the stability of the cosmological perturbation and to prevent the field from being a tachyon [14].

For the FRW metric, Eq. (9) should be expressed as

$$\left(-\partial_0^2 + \frac{\partial_i^2}{a^2} - 3H\partial_0\right) h_f = m_s^2 h_f, \tag{11}$$

where the term $-3H\partial_0$ gives a damping factor caused by the expansion of the universe. To illustrate the solution of Eq. (11), we take the de Sitter universe with a constant H. In this case, the solution is a damped plane wave

$$h_f = A(\vec{k})e^{-\frac{3}{2}Ht}exp\left(iq^\mu x_\mu\right), \tag{12}$$

where $q^\mu \equiv (\omega_m, \vec{k})$, $\omega_m = \sqrt{\vec{k}^2/a^2 + m_s^2 - \frac{9}{4}H^2}$ is the angular frequency and $A(\vec{k})$ is the amplitude. For simplicity and without loss of generality, we take $a = 1$ and neglect the damping effect as $\vec{k}^2/a^2 \gg H^2$. As a result, Eq. (9) leads to a simple plane wave solution

$$h_f = A(\vec{p})exp\left(iq^\mu x_\mu\right), \tag{13}$$

with $\omega_m = \sqrt{\vec{k}^2 + m_s^2}$. We can see that m_s is the cutoff frequency of the scalar mode of gravitational wave. For $\omega_m < m_s$, the wave vector becomes imaginary. The waveform is an exponential decay in distance, i.e., $h_f \propto \exp(-\vec{k}\cdot\vec{x})$. Thus, the scalar will not propagate in space

Journal of Physics: Conference Series **384** (2012) 012030 doi:10.1088/1742-6596/384/1/012030

below the cutoff frequency. The massive scalar mode will not propagate at the speed of light with the group-velocity

$$v_g = \frac{\left|\vec{k}\right|}{\omega_m} = \frac{\sqrt{\omega_m^2 - m_s^2}}{\omega_m}. \tag{14}$$

Note that the tensor mode in $f(R)$ is exactly the same as that in GR when a traceless gravitational wave propagates in a non-zero de-Sitter curvature R_d background.

Since the ΛCDM model can be viewed as a special case of $f(R)$ with

$$f(R) = R - 2\Lambda, \tag{15}$$

where Λ is the cosmological constant, the mass of the scalar mode is infinite, i.e., $m_s^2 = \infty$, which requires infinite large energy to excite the scalar mode. Clearly, there is no scalar mode in the ΛCDM model. Although the de Sitter curvature R_d is not zero, i.e.,

$$R_d = 4\Lambda \quad (\Lambda CDM), \tag{16}$$

the contribution from R_d is negligible because $\Lambda \approx H_0^2 \approx \left(10^{-33} eV\right)^2$, where H_0 is the present Hubble parameter.

In the exponential gravity model, the viable conditions are satisfied when $\beta > 1$ and $R_S > 0$ [8, 16]. The feature is that it is free from the fine tuning problem and it has only one parameter more than the ΛCDM model. According to the condition for the de-Sitter curvature in Eq. (4), R_d satisfies

$$\left(1 - \beta e^{-R_d/R_S}\right) R_d = 2R_d - 2\beta R_S \left(1 - e^{-R_d/R_S}\right). \tag{17}$$

Defining $x \equiv R_d/R_S$, Eq. (17) becomes

$$x = 2\beta - \beta e^{-x}(x+2). \tag{18}$$

The factor $e^{-x}(x+2)$ decreases very fast when $\beta > 1$, which is generally required by the viable condition for the exponential gravity. Therefore, we can obtain the asymptotic solution of x for a large β:

$$x = 2\beta \quad \text{for} \quad \beta \gg 1. \tag{19}$$

From Eq. (10), we derive the mass squared of the scalar mode in the exponential gravity as

$$m_s^2 = \frac{1}{3} R_S \left(\frac{e^{R_d/R_S} - \beta}{\beta} - \frac{R_d}{R_S}\right) = \frac{1}{3} R_S \left(\frac{1}{\beta} e^x - 1 - x\right). \tag{20}$$

Since in the large curvature regime $R/R_S \gg 1$, the theory will recover the cosmological constant model, R_S is roughly inverse proportional to β in the way that

$$\beta R_S \cong 2\Lambda = 9.94 \times 10^{-66} eV^2 \tag{21}$$

with the value of Λ obtained from WMAP 7 [17], SDSS 7 [18] and SCP Union2 observations [19]. Eq. (20) then can be approximated as

$$m_s^2 \cong \frac{2\Lambda}{3\beta} \left(\frac{1}{\beta} e^x - x - 1\right). \tag{22}$$

In Table 2, we show the exact m_s without any approximation for $\beta = 1.27$, 2, 3 and 4, respectively,

Table 2. Numerical results of the scalar mode mass m_s in vacuum with respect to different β in the exponential gravity model.

β	h	y_H^{ini}	Ω_m^0	$R_S\,(10^{-66}eV^2)$	$m_s\,(10^{-33}eV)$
4	0.7050	2.618	0.2761	2.452	24.36
3	0.7059	2.609	0.2758	3.263	11.39
2	0.7103	2.558	0.2738	4.824	5.069
1.27	0.7194	2.45	0.2701	7.39	1.86

where we have used the values of R_S obtained from in Ref. [20] under the constraints of WMAP 7, SDSS 7 and SCP Union 2 measurements.

When β is $\mathcal{O}(1)$, m_s is around $10^{-33}eV$. However, the cosmological observations do not give any significant upper bound on β. Thus, m_s could be arbitrary large in this case. As $\beta \to \infty$,corresponding to the ΛCDM model with $m_s \to \infty$, the scalar mode of gravitational wave vanishes.

In the Starobinsky Model, R_c is roughly the present cosmological density and λ and n are positive model parameters. From the solar system constraint and the bound on the violation of the equivalence principle, one gets $n > 0.9$ [21]. Since for $R \gg R_c$, the model will restore the ΛCDM model, we have $\lambda R_c \simeq 2\Lambda$, $R_d/R_c \simeq 2\lambda$ and $R_d \simeq 4\Lambda = 1.99 \times 10^{-65}eV^2$ when $\lambda \gg 1$.

We now consider gravitational waves in inner galaxy. In the presence of matter density, the scalar mode might not be able to exist in the viable $f(R)$ models. Consider a scalar mode of gravitational wave propagating within our Galaxy halo. The local homogeneous density of dark matter and baryonic matter is roughly $\rho \approx 10^{-24}g/cm^3$. If we take this matter density into our analysis, it will give a large contribution to the background curvature compared to the vacuum de Sitter curvature. (The ratio of the matter density ρ to de Sitter curvature R_d is about $\kappa^2\rho/R_d \simeq \kappa^2\rho/4\Lambda \approx 10^5$.) In this case, the condition for the background curvature R_0 in Eq. (4) should be modified as

$$f'(R_0)R_0 = 2f(R_0) - \kappa^2\rho, \tag{23}$$

where R_0 is the background curvature with matter. Note that for viable $f(R)$ models, the solutions to Eq. (23) can be approximated as $R_0 \simeq \kappa^2\rho$ at the high curvature regime.

In the case of the exponential gravity, Eq. (23) gives

$$x = 2\beta + r - \beta e^{-x}\,(x+2)\,, \tag{24}$$

where $x \equiv R_0/R_S$ and $r \equiv \kappa^2\rho/R_S$ are the ratios of the background curvature and matter density to R_S, respectively. Since $\beta R_S \cong 2\Lambda$ from (21), we find that the solution of Eq. (24) is extremely large,

$$x \simeq r \simeq \frac{\kappa^2\rho}{R_d/2\beta} \simeq 2 \times 10^5\beta, \tag{25}$$

which just leads to $R_0 \simeq \kappa^2\rho$. Thus, in the exponential gravity, the mass of the scalar mode will become an extreme in the galaxy region:

$$m_s \approx \sqrt{\frac{2\Lambda}{3\beta^2}e^{2\times 10^5\beta}} \approx \infty. \tag{26}$$

The corresponding cutoff frequency ω_m is also infinite. As a result, it is almost impossible to detect this scalar mode within our Galaxy under the exponential gravity scenario. Moreover, for any source that is massive enough to generate gravitational waves, we expect them to lay in the region with density higher than the baryonic/dark matter density $10^{-24}g/cm^3$. Therefore, the scalar mode will not have the chance to propagate from the source in the exponential gravity.

In the case of the Starobinsky model, the situation is quite different. The scalar mode of gravitational wave can have a light mass in the galaxy region. The minimum bound of the scalar mode mass is $m_s \gtrsim 10^{-24}eV$ when $\rho = 10^{-24}g/cm^3$. The corresponding cutoff frequency is quite small $f_m \gtrsim 10^{-9}$ Hz. This feature will allow the propagation of the scalar mode inside the galaxy. Hence, detecting the scalar mode in the Starobinsky model will be possible. In Fig. 1,

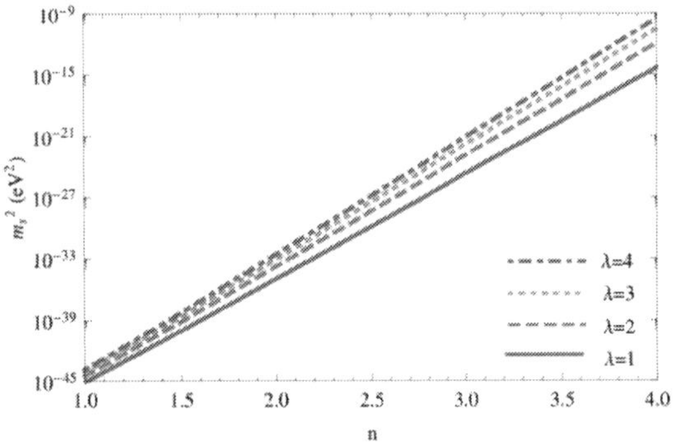

Figure 1. m_s^2 versus n in the Starobinsky model with matter density $\rho = 10^{-24}g/cm^3$ and $\lambda R_c \cong 2\Lambda$.

we depict the mass squared versus the model parameter n with different fixed values of λ, where we have used $\lambda R_c \cong 2\Lambda$. The scalar mode can still be very heavy when the index n goes to a large value, but the mass dependence on the parameter λ is not quite significant.

3. Conclusions

We have discussed gravitational waves in viable $f(R)$ theories. Using the weak field approximation on the field equation, we have confirmed that $f(R)$ will give an extra massive scalar mode besides the ordinary tensor mode in the standard GR. We have explicitly investigated the situations of the extra scalar mode of gravitational wave in the exponential gravity and Starobinsky models of the viable $f(R)$ gravity theories. In vacuum, we have shown that the typical mass squared of the scalar mode is in the order of the de-Sitter curvature $m_s^2 \sim R_d \approx 10^{-66}eV^2$ in both models. In the galaxy region, the galactic matter density of $\rho = 10^{-24}g/cm^3$ is about 10^5 larger than the de-Sitter curvature R_d in both models. In the exponential gravity, the mass of the scalar mode in galaxy is undetectable large. However, in the Starobinsky model, the mass can be much smaller with its lower bound in galaxy being about $10^{-24}eV$ (or 10^{-9} Hz). Therefore, it is possible to observe the scalar mode of gravitational wave in the Starobinsky scenario if there is an astrophysical source which generates this scalar mode.

Recently, there is an underway space-based gravitational wave probing experiment, the Laser Interferometer Space Antenna (LISA) [22], which is a proposed joint mission of the European Space Agency (ESA) and NASA. It will measure the low-frequency band (10^{-5} to 1 Hz) of gravitational waves with high signal-to-noise ratio. Since LISA will be located far from the Earth and other gravitational sources, the background curvature of it is very low compared to the ground-based experiments, which allows the propagation of the scalar mode in some viable $f(R)$. As a result, LISA has a great chance to direct detect not only the ordinary gravitational wave but also the clues of the deviation from Einstein's GR by analyzing the scalar mode behavior of gravitational wave. We note that other gravitational wave probes, such as ASTROD-GW [23] with the sensitivity in the $10^{-7} - 10^{-1}$ Hz band, may also detect the extra scalar mode. Finally, we remark that the scalar mode in a viable $f(R)$ cannot be observed by the ground gravitational searches due to the large background curvature.

Acknowledgments
We thank the hospitality of the Kavli Institute for Theoretical Physics China (KITPC) during the DSU workshop and the "Dark Matter and New Physics" program. This work was partially supported by National Center of Theoretical Science and National Science Council (NSC-98-2112-M-007-008-MY3) of R.O.C.

References
[1] Yang L, Lee C C and Geng C Q 2011 *JCAP* **1108** 029
[2] Nojiri S and Odintsov S D ECONF 2006 *C0602061* 06
[3] De Felice A and Tsujikawa S 2010 *Living Rev. Rel.* **13** 3
[4] Sotiriou T P and Faraoni V 2010 *Rev. Mod. Phys.* **82** 451
[5] Hu W and Sawicki I 2007 *Phys. Rev. D* **76** 064004
[6] Starobinsky A A 2007 *JETP Lett.* **86** 157
[7] Tsujikawa S 2008 *Phys. Rev. D* **77** 023507
[8] Linder E V 2009 *Phys. Rev. D* **80** 123528
[9] Chiba T 2003 *Phys. Lett.* B**575** 1 (2003)
[10] Nojiri S and Odintsov S D 2003, *Phys. Rev. D* **68** 123512; 2004 *Gen. Rel. Grav.* **36** 1765
[11] Capozziello S, Corda C, and De Laurentis M F 2008 *Phys. Lett. B* **669** 255; Corda C 2007 *JCAP* **0704** 009; 2008 *Int. J. Mod. Phys.* A**23** 1521; 2008 *Astropart. Phys.* **30** 209; 2010 *Eur. Phys. J. C* **65** 257
[12] Berry C P L and Gair J R 2011 *Phys. Rev. D* **83** 104022
[13] Naf J and Jetzer P 2011 arXiv:1104.2200 [gr-qc].
[14] Starobinsky A A 1980 *Phys. Lett. B* **91** 99
[15] Capozziello S, Stabile A, and Troisi A 2007 *Class. Quant. Grav.* **24** 2153
[16] Bamba K, Geng C Q, and Lee C C 2010 *JCAP* **1008** 021
[17] Komatsu E *et al.* (WMAP Collaboration) 2011 *Astrophys. J. Suppl.* **192** 18
[18] Percival W J *et al.* (SDSS Colaboration) 2010 *Mon. Not. Roy. Astron. Soc.* **401** 2148
[19] Amanullah R *et al.* (The Supernova Cosmology Project) 2010 *Astrophys. J.* **716** 712
[20] Yang L, Lee C C, Luo L W, and Geng C Q 2010 *Phys. Rev. D* **82** 103515
[21] Capozziello S and Tsujikawa S 2008 *Phys. Rev. D* **77** 107501
[22] Danzmann K and Rudiger A 2003 *Class. Quant. Grav.* **20** S1; see also http://sci.esa.int/lisa.
[23] Ni W T 2011 arXiv:1104.5049 [astro-ph.CO].

AUTHOR INDEX

Institute of Physics
Dirac House, Temple Back
Bristol BS1 6BE UK

CURRAN ASSOCIATES INC
proceedings.com